Carrier Air Group Commanders

Carrier Air Group Commanders

The Men and their Machines

Robert L. Lawson

Schiffer Military History
Atglen, PA

Acknowledgments

To acknowledge all of the wonderful people who contributed to this book is an impossible task, there simply were too many of you. So first of all, let me express a heartfelt thank you to everyone who assisted in any way to this effort; all of the former CAGs who kindly responded to my requests, squadron public affairs officers for sending photos, and especially to all of my airplane photographer friends who helped. I wish we could publish all of those great photos, but space just doesn't allow this.

At the risk of offending someone who will feel unappreciated by omission, I must name a few individuals whose help was outstanding. First, and foremost, thanks to my wife Sally, who patiently (sometimes more patiently than others) waded through my horrible handwriting to computerize the manuscript. If the Medal of Honor were awarded to wives, Sally would be a shoo-in. She faced the enemy with great intrepidity and valor, while upholding the finest traditions of the naval service.

Some others who deserve special mention (not in any hierarchy or sense of order) are: Pete Clayton, Barrett Tillman, Steve Millikin, Jan Jacobs, Hill Goodspeed, Harry Gann, Mike Grove, Doug Olson, Hal Andrews, Walt Boyne, RADM J.D. "Jig Dog" Ramage, and Roy Grossnick, along with his great staff at the Naval Aviation History Office. Also, the people of the Navy History Center, especially Bob Cressman and Chuck Haberlein.

To Mike Weeks, who compiled the CAG list appendix, your help has been beyond thanks and can never be repaid in full. Also recognized are: Don Montgomery, Bob Carlisle, John Elliott, Rick Burgess, CAPTs Rosario'~Zip" Rausa, Earle Rogers, and Bob Elder.

For those of you who contributed and don't find your name here, please know that your help is appreciated.

In the ten year gestation period of this book, I lost a few good friends, and this saddens me because they were some of the people for whom this book was written and dedicated. Farewell to CAPT Tommy Blackburn, VADM Bill Martin, CAPTs Bob Dose', Mac Wordell, Gerry O'Rourke, and Dave McCampbell. The service you gave to your country was incomparable.

Book Design by Ian Robertson.

Library of Congress Catalog Number: 00-100532

Printed in China.
ISBN: 0-7643-1035-6

We are interested in hearing from authors with book ideas on related topics.

Published by Schiffer Publishing Ltd.
4880 Lower Valley Road
Atglen, PA 19310
Phone: (610) 593-1777
FAX: (610) 593-2002
E-mail: Schifferbk@aol.com.
Visit our web site at: www.schifferbooks.com
Please write for a free catalog.
This book may be purchased from the publisher.
Please include $3.95 postage.
Try your bookstore first.

In Europe, Schiffer books are distributed by:
Bushwood Books
6 Marksbury Avenue
Kew Gardens
Surrey TW9 4JF
England
Phone: 44 (0)208 392-8585
FAX: 44 (0)208 392-9876
E-mail: Bushwd@aol.com.
Free postage in the UK. Europe: air mail at cost.
Try your bookstore first.

Contents

Prologue

Bob Lawson has gathered together in a unique and memorable fashion the tradition and the experiences of the spectacular individuals who represent and epitomize one of aviation's greatest challenging and rewarding commands. Naval Aviation, particularly carrier aviation, is in itself one of the world's most exhilarating careers in or out of the Armed Forces. The daily exercise of airmanship, courage, and unforgiving demands are common place in Naval Aviation. To be designated the leader of such a group of diverse aircraft and aviators, the Air Group Commander, or "CAG," is to be anointed as the best of the best!

The leaders that Bob has portrayed are only a few of these magnificent men who have historically led and continue to lead Naval Aviation to the peak of professional aviation. I am proud to have known and served with such men. Bob has done us all a great service by allowing us to share in some of the events during the best years of their lives.

Bob Lawson has accomplished the ultimate carrier "OK Three Wire" landing aboard with this fascinating insight into the history and exploits of one of the Navy's top fighting commands!

VADM Robert F. Schoultz, USN(Ret)

BIO: Vice Admiral Robert F. Schoultz

VADM Robert F. "Dutch" Schoultz spent his youth in Omaha, Nebraska. He entered the Navy as a student in the University of Nebraska's NROTC program and earned his wings and commission in September 1945.

Early in his career, Admiral Schoultz served in a variety of capacities in torpedo and attack squadrons and in USS Midway (CVA-41). In 1960, following duty on the Staff of the Commander in Chief, U.S. Pacific Fleet, he served as XO and then CO, Attack Squadron 93. During this tour, then-CDR Schoultz led his squadron on two Western Pacific deployments in USS Ranger (CVA-61). From August 1964 to December 1965, he was assigned to the Bureau of Naval Personnel, then joined USS Franklin D. Roosevelt (CVA-42) as Operations Officer, then Executive Officer.

CAPT Schoultz was Strike Warfare Officer and Assistant Chief of Staff for Plans on the Second Fleet Staff prior to assuming command of Arneb (LKA-56) in December 1969. He commanded Forrestal (CVA-59) from June 1971 to 1972. He then returned to the Bureau of Naval Personnel. In November 1973 CAPT Schoultz was promoted to flag rank and became Director of the National Military Command Center (J-3), Joint Chiefs of Staff in the Pentagon, then later served as Vice Director for Operations (J-3), Joint Chiefs of Staff until detaching in September 1976 to attend the Senior Officers Ship Material management Course, Idaho Falls, Idaho.

VADM Schoultz assumed command of Battle Force Sixth Fleet in April 1977. In August 1979, he reported to the OpNav Staff as Special Assistant to the Director for Command and Control (OP-094). He then assumed command of Naval Air Force, U.S. Pacific Fleet in January 1980. Prior to assuming duty as U.S. Commander Eastern Atlantic and Deputy Commander in chief U.S. Naval Forces, Europe 1 Feb 1985, he served as Deputy Chief of Naval Operations (Air Warfare).

VADM Schoultz retired in March 1987 and started a business consulting with aerospace corporations and other defense contractors. His consultations have been involved with the A-12, NATF, and AX-A/F-X programs, as well as AF-117X and JAST.

Vice Admiral Schoultz is married to the former Rosemary Parker of Falls City, Neb. They have two sons, Robert and Scott. VADM and Mrs. Schoultz reside in Falls Church, Va.

USN photo, courtesy VADM R.F. Schoultz, USN(Ret)

Introduction

The aircraft carriers of the U.S. Navy today are the single most powerful weapons of warfare that have been developed throughout the history of armed conflict. Since 1942 the carrier has been the manifest force for power projection and control of the seas. In the post-WWII period every president of the United States has ordered carriers to respond to crises and implement American foreign policy over the entire globe. The carriers have been sent as near as Cuba and Haiti to as far as Southeast Asia and the Middle East. At times, the mere presence of a carrier battle group has been sufficient to stabilize a volatile situation threatening a geographic area, or world peace in general.

Since WWII the U.S. Navy has been called upon to respond to more than 280 international crises. Aircraft carriers have been involved in more than 75 percent of those responses. However, it is aircraft, the carrier's "main battery," that make the carrier an effective tool of warfare, and it is naval aviators who fly these aircraft into combat. The men who lead these naval aviators are still respectfully called "CAG"—Commander, Air Group—even though the Navy terminology has changed to Air Wing.

The study of the U.S. Navy's air group and air wing commanders is a study of the history of carrier aviation in itself. This volume presents a history of the background of the establishment of the carrier air group commander billet and the attendant formal air groups. The book is divided into four sections; Section One presents an overview of naval aviation history from 1898 to 1922, when the U.S. Navy commissioned its first aircraft carrier, USS *Langley* (CV-1), with its wood-and-fabric biplanes. The following pages describe carrier aviation from *Langley* to the 90,000-ton nuclear supercarriers of today with their supersonic jets, one of which weighs nearly as much as *Langley's* entire original 34-plane complement. Although a brief overview of the development of U.S. carriers is covered, this is not a history of aircraft carriers, *per se.* For more definitive coverage of this subject readers will do well to refer to Norman Polmar's dated, but still wonderful, study "Aircraft Carriers" (Doubleday 1969), as well as Roger Chesneau's comprehensive "Aircraft Carriers of the World, 1914 to Present" (Naval Institute 1984) and Norman Friedman's technical study "U.S. Aircraft Carriers" (Naval Institute 1983).

Section Two covers the genesis and evolution of the air group/wing commander billet along with the development of the organizations they commanded. It contains profiles of the first CAGs, as well as their wing commander predecessors. Section Three has biographical sketches of more than 40 CAGs from WWII to the present. Their stories are presented by use of personal reminiscences and anecdotes, accompanied by photographs of the individuals. Their profiles give the reader a feel for the responsibilities, joys, and sorrows that accompany the assumption of the title: CAG. Additionally, the profiles give an insider's view of the U.S. Navy's carrier combat operation tactics during WWII, Korea, and Vietnam, while also covering major worldwide contingency operations in which carriers have been used since 1945, including Desert Storm.

The book's fourth section contains a selection of photographs and some profile drawings of the aircraft flown by CAGs from 1938 to the present. Heavy emphasis has been given to provide as many high-quality color photographs as could be located. This section should have appeal to both historians and enthusiasts in its coverage of the Navy's carrier aircraft during the past 70 years. Not only does it graphically present the U.S. Navy's aircraft, it also covers the evolution of carrier aviation itself.

Terms and designations used in this book appear according to the era in which they were used. Seemingly inconsistent or incorrect designations, such as *Forrestal* appearing as CVB, CVA, and CV-59, are a result of her reclassification from large aircraft carrier to attack aircraft carrier and, finally, aircraft carrier. All CVBs and CVs were reclassified CVA on 1 Oct 1952. Later, some CVAs became CVSs (anti-submarine support carrier), AVTs (auxiliary aircraft transport), or CVTs (training aircraft carrier). An exception to the CVA designation was the original *United States* (CVA-58) of the late '40s, which was the only carrier classified CVA for heavy carrier. The classification has not been used since, although the CVA designation was resurrected in 1952. In the 1970s (most during 1974), CVAs reverted to the CV classification when ASW forces were integrated into attack air wings.

As for air wings, originally they were called air groups, and on 1 Jul 1938 were named according to their ship; e.g., Saratoga Air Group. Previously, on 1 Jul 1937, the squadron designations were realigned to reflect the hull numbers of their carrier, such as *Saratoga's* (CV-3) VF-3, VT-3, VB-3, and VS-3. Although the air groups were often commonly referred to by number, numbered air groups did not exist until Carrier Replacement Air Group 9 was established 1 Mar 1942. By this time, air groups and individual squadrons began transferring between ships and groups according to wartime requirements. All prewar air groups were either disestablished or redesignated by 25 Sep 1943. On that date, Ranger Air Group became Carrier Air Group 4. Between March 1942 and 29 Jun 1944, the air group designation was CAG; then, on the latter date it was changed and broadened to include different carrier types. The designations then became CVBG, CVG, and CVLG, as well as CVEG for *Sangamon*-class escort carriers. The composite squadrons (VC) of the smaller CVEs retained the VC designation, but were reclassified as air groups. On 1 Sep 1948, all air groups became CVGs, regardless of the type carrier assigned. On 20 Dec 1963, air groups became air wings, and the designation changed to CVW.

Post-war replacement air groups were established 10 Mar 1958 and popularly called RAGs. Officially, they were RCVGs until 1 Apr 1963 when they became combat readiness air groups, but still designated RCVG. On 20 Dec 1963 they became RCVWs.

Antisubmarine carrier air groups came into being 1 Apr 1960 and were CVSGs. The CVSGs were phased out in the '70s; all were gone by 30 Jun 1973.

Regarding the air group/wing charts presented in the text; these charts reflect the composition (actual number of aircraft) of the units as of the date of the chart. They do not necessarily reflect the complement (number of aircraft authorized) of the unit. Complements have changed several times over the years to adapt the groups to the existing tactical requirements of the time, or to realign them according to the aircraft numbers and types available. The composition is fluid and will vary almost on a daily basis as a result of aircraft losses, transfers, or acceptances.

Generally, during the '30s, each of the carrier squadrons had a complement of 18 aircraft. Until 1930 *Lexington's* (CV-2) and *Saratoga's* air groups had one each fighting, bombing, scouting, and torpedo squadron in addition to their utility unit aircraft. In 1930 their composition increased to two fighting squadrons and eliminated the bombing squadron, which began to return in 1934. During 1931-34 the Marines came aboard with six-plane squadrons of Vought O2U and SU scoutbombers. Although the Marines continued to operate from carriers after 1934, it was only on temporary assignments until 1944. As the Navy continued to experiment with carrier tactics and its aircraft evolved, the air group compositions fluctuated until the late '30s when they stabilized on the original one squadron each of mission type.

Because of their smaller tonnage, *Ranger* (CV-4) and *Wasp (CV-7)* initially did not carry a torpedo squadron; instead, they had two scouting squadrons. With the advent of war, room was made for Douglas TBD Devastators and Grumman TBF Avengers in reduced strength on both carriers. Also, during early 1942, the air group compositions again began to fluctuate and continued to change throughout the war. On 1 Mar 1943 the carrier scouting squadrons were redesignated bombing squadrons and, during July, one bombing squadron was deleted from each group. A typical CV air group had 21 VF, 36 VSB, and 18 VTB aircraft authorized, indicating the emphasis on attack aircraft to strike the enemy. On 1 October the VF component was raised to 36 on the *Essexes*. CVL air groups were established at 12 VF, nine VB, and nine VT. In November the CVL groups increased to 24 VF, with nine VT for the remainder of the war. The SBD bombing squadrons were eliminated from the CVLs at that time.

On 10 Aug 1944, the CVGs were revised to 54 VF, 24 VB, and 18 VT. Four night fighters and two photo planes were included in the VF total. On 29 November, to counter the *kamikaze* threat, the fighter component was again increased, and the attack squadrons were decreased. The CVG composition became: 73 VF, 15 VB, and 15 VT, with two VF squadrons of 36 planes each and one for the air group commander. However, some CAGs continued to fly Avengers, usually dependent upon their previous flying backgrounds. Additionally, of the fighter total, two were VF(E) and four were VF(N) night fighters, along with two VF(P) photo planes. To help provide the increased fighter strength needed in the Pacific carriers, two Marine fighter squadrons (VMF-124 and -213) joined CVG-4 in *Essex* (CV-9), and the new VBF squadrons (bombing fighting) were formed in December.

Following the war, rapid demobilization caused an upheaval in the aviation units resulting in two complete reorganizations of naval aviation, one 15 Nov 1946 and another effective 1 Sep 1948.

In the 1946 reorganization, as related above, all carrier air groups were redesignated according to their ship classification. Additionally, carrier squadron mission types were limited to fighter and attack, eliminating the VBF, VB, and VT designations. The squadrons carried a suffix letter to indicate the type of air group to which they were assigned. An example of this reorganization is CVG-4, which originally began as Ranger Air Group and exists today as CVW-1:

WWII Designation	15 Nov 1946	2 Aug 1948
CVG	CVAG-1	CVG-1
VF-4	VF-1A	VF-11
VBF-4	VF-2A	VF-12
		VF-13 (established)
VB-4	VA-1A	VA-14
VT-4	VA-2A	VA-15

The September 1948 Naval Aviation reorganization resulted in all air groups becoming CVGs, with each squadron assuming a two- or three-digit sequential number without a suffix letter. The first one or two numbers of the squadron identified its parent air group, i.e., CVG-19's squadrons, for example, were designated VF-191, VF-192, VF-193, VA-194, and VA-195. On 20 July, CNO had directed that the CVG composition be changed to three VF and two VA squadrons. Also about this time, helicopters began carrier service in one-helo dets from their parent squadrons. The helo units were designated HU in 1948, and heavy attack VC units were redesignated VAH on 10 Nov 1955.

The 1948 directive resulted in a slight decrease in the squadron complements to allow for the new fighter squadron. It also established new dedicated mission carrier support units by converting some VC squadrons to VFN and VAN with night operation capabilities, as well as others to provide airborne early warning. These squadrons retained their VC designations until 1956, and usually operated with the CVGs in four-plane detachments. On 15 Jul 1956, the remaining VC squadrons became VA(AW), VF(AW), VAW, and VFP, according to their specific mission.

By the outbreak of the Korean War (it was a *war*, a *conflict* is what neighbors have over fences), some air groups had four VF squadrons—two flying jets and two flying F4U-4Bs in a fighter-bomber role—and only one VA squadron. Individual squadron complements were again at 18 aircraft. With the Navy now operating the large *Midway* class, along with the WWII-style as well as modernized *Essexes,* and with newer, more capable carriers on the horizon, the air group compositions began to vary in major degrees. These compositions are discussed in the text.

Another major change to the carrier air wings began 13 Nov 1980 with the establishment of Fighter Attack Squadron 125 (VFA-125), a replacement training squadron for the new McDonnell Douglas F/A-18A Hornet. On 25 Mar 1983 the VFA squadron designations were changed to strike fighter squadron, and in 1984 they began replacing Vought A-7 Corsair II squadrons in the fleet.

A note about the CAG profiles: Mention of individual awards has been limited to the Medal of Honor and the Navy Cross, the Navy's two highest decorations. Most CAGs end their careers with a chest full of ribbons. However, others, through no fault of their own and usually because of non-combatant assignments or the philosophy of their seniors during combat assignments, received fewer awards. By naming all of each individuals' awards it seemed to the author that it would imply an unfair competition. Thus, the decision was made to eliminate all but the two highest awards.

Robert L. Lawson
Dammeron Valley, Utah
July 1999

This book is sincerely dedicated to all those naval aviators who for the past seven decades have led their fellow aviators from the flight decks of the U.S. Navy's aircraft carriers—and who will forever be known as "CAG." Also, a special dedication to my mother, who has always been there with her love and encouragement.

1

A Historical Overview

"A plane carrier would carry 15 torpedo planes and, in my opinion, would be a menace to a whole division of battleships and in the same way a fleet of carriers could attack a place like Hawaii."

LCDR Edward O. McDonnell, USNR, Naval Aviator #16
arguing for carrier aviation before the General Board of the Navy, 1919.

The First Halting Steps

The birth of carrier aviation occurred on a blustery, forbidding day, 14 Nov 1910, when Eugene Ely, a civilian pilot flying a Curtiss "Hudson Flyer" pusher, flew from an 83-foot wooden platform specially constructed on the bow of USS *Birmingham* (CL-2). The armored cruiser was at anchor in Hampton Roads, VA, slowly making preparations to get underway for the takeoff demonstration.

Impatiently, Ely assessed the threatening weather situation and decided to launch immediately without benefit of "wind over the deck." The subsequent 57-foot takeoff run was spectacular, as the little craft dribbled off the bow, momentarily striking the water 37 feet below before continuing on two and one-half miles to land ashore at Willoughby Spit. Although a near disaster, this event was the first operation of an airplane from a ship.

Two months later, Ely made the first shipboard landing when he brought his Curtiss aboard another armored cruiser, USS *Pennsylvania* (ACR-4), 18 Jan 1911 at anchor in San Francisco Bay. After a brief celebration of the event, Ely's plane was gassed and "respotted" for launch. He then flew off *Pennsylvania* amid the shouts of a jubilant crew to land 10 miles away at the Tanforan aviation field south of San Francisco. It would be almost 12 years before another airplane would land aboard a U.S. Navy ship.

U.S. naval aviation has officially existed since 8 May 1911 when CAPT Whitaker I. Chambers signed requisitions for two Curtiss biplanes. This date is the recognized birthdate of naval aviation, although CAPT Chambers had reported for duty at the Navy General Board 1 Apr 1911 as Officer in Charge of Aviation. Since 1910, Chambers had been given the untitled job of handling all

Painting depicts Eugene Ely's 50 hp Curtiss Pusher dipping toward the water during his historic 14 Nov 1910 launch from *Birmingham* (CL-2) at anchor in Hampton Roads, VA. USN photo

Two months later Ely made the first shipboard landing aboard *Pennsylvania* (ACR-4) in San Francisco Bay. Acme Photo Service, courtesy Robert F. Dorr

Captain Whitaker I. Chambers, the Navy's first Director of Naval Aviation, spurred the Navy's early involvement with aircraft. His office was under a staircase. USN photo

matters relating to aviation. Although his "Office of Naval Aviation" was located under a staircase, Chambers, with no aviation experience, developed an avid interest in his task and became a forceful, enthusiastic advocate for developing a naval air arm. It was Chambers who, with political help, arranged the Ely demonstrations after much difficulty in persuading a reluctant Secretary of the Navy, George von L. Meyer. The Navy's interest in aviation dates back as far as 1898 when, at the recommendation of Assistant Secretary of the Navy Theodore Roosevelt, naval officers were assigned to sit on an interservice board investigating the feasibility of military applications for Prof. Samuel P. Langley's flying machine model, the Aerodrome.

During the years following the Wright brothers' successful first flight in 1903, the Navy assigned observers to various civilian air meets, as well as demonstrations conducted by the Wright brothers, as it watched with mild interest the growing new field of aviation. After observing the 1909 Rheims Aviation Meet, U.S. Naval Attaché at Paris, France, CDR Frederick L. Chapin, reported "the airplane would have a present usefulness in naval warfare, and that the limits of the field will be extended in the near future." Chapin also outlined two methods of operating airplanes from naval vessels: a catapult on battleships, or by building a wooden platform over the deck houses of auxiliary ships for takeoffs and landings.

On 23 Dec 1910, LT Theodore G. "Spuds" Ellyson was ordered to Glenn Curtiss' civilian aviation camp at North Island, San Diego, CA, to learn to fly. After successfully dodging North Island's jackrabbit population and undergoing further training at Hammondsport, NY, Ellyson received his pilot certificate 6 Jul 1911. On 1 Jan 1914 he was officially designated a "Navy Air Pilot," later changed to Naval Aviator #1. Also in 1911, LT John Rodgers was trained to fly at Dayton, Ohio, by the Wright brothers to become Naval Aviator #2, and LT John H. Towers joined Ellyson at Hammondsport to become #3.

Gradually, more naval officers were ordered to flying duty. Meanwhile, in September 1911, the Navy opened its first air facility at Greenpoint, MD, near the Naval Academy. In 1914 this unit moved to Pensacola, FL, to establish the first naval air station. As a result of CAPT Chambers' perseverance and firm resolve, naval aviation was finally launched, haltingly, but airborne nonetheless.

In January 1913, seeking more ways to get aviation involved with the fleet, Chambers ordered the entire neophyte Naval Aviation Unit from Annapolis to the Atlantic Fleet training base at Guantanamo, Cuba. For eight weeks the aviators flew support missions, such as scouting, mine spotting, and looking for submerged submarines. Led by John Towers, the unit's aviators also conducted demonstrations and took fleet personnel for airplane rides in an effort to demonstrate various applications that aviation might bring to the fleet.

Expanding its interest in shipboard applications for aviation, in 1912 the Navy had begun developing a method to operate aircraft from its ships. LCDR Henry C. Mustin made the first ship catapult launch 15 Nov 1915 in the Curtiss AB-2 flying boat from USS *North Carolina* (ACR-12). On 12 Jul 1916, LT Godfrey De C. Chevalier was catapulted from the battleship to complete calibra-

LT Theodore G. "Spuds" Ellyson (third from left) with (from left) Army 2nd LTs G.E.M. Kelly, John C. Walker, Jr., and 1st LT Paul W. Beck at Glenn Curtiss' first military aviation training school at North Island, San Diego, CA, in 1911. Ellyson became Naval Aviator No. 1 (NA #1), and became a leading proponent of carrier aviation. He died in 1938 during a cross-country flight while serving as executive officer of *Lexington* (CV-2). USN photo

tion of the first catapult designed for shipboard use, making *North Carolina* the first U.S. Navy ship capable of operating airplanes.

Flying airplanes and experimenting with them was one thing; getting them into the fleet was another. Although the Navy conducted aviation combat operations from battleships at Veracruz, Mexico, in 1914 and from shore bases during WWI, the aircraft and personnel were not fleet units. On 3 Feb 1919, Fleet Air Detachment, Atlantic Fleet was established under the command of CAPT G.W. Steel, Jr., to become the Navy's first aviation command under a fleet organization. Shortly thereafter, the concept of individual aviation squadrons began to evolve as permanent fleet units. As early as 5 Feb 1919, the Director of Naval Aviation Weekly Report shows the existence of numbered squadrons land-based at various naval air stations. They are referred to merely as "Squadron One," "Squadron Two," etc. In the excellent reference work, "*The Dictionary of Naval Aviation Squadrons, Vol. 1*," produced by Roy Grossnick, Naval Aviation Historian, and published by the Naval Historical Center, the genesis of designated naval aviation fleet squadrons is cited as follows:

> ...General Order Number 533 (Series 1913) of 12 July 1920 "provided for the organization of naval forces afloat into the Atlantic, Pacific and Asiatic Fleets and for the formation of type forces within each fleet, designated as Battleship, Cruiser Destroyer, Submarine, Mine, Air, and Train." Aviation was now a distinct part of the fleet organization...
>
> ...Squadron designations were not immediately identified in the fleet organization after General Order 533 was issued. The first known reference to a squadron-like organization in the fleet appears in September 1920 when Airboat Divisions 1 and 2 of Air Force Atlantic Fleet are identified during their visit to Annapolis. In August 1921, a reference was made to Torpedo Plane Squadron 5...

General order 541 of 17 July 1920 had also established aircraft type classes as V for heavier-than-air types and Z for lighter-than-air. Heavier-than-air types were sub-designated, such as VF (fighting plane), VO (observation plane), VT (torpedo bombing plane), VS (scouting plane), etc. Torpedo Plane Squadron 5 is the first squadron in known reference sources identified under this system. Established in 1919 as an element of Fleet Air Detachment Pacific, VT-5 maintained a more than 80-year lineage to Fighter Squadron 14 of the 1990s to claim title as the Navy's oldest continuous squadron.

Launching the Carriers

The performance of naval aviation during WWI helped spark an increased interest in aviation by naval leaders and manifested its potential in the Navy's future. Although naval aviation had grown to 6,716 officers and 2,107 USN/USMC aircraft by the end of the war, demobilization reduced the force to 620 pilots and 1,205 combat aircraft by 1 Jul 1920. The aircraft numbers would continue to diminish until 1925, with a low of 491 combat planes.

After the war, a small group of naval aviators, primarily LCDR Godfrey De C. "Chevy" Chevalier (NA #7), LT A. Melville "Mel" Pride (NA #1119), and led by LCDR Kenneth "Ken" Whiting (NA #16), was the driving force behind the Navy's entry into carrier aviation. Mel Pride was a Massachusetts Institute of Technology reserve officer, while the other men were Naval Academy graduates. During WWI Chevalier had observed Royal Navy efforts to operate aircraft from ships and foresaw the application of a ship dedicated to launching and recovering aircraft. The naval aviators were successful in persuading the Navy to produce an experimental aircraft carrier and, in June 1919, the General Board of the Navy recommended that a Naval Air Service be established and that the collier USS *Jupiter* (AC-3) be converted to a temporary aircraft carrier for experimental work. Additionally, the Board desired construction of wooden platforms on battleship gun turrets from which to fly aircraft for scouting and gunnery spotting missions. The Naval Appropriations Act of July 11, 1919, authorized funds for *Jupiter's* conversion to a carrier. On 1 July, the secretary of the Navy had authorized installation of aircraft launching platforms on top of two main turrets in eight battleships.

During *Jupiter's* conversion, development of an aircraft arresting system was conducted at NAS Hampton Roads under the guidance of LT Fred W. "Horse" Pennoyer, a naval construction engineer who would later undergo flight training. After much experi-

LCDRs Virgil C. "Squash" Griffin (NA #4) (left) and Godfrey de C. "Chevy" Chevelier (NA #7) on board *Langley* in November 1922. Griffin made *Langley's* first takeoff (17 Oct 1922) and Chevalier her first landing (22 Nov 1922). Chevalier would die two months later during a cross-country flight. Griffin served in aviation until his retirement in 1947. USN photo

LT A. Melville "Mel" Pride (third from left), in 8 Sep 1928 photo with other *Saratoga* officers, (from left) LTs C.A.F. Sprague, B. Rhodes, and L.D. Webb. Mel Pride (NA #1119) was a member of the first group of naval aviators to urge the Navy to develop aircraft carriers. He conducted most of the flight experiments during the design of carrier landing arresting systems. Pride continued a distinguished 42-year naval career, retiring in 1959 with the rank of four-star admiral. USN photo, courtesy CAPT W.E. Scarborough, USN(Ret)

mentation with axle hooks and fore-and-aft wires to guide the arresting hook, and athwartship wires attached to weights to retard aircraft during landing, a similar arresting device called the "trap" was installed on the ship.

On 20 Mar 1922 ex-collier *Jupiter* was recommissioned as USS *Langley*, under her executive officer CDR Kenneth Whiting. *Langley* was originally classified OCV-1 until 6 Jun 1928 for aircraft carrier, second line, as opposed to CV for aircraft carrier, first line. No record of actual use of the OCV designation has been found, and traditionally, *Langley* has always appeared with CV-1 as the first carrier. Chevalier was her first air officer and, during the ship's conversion, oversaw the training of 15 pilots assigned to *Langley*. During training, the pilots, all new to landplanes, practiced touch-and-go landings on a 100-foot platform built over a coal barge. Pennoyer continued refining *Langley's* arresting gear design, while Pride assisted and conducted the aircraft experimental work for the arresting gear.

Langley's first "air group" was the Langley Detachment, consisting of a few Aeromarine 39-B trainers and Vought VE-7SF fighters, all fitted with arresting devices for carrier landings. Except for the three *Langley* pioneers, the detachment pilots were all newcomers to landplane aviation, recently trained by the Army. With only 314 aviators as of 1 Jul 1922 and little need for landplane pilots, the Navy was at that time still training only in seaplanes.

Nearly 12 years after Ely's shipboard demonstrations, the U.S. Navy began its first flight operations from an aircraft carrier. On 17 Oct 1922, while *Langley* was at anchor in the York River near Yorktown, VA, LCDR Virgil C. "Squash" Griffin, Jr. (NA #41), flew VE-7SF A-5932 from her 523-foot flight deck to launch nearly eight decades of unparalleled naval history. Griffin landed at Hampton Roads. Nine days later, *Langley* was underway in the York River. Chevalier flew out from Hampton Roads in a 39-B (A-606) to make the first of more than two million traps—carrier arrested landings—by naval aviators who were to follow, on board nearly 150 U.S. Navy aircraft carriers. Chevalier, however, would never see the fruits of his labor; he died 14 November from injuries received two days earlier in a non-carrier related crash in bad weather.

Langley spent her first months conducting experimental operations in Chesapeake Bay before moving to Pensacola in January 1923 for the winter. In both areas her planes were based ashore. During the year more service-type aircraft joined the Langley Det, such as the Curtiss-built TS-1 fighter and Vought UO-1 scoutplane. The TS-1 was designed by the Navy, and became the first built-for-the-purpose carrier plane. All pilots first qualified in the 39-B trainer before moving on to the faster service types.

With the advent of spring *Langley* returned to Norfolk for modification and operations. She became somewhat of a "show-

CDR Kenneth "Ken" Whiting (NA #16), *Langley's* first executive officer, was the driving force behind the movement to bring carrier aviation to the U.S. Navy. Whiting died on active duty in 1943 as a captain. USN photo

Mel Pride launches in a Nieuport 28 from a gun turret platform on board the battleship *Arizona* (BB-39) in 1920. The Navy's initial shipboard aviation operations began in 1912 with catapult experiments, and later with platform-launched fighters and scoutplanes. USN photo, courtesy F.C. Dickey Jr.

During *Jupiter's* conversion to an aircraft carrier, development of an aircraft landing arrestment system was conducted at NAS Hampton Roads, VA, on a wooden turntable. USN photo

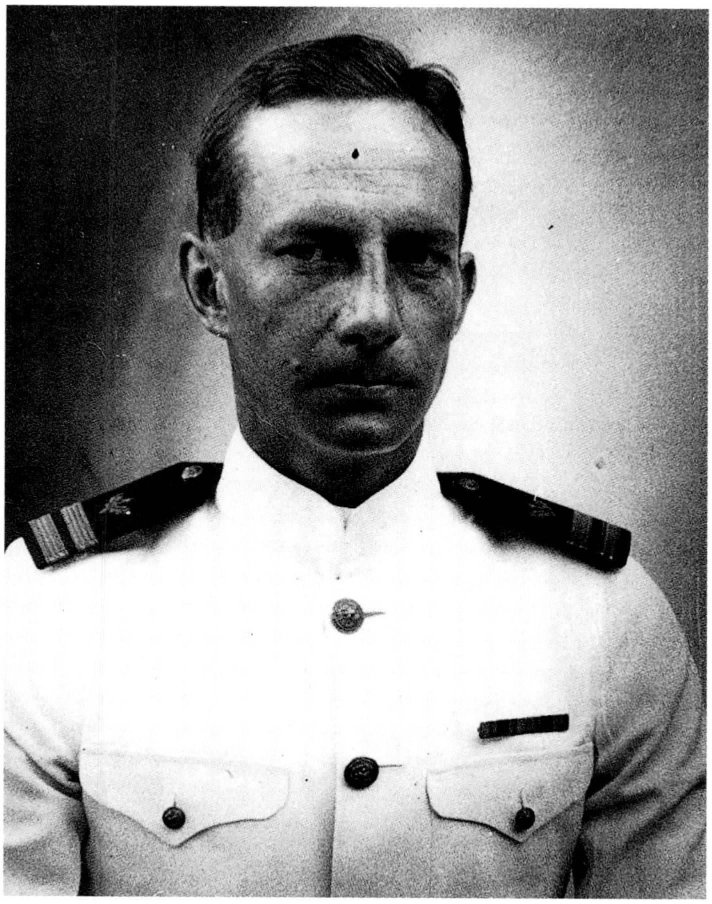

LT Frederick W. "Horse" Pennoyer, a naval construction engineer, helped design *Langley's* (CV-1) arresting gear. In 1923 he underwent flight training at Pensacola to become the Navy's only dual-qualified engineer/aviator. USN photo, National Archives PN-56270-8-23

LCDR Godfrey De C. Chevalier makes *Langley's* first landing in an Aeromarine 39-B trainer 26 Oct 1922. USN photo, courtesy Naval Aviation News

Langley's early flight operations were conducted at anchor in Chesapeake Bay. Her pilots first qualified in the Aeromarine 39-B (seen here landing), then advanced to the front-line types. USN photo

Vought VE-7SF carrier-modified landplane fighter lands in *Langley's* arresting gear consisting of fore-and-aft wires and crossdeck pendants. The plane has early-style axle hooks and trapeze tailhook. Fore-and-aft wires were discarded from carriers in 1929. USN photo, courtesy National Air and Space Museum

Originally designed by the Naval Aircraft Factory, the TS-1 was built by Curtiss and was the first U.S. Navy aircraft specifically designed for carriers. This TS-1 was assigned to Fighting Plane Squadron One at NAS San Diego, CA (North Island), during 1923-27. Most fleet fighters and torpedo bombers of the early '20s could be used as carrier planes, landplanes, or seaplanes, merely by changing the landing gear equipment. USN photo, courtesy W.L. Swisher

boat" during a public relations cruise, as her pilots demonstrated landings while she was anchored off New York and Boston. The cruise precipitated a schism in the already tenuous relationship between *Langley's* non-aviator commanding officer CAPT H.R. "Stiffy" Doyle (who had not felt it important enough to attend her commissioning ceremony) and XO Kenneth Whiting. Whiting rightly felt that Doyle had no understanding of aviation or aviators' problems. Consequently, he traveled to Washington to help convince politically influential friends to pass a law requiring that future commanding officers of aviation ships be naval aviators. Whiting's actions in this matter would have a profound effect on future carrier operations. During WWII President Franklin D. Roosevelt, a former assistant secretary of the Navy, further ordered that all carrier task forces be commanded only by naval aviators.

In January 1924 *Langley* conducted her first fleet operations. She joined the Battle Fleet Pacific and the Scouting Fleet Atlantic off Panama to engage in Battle Problem II, the second of a long series of annual fleet exercises to be held until WWII. The 1924 Battle Problem was an offensive/defensive exercise involving an attack on the Panama Canal. By this time "The Covered Wagon," as she was affectionately known, had added all current carrier aircraft to her complement.

Douglas' DT-2 torpedo bomber has the distinction of flying the first tactical missions from U.S. Navy carriers. During the January 1924 Fleet Battle Problem off Panama, *Langley's* innovative XO, CDR Ken Whiting, successfully launched two camera-equipped DT-2s to simulate an attack against land targets. USN photo, courtesy CAPT J.P. Cartwright, USN(Ret)

Dropping her experimental status in 1924, *Langley* joined the Battle Fleet on the West Coast, and by May 1928 was operating with her full complement of 34 airplanes. Although not yet termed, this was the beginning of the Navy's air groups. USN photo/National Archives 80-G-424475

Still in an experimental category, *Langley* was not assigned to either fleet and was to be involved only as an observer of the event. Ever the innovator, Whiting launched two Douglas DT-2 torpedo planes carrying crew members armed with cameras to convince the naval hierarchy that carrier aircraft were capable of tactical missions against land targets. The aircraft were flown by LT Stanton H. Wooster with Aviation Machinist Mate Chief James H. Densback, and LT Jackson R. Tate with an unidentified crewman. Further launches of *Langley* aircraft against the Scouting Fleet at Colón-Cristóbal were made to impress upon the fleet commanders the vulnerability of ships in port. The commanders were not impressed and felt that the airplanes were only "stunting." Not until the 7 Dec 1941 Japanese attack on Pearl Harbor would all U.S. naval officers finally recognize the devastating potential of carrier aircraft against an immobile fleet trapped in port.

Returning to Pensacola during spring 1924, *Langley* continued with her experiments and training. Whiting, Griffin, Pride, and Pennoyer (now a naval aviator) were ordered to Washington for further assignment as ship's company for *Saratoga*, which, along with her sister battle cruiser, *Lexington*, was undergoing conversion to become the next carriers. In the fall, *Langley* passed through the canal she had successfully "attacked" and joined the Battle Fleet on the West Coast 29 November. Having shed her experimental status, she was now a fully operational fleet unit. She became flagship of Commander, Aircraft Squadrons, Battle Fleet, CAPT Stanley E. Maser.

Langley operated in the California and Hawaiian areas for the next 12 years, while also participating in fleet operations in the Caribbean, as well as making visits to the East Coast. During January 1925, CV-1 began her first operations with a fleet squadron, Fighting Squadron Two. The next year, LCDR Gerry Bogan, CO VF-1B, and LCDR Daniel W. Tomlinson, CO VF-6B, led a flight

During 1936-37 *Langley* was converted to a seaplane tender, resulting in the removal of a portion of her flight deck. Reclassified AV-3, she served with the Scouting Fleet until the outbreak of WWII. She was sunk off Java 27 Feb 1942 while ferrying Army P-40 fighters to Tjilatjap. USN phot

Saratoga **(CV-3) at sea with her air group shortly after her 16 Nov 1927 commissioning. Converted from partially constructed battle cruisers,** *Sara* **and** *Lexington* **(CV-2) were giants of their time at 33,000 tons empty displacement and with 866-foot flight decks. USN photo**

of 31 F2B-1s from their squadrons 6 Apr 1928 to land aboard *Langley* in Coronado Roads off North Island. This was the first time an "air group" flew aboard a U.S. carrier at sea. By 1928 *Langley* was operating with four fleet squadrons, which combined with her utility unit for a designed total of 34 airplanes. On 25 October 1936, as a result of the naval treaty limitations on aircraft carrier tonnage, she began conversion to a seaplane tender. Part of her flight deck, which had been the site of so many innovations for naval aviation, was removed. In April 1937, reclassified AV-3, she became a unit of the Aircraft Scouting Force, with which she operated until the outbreak of WWII.

On 8 Dec 1941 (Far East time), she was at anchor off Cavite in the Philippines during the Japanese attack. Moving to Australia, she participated in anti-submarine operations until 11 Jan 1942. On 22 Feb *Langley* departed Freemantle en route to Tjilatjap, Java, to deliver P-40 fighters. On the morning of the 27th she was unsuccessfully attacked twice by Japanese bombers. However, a third assault produced five hits on the Covered Wagon, and she was abandoned to be ultimately sunk by torpedoes from her escorting destroyers. Thus came to a tragic close the career of the ship that took U.S. naval aviation to sea.

Lexington and Saratoga

Following WWI, there was a growing tide of isolationism in the U.S. This trend, coupled with a desire in this country and other nations to reduce government spending, prompted President Warren G. Harding to invite representatives from Britain, Japan, France, and Italy to participate in a disarmament conference held in Washington, DC, beginning in November 1921. On 6 Feb 1922 a naval limitation treaty was signed, which established a 10-year moratorium on building capital ships and allowed the U.S. to build a series of aircraft carriers not to exceed a total of 135,000 tons. Britain was allowed the same tonnage, Japan 81,000 tons, with France and Italy 60,000 tons each. No individual new carrier was to exceed 27,000 tons light, with the exception that each country was allowed two existing or already under construction carriers of up to 33,000 tons

Lexington **(CV-2) joined her sister ship** *Saratoga* **14 Dec 1927. Both ships were capable of carrying 120 aircraft, but normally operated with 70-80. USN photo**

each. These parameters, coupled with the London Naval Treaty of 1930, would dictate the design and capabilities of U.S. carriers until expiration of the latter treaty in 1936.

On 1 Jul 1922 Congress authorized funds to convert two battle cruisers, on which construction had been started in February but halted at about one-third completion, to be converted to aircraft carriers. These ships became the 33,000-ton *Lexington* (CV-2) and *Saratoga* (CV-3), and they would introduce the air group concept to naval aviation. The new carriers were giants of their time, with 866-foot flight decks.

Saratoga, 35.4 percent complete at start of conversion, was the first of the two sister ships to reach the fleet, commissioned 16 Nov 1927, with *Lexington* (33.8 percent at conversion) following 14 December. Both ships were capable of carrying as many as 120 aircraft, including the ship's utility unit and spares stowed in the hangar bay overhead. However, this number was not practical; Lex and Sara normally operated prior to WWII with four squadrons plus the utility unit, for a total of 75-80 aircraft. Over the years the total number of aircraft and squadrons operated by *Lex* and *Sara*, as well as future classes of carriers, would vary according to the size of the class, current aircraft, and air group composition. Interestingly, the average number of aircraft assigned for large carriers has remained fairly close to the first air group totals.

In mid-1929, the first complete year of *Lexington's* and *Saratoga's* fully operational fleet status, their air group compositions were:

Saratoga
June 1929

VF-1B	19 F2B-1, 1 FU-2
VB-2B	18 F3B-1, 1 FU-2
VS-2B	12 O2U-2
VT-2B	27 T4M-1
Total	78
Utility Unit	2 FU-2, 3 02U-1, 1 O2U-2

Lexington
June 1929

VF-3B	12 F3B-1, 1 FU-2
VB-1B	15 F3B-1
VS-3B	12 O2U-2
VT-1B	28 T4M-1
Total	68
Utility Unit	1 UO-1, 3 O2U-1

Ranger

The U.S. Navy's first designed and built-from-the-keel-up carrier was *Ranger* (CV-4), which joined the fleet in 1934. Commissioned 4 June, she was a miniature 14,500 tons (700 tons more than her planned 13,800) and had a 710-foot flight deck. Her size was dictated by plans of the General Board of the Navy to build five carriers in five years and remain within the 69,000-ton limitation remaining from the disarmament treaty authorization. Additionally, in the late '20s, carrier design philosophy slanted toward smaller carriers and, with *Lexington* and *Saratoga* not in the fleet until after *Ranger's* size had been decided, "lessons learned" from the new ships could not influence the thinking of naval planners.

Ranger began full fleet operations in 1935 with her air group that had 98 aircraft assigned. However, in reality, she operated more with her design capability of 75. As a result of her size, she did not have the larger torpedo bomber aircraft that Lex and Sara had. In mid-'35 her air group composition was:

Ranger
June 1935

VF-3B	18 F2F-1, 19 F4B-1,1 O3U-1, 1 SU-2, 1 SF-1
VB-3B	18 BG-1
VB-5B	15 BF2C-1,1 O2U-2, 1 SF-1
VS-1B	20 SU-4
Total*	76
Utility Unit	3 O3U-3, 2 JF-1

*(At this time, VF-3B was transitioning from Boeing F4B-1s [not included in total] to the Grumman F2F-1, resulting in an assignment of two squadrons' worth of airplanes. Additionally, as seen above, some squadrons had a few off-type "hacks" assigned for utility and cross-country navigation purposes that were normally left ashore when the squadron went to sea.)

With a size comparable to the later WWII escort carriers, *Ranger* was prevented from operating with the Fast Carrier Task Forces of

***Ranger* (CV-4) was the U.S. Navy's first built-from-the-keel-up carrier. Naval treaty agreements limited her to 14,500 tons with a 710-foot flight deck. She joined the fleet 4 Jun 1934 and was capable of operating 75 aircraft of the time. USN photo/National Archives 80-G-43167**

Name-ship of her class, *Yorktown* (CV-5) was the first of three ships of her class weighing in at 19,900 tons. She normally operated 60-70 aircraft from her 802-foot flight deck. USN photo

1943-45. After participating in the pre-war Neutrality Patrol mission and several Army aircraft ferry missions to Africa, she conducted combat operations during the invasion of North Africa in November 1942 and at Vestfjord, Norway, in October of the following year. On 3 Jan 1944 she was designated a training carrier and operated as such off the East Coast until reporting to San Diego 25 Jul 1944. She remained in a training status on the West Coast until the end of the war, when she returned to the East Coast and continued qualifying pilots. After a total of 92,262 landings, she was decommissioned 18 Oct 1946 and scrapped. Her air group composition during the North African and Norway campaigns reflect the tremendous evolution of carrier aircraft capabilities between 1935 and 1942, as the Navy moved from all-biplane air groups to all-monoplane:

Ranger Air Group		Ranger/CVG-4	
North Africa—Nov 42		Norway—Oct 43	
CDR David B. Overfield, CAG		CDR Joseph A. Ruddy Jr., CAG	
VF-9	27 F4F-4	CAG	1 TBF-1
VF-41	28 F4F-4	VB-4	30 SBD-5
VS-41	18 SBD-3	VF-4	23 F4F-4
VT-4*	10 TBF-1	VT-4	9 TBF-1
Total	83	Total	63

*(VT-4 remained in U.S., one TBF assigned to CAG accompanied Ranger Air Group to Africa.)

And Then There Were Eight

Early operations of *Lexington* and *Saratoga* convinced the General Board of the need for carriers larger than *Ranger*, even before she was commissioned. By 1932, it recommended that future carriers be a minimum of 20,000 tons light. Consequently, two ships,

***Enterprise* (CV-6) was commissioned 12 May 1938. She would become the most decorated carrier of WWII, while participating in nearly every battle of the Pacific war. USN photo**

Last of the *Yorktown*-class, *Hornet* (CV-8) was doomed to have the shortest service life. Commissioned 20 Oct 1941, she was sunk 26 Oct 1942 during the Battle of Santa Cruz. USN photo

Yorktown (CV-5) and *Enterprise* (CV-6), were authorized. *Yorktown* was commissioned 30 Sep 1937 at 19,900 tons with an 802-foot flight deck, as was *Enterprise*, which followed 12 May 1938. *Hornet* (CV-8), also a *Yorktown*-class carrier, came after the naval treaty limitations had expired and joined the fleet 20 Oct 1941 slightly larger at 20,190 tons and with an 814-foot flight deck. The design air group capacity for these ships totaled 96 airplanes, but in reality they normally carried 60–70. Their earlier air group complements were:

Yorktown Air Group January 1939		Enterprise Air Group June 1939	
LCDR Miles R. Browning, CAG		LCDR Giles E. Short, CAG	
VB-5	18 BT-1, 1 03U-3	VB-6	18 BT-1
VF-5	18 F2F-1, 1 SBC-3	VF-6	18 F3F-2, 2 SBC-3, 2 SU-2
VS-5	18 SBC-3	VS-6	18 SBC-3
VT-5	18 TBD-1	VT-6	18 TBD-1
Total	74	Total	76
Utility Unit	2 J2F-2, 3 O3U-3 1 SBC-3	Utility Unit	1 J2F-1, 1 J2F-3, 3 03U-3

Hornet Air Group April 1942 (actual strength) LCDR Stanhope C. Ring, CAG	
VF-8	18 F4F-3
VB-8	18 SBD-3
VS-8	17 SBD-3
VT-8*	9 TBD-1, 14 TBF-1
Total	62

*(VT-8, transitioning to TBFs, was operating as a split unit with its *Avengers* at Norfolk.)

Wasp (CV-7) was built under the constraints of the remaining tonnage left in the naval disarmament authorization. As a result, in terms of size, she was a virtual throwback to the *Ranger* design, displacing only 14,700 tons light, with a 727.5-foot flight deck when commissioned 25 Apr 1940. Her design was compromised in order to give her maximum aircraft capacity, which resulted in lesser armor and torpedo protection. These deletions would ultimately contribute to her loss from Japanese submarine torpedoes off Guadalcanal 15 Sep 1942. Wasp Air Group was authorized the following units and aircraft as of 31 Dec 1941.

Wasp Air Group 31 Dec 1941 LCDR Wallace M. Beakley, CAG		
VF-71	18	F4F-3
VF-72	18	F4F-3, 2 SB2U-2, 1 SNJ-3
VS-71	4	SB2U-1, 2 SB2U-2, 2 TBD-1
VS-72	18	SB2U-2
Total	65	
VT-7*	6	TBF-1
Utility Unit	1 1	J2F-2, 1 J2F-3, 1 O2SU-1, SON-1, 1 SB2U-2 (CAG)

*(First TBFs accepted in the fleet 25 Mar 1942 and did not operate from *Wasp* until August 1942.)

The Magnificent Essexes

The ships that truly defined carrier aviation and were the primary instrument in winning the Pacific sea war were the *Essex*-class carriers. The lead ship of the class, which ultimately numbered 26 hulls (two were not completed), was authorized at the same time as *Hor-*

Treaty limitations caused *Wasp's* (CV-7) design size to be held at only 14,700 tons, a virtual throwback to *Ranger*. Weight-saving compromises in her design contributed to her loss after being torpedoed off Guadalcanal 15 Sep 1942. USN photo

The ships that defined carrier aviation were the 24 *Essex*-class ships built between 1941 and 1950. Their flight decks varied from 860 to 878 feet, and they were designed to carry as many as 100 aircraft. USN photo/National Archives 80-G-68097

Franklin (CV-13) was heavily damaged by explosions set off on her hangar and flight decks by bombs from two Japanese aircraft off Kyushu 19 Mar 1945. Survival after the severe damage sustained by *Franklin* and sister ship *Bunker Hill* (CV-17) attest to the design quality of the *Essex*-class, as well as to the perseverance of their crews. USN phot

net under the treaty limit. World events and the treaty expirations soon allowed a change in *Essex's* original design size of 20,100 tons to 27,100 at commissioning. Modifications of design and alterations changed the characteristics of the ships that followed, resulting in flight deck lengths of 860-878 feet. The measure of the true value and efficient design of the *Essex* class is the fact that, although modernized, this class would operate in front-line service from 1942 to 1976. One ship, *Lexington* (CV-16), served from 17 Feb 1943 until 8 Nov 1991, her final 30 years as the training carrier at Pensacola. Although several *Essexes* were severely damaged during WWII, none was sunk—another tribute to their design, as well as their crews.

Air group composition is dictated by size of aircraft and design of the ship (the reverse is also true), as well as tactical requirements of the current situation of the carrier. Although originally designed to carry a 74-plane air group, wartime needs required a variety of *Essex*-class compositions. A typical early *Essex*-class air group is shown here when the emphasis was placed on attack aircraft for strikes against enemy targets. Later, after the October 1944 introduction of the *kamikaze* threat, the emphasis was shifted to fighters for fleet protection, resulting in an increase in the total number of aircraft. Additionally, the strike-fighter concept was strengthened (begun by early dive-bombing experimentation in the late '20s) by bringing Vought F4U Corsairs to the carriers as fighter-bombers with the newly formed bomber fighter (VBF) squadrons. In fact, fighter squadrons always had an additional limited bombing mission, but not until the advent of VBF squadrons did this role equal the fighter mission.

Antietam (CV-36) enters Yokosuka Harbor, Japan, during her 1951-52 Korean War deployment with CVG-15. During the Korean War, the *Essex*-class carriers carried more than 100 aircraft, but could not effectively operate that number. Consequently, some aircraft were left in Japan until combat losses made room for them. PH1 Steve Kanyusik

Yorktown (CV-10)/CVG-5	
Nov 1943	
CDR Charles L. Crommelin, CAG	
VF-5	36 F6F-3
VB-5	36 SBD-5
VT-5	18 TBF-1
Total	90

Essex (CV-9)/CVG-83	
Mar 1945	
CDR Harmon T. Utter, CAG	
VF-83	34 F6F-5/3/N/P
VBF-83	36 F4U-1D
VB-83	15 SB2C-4
VT-83	15 TBM-3
Total	100

As previously mentioned, the *Essexes* would serve in a combatant status from 1943 to 1976. During that time, U.S. Navy carrier aircraft that were routinely operated from their decks would evolve from the 380-mph Grumman F6F *Hellcat* to the Vought F-8 *Crusader* with a maximum speed in excess of 1,100 mph. The *Hellcat* weighed 9,238 pounds empty, and the F-8 weighed in at 17,836 pounds empty. Landing approach speeds almost doubled, from around 85 mph to near 155. In attack aviation, the Douglas SBD-5 *Dauntlesses* of 1943 were 6,675 pounds empty, and the Douglas A-3B *Skywarriors* of 1976 were 39,409 pounds. Their maximum gross weight was 82,000 pounds, nearly the weight of nine empty Hellcats. Yet, the *Essexes* handled them all. True, it took major modernization alterations to keep the old girls up to date with the new aircraft, but the fact that the ships were able to accept the modernizations proves the efficacy of the basic design begun in the late '30s.

"Combustible, Vulnerable, and Expendable"—the CVEs

As early as 1927 LCDR Bruce G. Leighton (NA #40) had predicted in a study of "light carriers" virtually every application that would be assigned to the escort carriers of WWII. The impetus, however, that brought about the "jeep" carriers was the need for a small transport carrier to ferry aircraft overseas. Between 2 Jun 1941, when *Long Island* (AVG-1), the first escort carrier, was commissioned, and 15 Jan 1946, when *Palau* (CVE-123), the final escort carrier to enter service was commissioned, 86 CVEs joined the fleet in six different classes. Their sizes varied from *Long Island* (a converted C-3 cargo ship, *Mormacmail*) at 7,886 tons with a 418-foot flight deck, to the built-as-carriers *Commencement Bay*-class at 18,908 tons with 501-foot flight decks.

The CVE's (originally AVG until 20 Aug 1942, then ACV until 15 Jul 1943) aircraft transport role quickly expanded to include such missions as anti-submarine patrol, convoy escort, scouting, and reconnaissance, as well as offensive missions against enemy ships and land bases. The diminutive warriors earned their status as full-fledged members of the carrier team, fighting alongside their bigger *Essex*-class counterparts.

Although sometimes morbidly referred to by their crews as "Kaiser Coffins" because of their light construction, CVEs proved their combat effectiveness, initially in the Atlantic and Caribbean, and are generally credited with breaking the back of the German U-boat Wolfpacks. Besides their convoy escort/ASW role, the escorts participated in the November 1942 North African invasion and the August 1944 Southern France campaign. The four *Sangamon*-class CVEs, converted from fleet oilers, operated off North Africa with *Ranger* during the first days of Operation Torch, three of them in an offensive role covering the invasion forces with TBFs, SBDs, and F4Fs. The fourth carrier, *Chenango* (CVE-28), ferried Army P-40s. Later, during the Southern France invasion, *Tulagi* (CVE-72) and *Kasaan Bay* (CVE-69) helped spearhead the Allied assault. In the ASW category, Atlantic Fleet CVE task groups were credited with participating in the sinking of 55 Axis submarines, 30 by aircraft alone. These totals include one Vichy French boat and one unfortunate Japanese transport submarine sunk while ferrying material to Germany.

In the Pacific, the escort carriers' combat operations ranged from the Solomons in August 1942 to the Japanese homeland in 1945. Although the Pacific submarine threat was less severe than in the Atlantic, Pacific Fleet CVE task groups sank 11 Japanese boats. Their aviators flew alongside the more heralded Fast Carrier Task Force aircrews throughout the Pacific War, performing the same missions. The epic story of Taffy Three off the Philippines in October 1944, where a tiny CVE task unit fought a major surface engagement with Japanese battleships and cruisers, is legendary. (See "*The Little Giants*" and "*Hunter-Killer*," Wm. T. Y'Blood, Naval

Born from a need for a fleet of small transports, the CVE escort carriers proved as valuable as their larger contemporaries in a variety of missions. *Santee* (CVE-29), with VGF-29 F4F-4s, along with VGS-29 TBF-1s and SBDs, was one of 86 CVEs to serve with the U.S. Navy in six different classes. USN photo

Institute Press 1987 and 1983.) CVE pilots were also credited with 565 1/2 aerial victories, all but 14 in the Pacific. Five pilots earned the rank of ace. Six escort carriers were lost to enemy action during the war, *Bismarck Sea* (CVE-95), *Block Island* (CVE-21), *Gambier Bay* (CVE-73), *Liscome Bay* (CVE-56), *Ommaney Bay* (CVE-79), and *St. Lô* (CVE-63).

As stated, the CVE program began as an aircraft ferry system, but the urgent need for protection of the Atlantic convoys almost immediately brought squadron assignment to the carriers. Initially, they did not carry air groups as such, nor did they have an air group commander. Early ideas for the ships even included assigning blimps to them, but the plan was quickly discarded in favor of traditional carrier aircraft types. The first squadron assignments were not systematic, and varied according to the envisioned employment of the carrier at the time. On 20 Mar 1941, *Long Island's* first squadron, VS-201, was formed with nine Curtiss SOC-3A *Seagulls*, one SOC-1A, and seven Brewster F2A-3 Buffalos aboard her by February 1942. CVE aircraft complements varied from 21 for *Long Island* to 33 for the *Commencement Bay*-class ships that, by the end of the war, were operating with Marine air groups consisting of one fighter squadron and a TBM squadron.

The designation of CVE air groups is somewhat confusing. Essentially, some CVE squadrons in existence (designated VGS in 1942) became composite squadrons (VC) 1 Mar 1943 and were considered air groups. They carried this designation until all were disestablished after the war. Again, depending on the class of ship, composite air group size varied, but the aircraft types assigned were fairly consistent. A typical Atlantic operational ASW composite air group in early 1943 was VC-9 assigned to *Bogue* (CVE-9), with 12 F4F-4s and eight TBF-1 *Avengers* embarked. Although considered as one unit, the VC outfits really operated as two squadrons, because of the differences in aircraft types and missions.

Other escort carrier squadrons were formed into escort carrier air groups (CVEG), the first of which was *Sangamon's* (CVE-26) Escort Carrier Air Group 26, with VC-26's seven SBD-3s and eight TBF-1s, along with VF-26's nine F4F-4s. In the beginning, the escort carrier air group number coincided with the carrier's hull number, but this system did not last long because of operational necessity requiring transfer of air groups from one ship to another. However, the CVE air group compositions remained relatively consistent until late 1944, with the exception of the SBDs being replaced by additional TBFs/TBMs. An increase in the CVE groups occurred in fall 1944. By 7 Apr 1945, *Anzio* (CVE-57) was in the Western Pacific with VC-13 embarked, consisting of 16 FM-2 *Wildcats* and 12 TBM-3s, while the larger *Santee* (CVE-29) had VC-24 embarked with 17 F6F-5s and one F6F-5P of VF-25, as well as VT-25's 11 TBM-3s and one TBM-3P. Also in April 1945, newly formed Marine carrier air groups began showing up on the big *Commencement Bays*; the first time in history for all-Marine carrier air groups. Originally authorized at 18 VF and 12 VT, typical was the new *Gilbert Islands* (CVE-107) with MCVG-2's complement:

7 April 1945
LCOL William R. Campbell, CAG

VMF-512	4	F3A-1
	2	FG-1A
	23	F4U-1D
VMTB-143	22	TBF-1
	2	TBM-3
Total	53	

The CVEs continued their ASW service after WWII, helping develop new aircraft for the mission and pioneering the development of helicopters aboard carriers. The Marines rejoined them in 1951 to fight in Korea.

The CVLs

The early-war requirement for more carriers to quickly join the fleet brought about the conversion of nine *Cleveland*-class light cruisers to the *Independence*-class carriers (CV until 15 Jul 1943, then CVL, for small carrier, not light carrier as usually seen). Following the war, two larger designed-as-carrier CVLs, *Saipan* (CVL-48) and *Wright* (CVL-49) joined their surviving predecessors. One ship, *Princeton* (CVL-23), was sunk as a result of a dive bomber attack

***Independence* (CVL-22) was the first of nine WWII carriers built from cruiser hulls. VC-22's SBDs, seen here 15 July 1943, were replaced by TBF-1s by September. USN photo, courtesy A.D. Baker**

Two larger, designed-and-built-as-CVLs were commissioned after the war. *Wright* (CVL-49), with VMA-211 embarked, is anchored in Guantanamo Bay, Cuba, during Jan 1954. Michael J. Kolaska

24 Oct 1944 off the Philippines, the only Fast Attack Carrier lost during WWII. (Some of the prewar carriers sunk before the Fast Attack Carrier terminology evolved would have qualified in this category.) The *Independence*-class CVLs displaced 10,662 tons light and had a 552-foot flight deck, while the *Saipan*-class displaced 14,500 tons with 611-foot flight decks. The *Independences,* however, had a pilot's attention-getting 73-foot width, seven feet narrower than the most narrow CVE!

The cruiser-conversion CVLs came about as a result of prodding by President Franklin D. Roosevelt, who also fostered the CVE program. As a result of his former service as Assistant Secretary of the Navy, Roosevelt retained an active interest in naval affairs throughout his presidency. Name-ship of the class, *Independence* (CVL-22), was commissioned 1 Jan 1943 and deployed to the Central Pacific in July 1943 with CVLG-22, comprised of:

Independence (CVL-22)/CVLG-22
LCDR James McC. Peters, CAG
July 1943

VF-22	12	F6F-3
VC-22	9	TBF-1
	9	SBD-5
Total	30	
VF-6 DET	12	F6F-3 (temporarily assigned)

The SBDs were soon replaced on the CVLs by TBF-1s. On the CVs, they were converted to Curtiss SB2C Helldiver squadrons, much to the chagrin of SBD aircrews. At war's end, *Independence, Belleau Wood* (CVL-24), *Cowpens* (CVL-25), *Monterey* (CVL-26), *Bataan* (CVL-29), and *San Jacinto* (CVL-30) were all on station in WestPac. The remaining two sisters, *Langley* (CVL-27) and *Cabot* (CVL-28), had recently returned home from the combat area. *Monterey's* CVLG-34 was typical at that time with:

Monterey (CVL-26)/CVLG-34
Aug 1945
LCDR Robert W. Conrad, CAG

VF-34	24	F6F-5
	1	F6F-5P
VT-34	8	TBM-3E
	1	TBM-3P
Total	34	

As with some of the CVEs, a few CVLs served in postwar ASW development roles, and one, *Bataan*, served in Korea with Marine squadrons.

The Midway Class

Design of the *Midway*-class carriers began as early as 1940, with discussions leading to nine proposals for a new follow-on carrier to

***Midway* (CVA-41) at anchor during her 1954 Mediterranean cruise with CVG-6. Design of the *Midway* class began in 1940; they were the first U.S. carriers to have armored flight decks. CAPT. J.A. Overn, USN(Ret.)**

the *Essexes*, none of which had yet been built. Early-war carrier losses and damage led to revisions of the design selected, a proposal that resulted in the building of the largest carrier in history at the time. The *Midways* were also the U.S. Navy's first carriers with armored flight decks, a result of the survival of HMS *Illustrious*, whose armored flight deck saved her from destruction by German dive bombers in June 1941. Six ships of the class were authorized (CVBs 41-46), but three were canceled by the end of the war. Contract for construction of *Midway* was let 7 Aug 1942.

Midway (CVB-41), the name-ship and first of the three CVBs (for large aircraft carrier, not battle carrier, as often reported), was commissioned 10 Sep 1945, eight days after Japan's signing of unconditional surrender on board *Missouri* (BB-63) in Tokyo Bay. The three ships of the class—*Midway, Franklin D. Roosevelt* (CVB-42), and *Coral Sea* (CVB-43)—varied somewhat in design details. *Coral Sea*, not commissioned until 1 Oct 1947, had the advantage of design modifications resulting from lessons learned after major battle damage sustained in 1944-45 by *Franklin* (CV-13) and *Bunker Hill* (CV-17). At commissioning, they displaced 48,000 tons light and had 924-foot flight decks. As with the *Essex* class, during their service life, the *Midways* would receive major modifications to keep them abreast of the newer, faster, and heavier aircraft that were produced during their service careers. Eventually, *Midway* would put on a matronly 13,000 tons, mostly as a result of her angle deck and increased flight deck area.

Wartime deployment design complement for the *Midway*-class air groups was 136 aircraft, a vast increase over the *Essexes*. However, in reality, they rarely operated with more than 110 or so. (*Midway* did have 121 during her shakedown cruise.) By the time the CVBs became fully operational in the late '40s, their air group compositions were essentially the same as *Midway's* CVBG-1 on 1 Feb 1948:

Midway/CVBG-1 1 Feb 1948
CDR R. Emmett Riera, CAG

CAG	1	F4U-4B
VF-1B	32	F4U-4B
VF-2B	32	F4U-4B
VA-1B	24	AD-1
VA-2B	24	AD-1
Total	113	

Although the increased size of an aircraft carrier allows it to carry more aircraft, other operational limitations affect the numbers. The location of catapults, as well as elevators, coupled with the restrictions of moving aircraft on the hangar and flight decks, reduce the actual number of aircraft that can be handled at one time. Thus, the *Midway* class could operationally handle only about 10 more aircraft during flight ops than could the *Essexes*.

Midway was the first and last of her class, surviving until decommissioned 11 Apr 1992; *Franklin D. Roosevelt* preceded her 1 Oct 1977, and *Coral Sea* 30 Apr 1991. During 7 Oct 1990–17 Apr 1991 *Midway* and her Air Wing Five, led by CAPT James M. Burin, flew missions against Iraq from the Persian Gulf in Operations Desert Shield and Desert Storm. CVW-5's composition reflects the advances of carrier aircraft and technology during *Midway's* service life.

Midway/CVW-5 Jan 1991
CAPT James M. Burin, CAG

VFA-195	12	F/A-18A
VFA-151	12	F/A-18A
VFA-192	12	F/A-18A
VA-185	8	A-6E/KA-6D
VA-115	8	A-6E/KA-6D
VAW-115	4	E-2C
VAQ-136	4	EA-6B
HS-12	6	SH-3H
Total	66	

Once the Navy's largest carriers and, for the first decade of their lives, the world's largest warships afloat, by the 1990s the *Midways* had become the Navy's smallest carriers. *Midway* was technically able to operate the large F-14 *Tomcat*, but could not do so efficiently. Although some limited F-14 operations were conducted, *Tomcats* were never assigned to *Midway*-class carriers. However, Midway served her country well for nearly 50 years.

The Essex-Class Conversions

By late 1949, Chief of Naval Operations ADM Forrest P. Sherman had authorized a plan to modernize the Navy's *Essex*-class carriers. During the three years of the Korean War (June 1950–July 1953),

During its 47-year career, the *Midway* class received major improvements, making each ship virtually one-of-a-kind. *Midway*, with CVW-5 embarked, is off Japan 18 Aug 1984 with a new bow, flight deck, and matronly 13,000 tons more weight. Bob Lawson

Construction of *Oriskany* (CV-34) was suspended at the end of WWII. Completed in 1950, she was the first *Essex*-class with Project 27A improvements. Bud Sheehan

the active carrier strength rose from 15 of all types (3 CVBs, 4 CVs, 2 CVLs, and 6 CVEs) to 39.

The basic Ship Characteristics Board (SCB) 27 program had been planned as early as 1946 for *Oriskany* (CV-34), construction of which had been suspended in 1945. On 1 October 1947, construction was resumed under Project 27A, and she was commissioned 14 Oct 1950. The major improvements provided by 27A were new Mk 5 arresting gear and more powerful H-8 hydraulic catapults. The flight deck was strengthened to handle the new heavy attack aircraft with planned weights of 52,500 pounds. Elevator capacity was increased to handle 30,000 pounds, and fuel storage was expanded to service the thirsty new jets. The hull was blistered (widened) to accommodate the 6,000-ton increased displacement. Additionally, improved radar provided increased self-defense capabilities. The most visible change was the redesigned island and removal of the twin 5-inch gun mounts from the flight deck. The loss of these mounts was partially compensated by the addition of four single 5-inch mounts in sponsons added to the starboard side to maintain adequate anti-aircraft protection. On the flight deck, small retractable jet-blast deflectors were added behind the two catapults to protect flight deck crews.

Nine *Essexes* received the SCB 27A refit, all completed by November 1953. By 1957, a total of 15 *Essex*-class carriers would receive one of two modernizations: 27A or 27C.

During the late '40s and early '50s, the British, who for three decades had led the world in many aircraft carrier innovations, had been working on three major developments that would virtually save the future of carrier aviation. Jet aircraft were beginning to enter naval service, and their increased landing speeds, as well as operating weights, were threatening to bring landing accident rates to unacceptable levels. Three British programs adapted by the U.S. Navy reversed this trend—steam catapults, the mirror landing system, and the angled deck. *Hancock* (CVA-19) (CVs and CVBs became CVA "attack carriers" 1 Oct 1952) was selected for the first U.S. carrier steam catapult installation. *Intrepid* (CVA-11) and *Ticonderoga* (CVA-14) followed, the only three ships to have C-11 steam catapults as straight-deck carriers under the new Project 27C program.

Following tests during May 1951 aboard *Midway* with a painted simulated angled deck, *Antietam* (CV-36) entered the yards to receive an experimental "canted deck" (changed to angled deck by

The introduction of new, faster jets to carrier aviation in the early '50s brought an alarming increase in landing accidents. Amazingly, test pilot CDR George C. Duncan survived this ramp strike aboard *Midway* (CVB-41) on 23 Jul 1951 to become CAG 5 during 1953–54. USN photo, National Archives 80-G-432165

Serving during 1950–76, *Oriskany* received further modifications, allowing her to operate the latest jet aircraft. CDR Pete Clayton, USN (Ret)

CNO directive in 1955). Successful completion of *Antietam's* test program 1 July 1953 led to a third modernization program, Project 125, to further improve the *Essexes*.

Eight of the nine Project 27A carriers received angled decks and enclosed "hurricane" bows under the 125 program. *Lake Champlain* (CVA-39) remained a 27A until her decommissioning in 1966, the "last of the straight decks." Additional Project 125 alterations included the removal of the aft centerline elevator and installation of a new elevator on the aft starboard deck-edge. The 125 program did not provide for steam catapults, leaving these ships with the less powerful and somewhat less reliable hydraulic H-8s. This factor contributed to the eight 27A carriers ultimately being reclassified CVS—anti-submarine warfare support carriers—to operate prop-driven Grumman S2F Trackers and helicopters.

The final modernization of the *Essexes*, or "27 Charlies," as they were popularly called in the fleet, provided all of the refinements of the 27A and 125 programs, plus C-11-1 steam catapults. *Lexington, Bon Homme Richard* (CVA-31), along with *Shangri-La* (CVA-38) all received the 27C alterations in one overhaul. By 1957 *Intrepid, Tico*, and *Hancock* had also completed the 125 program. *Oriskany* became "one-of-a-kind" when she underwent a special 125A modernization during 1958-59 that essentially gave her all the elements of the 27C conversion with further upgraded features. In 1955, the first U.S. Navy mirror landing system was installed on board *Bennington* (CVA-20) for evaluation. Ultimately, all CVAs and CVSs received the mirror landing system, or its Fresnel lens optical landing system successor.

The three *Midway*-class carriers received extensive modernizations similar to the *Essex*-class conversions. In 1954, FDR began Project 110, a program similar to SCB 27C that added a third catapult in the angle deck. *Midway* followed in 1955. *Coral Sea* was the only carrier to receive a SCB-110A, which virtually rebuilt the ship. *Midway* was also essentially reconstructed during 1966-70. Cost overruns on *Midway* sealed *Roosevelt's* fate for early retirement when she received only limited modernization in 1969. After these conversions, each of the *Midways* was essentially a one-class ship.

***Antietam* (CVA-36) received an experimental angled deck in 1953, a feature added to other *Essexes* under Projects 125 and 27C. Never modernized, *Antietam* was reclassified as an antisubmarine warfare support carrier and went to Pensacola as the training carrier 1957-62. USN photo**

Although receiving the original Project 27 modernization, *Lake Champlain* (CVA-39) was not further modified and became the last of the straight-deck *Essexes* as CVS-39. CAPT Ian F. Brown, USN(Ret)

Intrepid (CVS-11) underway 10 Jan 67 with 27C steam catapults and aft starboard deck-edge elevator modifications. She was one of three straight-deck "27 Charlies," receiving P125 improvements in a second fitting. High tempo Vietnam war ops are evident, as *Intrepid* is shown during workups for her April 1967 Vietnam deployment having only returned from there 21 Nov 66. Although reclassified CVS, she operated as an attack carrier. USN photo

Eight *Essexes* received 27A and 125 modernizations in two fittings. *Hornet* (CVS-12) shows the angled deck and enclosed hurricane bow of the 125 program. However, she and seven sisters did not get C-11 steam catapults (note absence of extended bridle arresters on bow) of the 27C program, thus causing their ultimate reclassification as CVS ASW carriers, unable to operate the later, heavier jet aircraft. Bob Lawson

A comparison of the standard *Essex*-class complement at the outbreak of the Korean War and that of a modernized *Essex* during the Vietnam War is an interesting study of the changes in carrier aviation accompanying the furious pace of technology during the '50s and '60s.

Valley Forge (CV-45)/CVG-5
June 1950
CDR Harvey P. Lanham, CAG

		CAG Unit	
VF-51	18 F9F-3	3	F4U-5N
VF-52	18 F9F-2	2	F4U-5P
VF-53	18 F4U-4B	2	AD-3N
VF-54	18 F4U-4B	1	AD-3Q
VA-55	15 AD-4	3	AD-3W
		3	AD-4Q
		1	HO3S-1
Total		102	

Bon Homme Richard (CVA-31)/CVW-5
Jan 1968
CDR D. Bruce Miller, CAG

VF-51	12 F-8H	VFP-63 Det 31	4 RF-8A
VF-53	12 F-8E	VAW-111 Det 31	3 E-1B
VA-93	14 A-4F	VAW-13 Det 31	3 EKA-3B
VA-94	14 A-4E	HC-1	3 HU-2C
VA-212	14 A-4F		
Total	79		

Another interesting comparison is the number of aircraft actually operated and method of utilization. During Korea, the *Essexes*

Lexington **(AVT-16) steams in the Gulf of Mexico as she qualifies student naval aviators 14 Jul 1982. "The Blue Ghost" of WWII fame served from 1943 until 8 Nov 1991; her final 30 years as the training carrier at Pensacola. Bob Lawson**

could not effectively operate 102 aircraft; consequently, some planes were dropped off in Japan to return aboard after combat losses made room for them. In Vietnam the air wings deployed with one extra fighter and two bombers per squadron to offset combat losses. As during Korea, some aircraft were left with beach dets, or were in maintenance at NAS Cubi Point, P.I., reducing the actual number on board.

The CVS and Carrier Antisubmarine Air Groups

Sea-based airborne antisubmarine warfare came into being with the creation of the escort carrier in 1940. It was initially conceived as an anti-air platform to provide fighter protection for North Atlantic convoys in mid-ocean. It was ultimately adapted by the U.S. Navy to a Hunter-Killer (HUK) role in the Atlantic, and to a lesser degree, in the Pacific. The importance of ASW carriers had a resurgence in the immediate postwar years, as relations with the Soviet Union rapidly deteriorated and its submarine threat increased.

In mid-1946 two *Commencement Bay*-class CVEs were assigned to ASW duties, one on each coast. In the Pacific, Carrier Escort Air Group 41 (CVEG-41) was established 26 Mar 1945. It was composed of VF-41 and VT-41, then redesignated CVEG-1 (VF-1E/VA-1E) on 15 Nov 1946. The ASW group operated from *Badoeng Strait* (CVE-116). In the Atlantic, CVEG-42 was established 19 Jul 1945, made up of VF-42 and VT-42. It was redesignated CVEG-2 (VF-2E/VA-2E) on the same date as its Pacific counterpart. CVEG-2 operated from *Sicily* (CVE-118). By late 1947 *Mindoro* (CVE-120) and CVEG-3 (VA-3E/VF-3E) had joined the East Coast ASW force. On 1 Sep 1948, CVEG-1, -2, and -3 were redesignated VC-21, -22, and -23, respectively.

The original aircraft complement for the CVEGs was eight VF (F6F-5N) and 12 VT (TBM-3E), but was modified to 18–20 VA and six VF by September 1947. Early ASW exercises proved the VF aircraft were not contributing tactically to the ASW mission, consequently, on 1 Sep 1948, the composite squadron (VC) antisubmarine designation was established, built around a squadron of 18 TBM-3Es (later replaced with ASW-modified TBM-3S and-3Ws). By 1949 there were seven active-duty ASW VC squadrons, and in April 1950, these units were redesignated air antisubmarine squadrons (VS).

In the 1950s sea-based airborne ASW underwent many changes. The CVEs were replaced by the *Essex* -class CVS. The 1951 Grumman AF-3S/-3W hunter-killer Guardians that replaced the old Avengers were in turn replaced by the single-package search-and-attack Grumman S2F Tracker in 1954. The following year, the Sikorsky HO4S (UH-19) ASW helicopter was introduced with dunking sonar that led to the HSS-1 (SH-34) Seabat.

By the end of the '50s, the ASW forces were built around nine *Essex*- class CVSs. Their air groups consisted of one each VS (20 S2F) and HS squadrons (16 HSS-1), along with a detachment of four AD-5W early warning aircraft, plus a possible detachment of fighters (usually Douglas A4D Skyhawks).

In 1960 the carrier ASW forces were reorganized to create nine deployable air groups, consisting of a CAG staff, two VS, and one HS squadron, along with a VAW detachment. Each carrier ASW group was assigned to a particular CVS. To double the number of VS squadrons in accordance with the CVSG organization, each large 20–24-plane squadron was split in two, creating two 10-plane squadrons and more command billets. HS squadrons normally operated 15 HSS-1s, along with a four-plane AD-5W VAW detachment.

During the last half of the '60s, the CVS HUK group began to decline as the Navy's primary ASW system. The S-2s and CVSs were beginning to show their age, as well as reaching the limits of their growth potential. It was decided in 1967 to reduce the CVS forces and develop a jet aircraft replacement (VSX) for the aging S-2. One-by-one, the CVSGs and their carriers began to disappear.

Three ships received the 27C and 125 improvements in one fitting: ***Shangri-La*** **(CVA-38), seen here 11 Feb 1970 off Puerto Rico with CVW-8 embarked,** ***Bon Homme Richard*** **(CVA-31), and** ***Lexington*** **(CVA-16). USN photo, F. Childre**

By May 1970, three of the four remaining Pacific Fleet CVSs were decommissioned, leaving *Ticonderoga* (CVS-14) as the only PacFlt ASW carrier. In the Atlantic, only *Wasp* (CVS-18) remained.

The early '70s was a period of transition for carrier-based ASW. The CVS had gone away more rapidly than expected, and the attack carrier, which had depended on it, had no indigenous airborne ASW capability. To overcome this problem, the flexible air wing was developed in the CV concept, resulting in ASW aircraft being integrated into attack air wings.

During 1971 units of CVSG-54 on the East Coast, along with squadrons from the West Coast CVSG-53, deployed with CVW-3 in *Saratoga* (CV-60) to test the integrated air wing concept. On 30 June 1973 the three remaining CVSG staffs were disestablished. On the West Coast, VS-33, -37, and -38 were assigned to CVW-11 in the first PacFlt CV, *Kitty Hawk* (CV-63). The CV air wing tests were successful, and gradually the carriers and air wings were changed to the new format. Lockheed S-3A Vikings began replacing the S-2s in the mid-'70s, and the SH squadrons began transitioning to the Sikorsky SH-60 Sea Hawk in 1989. Early flexible air wing deployments sometimes resulted in sending one squadron ashore overseas during the cruise because of overcrowding in the ship.

The CVSGs that had operated from mid-1960 to mid-1973 were organized and operated along the same lines as attack air wings. They had a senior commander as CAG, usually a VS pilot. By the late '60s, as seen in the accompanying chart, the CVSG strength had dwindled. The S-2s were aging, along with their carriers, and would soon be replaced by the jet-powered Vikings as their squadrons were integrated into the CVWs. On 1 Apr 1993, to reflect their expanded mission beyond ASW work, the VS squadrons were renamed "sea control squadrons," but retained the VS designation. During the 1990s, the VS squadrons added the all-important role of aerial refueling for the air wings to their mission.

USS H*ornet* (CVS-12)/CVSG-57
30 Sep 1968–13 May 1969
WestPac/Vietnam Cruise
CDR Raymond H. Barker, CAG

VS-35	8 S-2E
VS-37	9 S-2E
HS-2	15 SH-3A
VAW-111 Det 12	4 E-1B
CVS-12 COD	1 C-1A
Total	37

The Age of the Supercarrier

Postwar demobilization and Defense Department budget cuts precipitated fierce inter-service competition for funds. A major argument ensued between the Navy and the newly established Air Force over roles and missions. The Air Force contended it should control all strategic air missions, and there was little need in the future for aircraft carriers; instead, the highly vaunted B-36 could handle the missions. Secretary of Defense Louis A. Johnson and President Harry S Truman sided with the Air Force in the debate, which culminated in 1949-50 and resulted in the "Revolt of the Admirals." Cancellation of a new carrier, *United States* (CVB-58), led to the resignation of Secretary of the Navy John L. Sullivan and the eventual firing of CNO ADM Louis E. Denfeld.

Artist's rendering of *United States* (CVB-58) shows flush-deck design, along with four catapults and elevators of the doomed "supercarrier." Inter-service bickering and an unsympathetic Truman administration led to her cancellation five days after her 18 Apr 1949 keel laying. Bruno Figallo

The "supercarriers" joined the fleet 1 Oct 1955 with the commissioning of *Forrestal* (CVA-59), a 60,100-ton giant with a 1,018-foot flight deck. Three similar sister ships followed 1956-59; they normally operated approximately 75 aircraft in the early years. *Forrestal's* original two-mirror landing system has been circled for identification. McDonnell Douglas

***Kennedy* (CVA-67), with CVW-1 embarked, steams through the Med during her 1971 maiden deployment. Although technically an "improved *Forrestal*," her design makes her another virtual one-of-a-kind. JFK was the U.S. Navy's last conventionally-powered attack carrier. USN photo**

Construction of a new "supercarrier" had been authorized by Congress in July 1948, and her keel was laid 18 Apr 1949. Five days later she was canceled by Secretary Johnson, virtually sounding a death knell for naval aviation. *United States* (CVA-58) was designed at 66,850 tons with a 1,034-foot straight deck. She was to have four steam catapults (two angled port and starboard) with four elevators—three aligned along the flight deck periphery with one on the fantail. Essentially a flush deck design, she would have had a small island on the starboard side that would be lowered during flight operations.

The outbreak of the Korean War in June 1950 caused a realignment in Pentagon thinking about the value of aircraft carriers and the roles for which they were needed. A new carrier design was approved with even more refinements, and the keel for *Forrestal* (CVB-59) was laid 14 July 1952. She, and all future "big deck" carriers, incorporated the four-catapult, four-elevator capability into her design. All elevators were placed alongside the deck edge, and a traditional island was installed on the starboard side of the flight deck. *Forrestal* was the first carrier designed for jet operations, and was commissioned CVA-59 on 1 October 1955. She was followed by sister-ships *Saratoga* (CVA-60) in 1956, *Ranger* (CVA-61) in 1957, and *Independence* (CVA-62) in 1959. The *Forrestals,* the first of the supercarriers, had 1,018-foot flight decks and displaced 60,100 tons light—fully loaded they displaced 79,000-80,000 tons.

Three more "improved *Forrestals*," incorporating design changes to handle the ever-increasing technology of carrier aviation in the '50s and '60s, entered service as the *Kitty Hawk*-class (CVA-63). *Constellation* (CVA-64) and *America* (CVA-66) were officially *Kitty Hawk*-class ships, but varied so much in design from CVA-63 to each warrant a one-ship class status. *Kitty Hawk* was commissioned 9 June 1961, and *John F. Kennedy* (CVA-67) reported to the fleet 7 Sep 1968. She incorporated a new hull design, as well as arrangement improvements in her design, and represents a class of her own. After much controversy over her propulsion plant, *Kennedy* was the last conventionally powered attack carrier built. The debate over conventional vs. nuclear-powered carriers continues, and will probably exist as long as carriers are built.

Joining the fleet 25 Nov 1961, *Enterprise* (CVA(N)-65) was the Navy's first nuclear carrier. Her four-and-one-half acre, 1,079 foot flight deck is the largest of all carriers. USN photo PH3 J.B. Avera

Possibly the ultimate in aircraft carrier design, the nine ships of the *Nimitz*-class in service, authorized or under construction, may be the last of the Navy's (or the world's) first-line combat carriers. *Carl Vinson* (CVN-70), with CVW-15 embarked, plows through the Pacific 14 Dec 1986. USN photo PH2 Vaughn Naegel

Forrestal/CVG-1 30 Jun 1957
CDR William C. Bryan, CAG

VF-14	12	F3H-2N
VF-84	14	FJ-3M
VA-76	12	F9F-8B
VA-15	14	AD-6
VAH-1	12	A3D-1
VAW-12 Det 42	4	AD-5W
VA(AW)-33 Det 42	4	AD-5N
VFP-62 Det 42	2	F2H-2P
HU-2 Det 42	1	HUP-2
Total	75	
Aircraft assigned to CVA-59	1	TF-1, 1 AD-4Q

Entering the Nuclear Age

Studies into a nuclear-powered aircraft carrier were begun as early as 1949. After several years of development of nuclear reactors conducted both by the Navy and a civilian ship reactor program, the Navy's propulsion plant was selected. *Enterprise* (CVA(N)-65) was ordered in 1956 as a Fiscal Year 1958 building authorization. Commissioned 25 Nov 1961, with eight nuclear reactors, she succeeded the *Forrestals* as the world's largest warship at 92,900 tons full load. Her flight deck was an immense four and one-half acres, with a length of 1,079 feet, and is still the largest deck of all carriers. Her four C-13 steam catapults allowed her to launch aircraft weighing 78,000 pounds—a total more than that of *Langley's* entire 34-plane complement. Her original reactors gave her the ability to steam 400,000 nautical miles unrefueled at 20 knots. Rapid advances in nuclear reactor technology, however, dictated that *Enterprise* would become a one-of-a-kind ship. Her mid-Vietnam War complement indicates her fighting capabilities.

Enterprise/CVW-14 Jan 1968
CDR Ernest E. Tissot, CAG

VF-92	12	F-4B
VF-96	12	F-4B
VA-113	14	A-4F
VA-56	14	A-4E
VA-35	12	A-6A
RVAH-1	4	RA-5C
VAW-112	3	E-2A
VAW-13 Det 65	3	EKA-3B
VAH-2 Det 65	2	KA-3B
HC-1 Det 65	4	UH-2C
Total	80	

Again, additional aircraft were carried to replace combat losses.

Perhaps the ultimate in aircraft carrier design is the *Nimitz*-class. Some of the refinements in *Kennedy's* design were used in *Nimitz* (CVN-68), and because of the long delay between the completion dates of each carrier in this class, each has significant changes. For example, CVN-71 and subsequent ships have 3,000 tons of magazine protection fitted. *Nimitz* displaces 93,900 tons full load with a 1,092-foot flight deck. Her range at 20 knots is an incredible 1,000,000 nautical miles. Technological developments allow the *Nimitzes* to have only two reactors, thus allowing room for additional magazine capacity. The *Nimitz* class was designed to carry 89 aircraft, but later events would throw air wing compositions into turmoil.

Nimitz was laid down 22 June 1968 at Newport News, VA, and commissioned 5 May 1975. She was followed by *Dwight D. Eisenhower* (CVN-69) in 1977, *Carl Vinson* (CVN-70) in 1982,

Theodore Roosevelt (CVN-71) in 1986, *Abraham Lincoln* (CVN-72) in 1989, and *George Washington* (CVN-73) in 1992. The latest carriers *John Stennis* (CVN-74) and *Harry S. Truman* (CVN-75) joined the fleet in December 1995 and July 1998, respectively. One more *Nimitz*-class, *Ronald Reagan* (CVN-76), has been authorized by Congress and is scheduled for completion in 2003. A final *Nimitz*, CVN-77, is planned for completion in 2008. A completely new carrier design (CVX) is under study as replacement for the *Nimitzes* in the new millenium.

Redefining the air wings for the next century

During the late 1980s, the composition of carrier air wings underwent a great deal of experimentation that resulted in several options for the number and types of aircraft to be deployed. At one time, there were five different types of air wings operating aboard carriers. However, the end of the Cold War precipitated draconian cuts in the defense budget in the early 1990s and left an uncertainty about the future composition, as well as number of air wings for the Navy. In 1994, the authorized air wing level was cut to 10 active and one Naval Air Reserve Force wing. The carrier force was reduced to one Reserve and 11 fleet carriers. One fleet carrier is normally inactive, undergoing overhaul, leaving a maximum of nine deployable ships at any one time—a level similar to that of first-line carriers at the outbreak of WWII and the Korean War. An offsetting asset was the fact that the Navy was entirely equipped with big-deck carriers: five conventionally powered and six nuclear.

After operational deployment evaluations of several air wing compositions, the Navy settled on a "standard" air wing with approximately 50-54 tactical and 27 support mission aircraft, for a total of 77-81. A typical air wing of the mid-late 1990s was designed as follows:

- 10-14 F-14 *Tomcat* fleet defense fighters with a limited attack role. F-14s are scheduled for replacement by F/A-18E/Fs by 2010.
- 30-36 F/A-18 *Hornet* strike fighters, specifically designed for dual attack and fighter missions.
- 14 A-6E *Intruder* medium attack deep penetration bombers. (Retired in 1997; replaced by F/A-18Cs.)
- 4 E-2C *Hawkeye* airborne early warning and tactical control aircraft.
- 5 EA-6B *Prowler* electronic warfare aircraft.
- 8 S-3B *Viking* ASW aircraft with dual tanker mission.
- 6 SH-60F *Seahawk* ASW helicopters.
- 2 HH-60H *Seahawk* combat rescue helicopters.

Additionally, each air wing included two ES-3A electronic intelligence-gathering versions of the *Viking*, as well as two C-2A *Greyhound* carrier onboard delivery transport aircraft.

Although the average of 50-54 tactical aircraft is a considerable reduction in number from recent decades, the addition of newer "smart" weapons is theorized to offset the deficiency in tactical aircraft available to the battle group commander. Additionally, new, larger versions of the Hornet, the F/A-18E and F models, are to replace the F-14 and older Hornets early in the next decade. In a cost cutting effort, studies are also underway for other new versions of the Hornet to replace most if not all of the remaining fixed-wing aircraft types in the air wings, resulting in virtually a one aircraft type force. Only the next war will confirm or disprove the viability of these theories.

Massive carrier commitments during the Gulf War of 1991 and later concurrent world crises of the late 1990s, such as Iraq and Yugoslavia, indicated the U.S. Navy's carrier force was stretched to its limits to provide necessary support for the nation's foreign policies. In order to support simultaneous combat operations in both the Middle East and the Balkans, the Navy was forced to alter its normal operations of having carriers continually on station in various parts of the world, such as the Mediterranean, western Pacific, and Persian Gulf. With its current force level, it is questionable if the Navy can provide adequate sustained carrier power in the new century if the nation is involved in more than one major, or two moderate combat operations at one time.

2

Evolving the System

"The art of war is of vital importance to the state. It is a matter of life and death, a road either to safety or to ruin. Hence under no circumstances can it be neglected.
...The Commander stands for the virtues of wisdom, sincerity, benevolence, courage, and strictness."

Sun Tzu, ca. 500 B.C.

The Process

CAG—Commander, Air Group—is one of the most revered and sought-after titles in U.S. naval aviation. Since its rudimentary beginnings in the mid-1920s, approximately 1,000 men have held the honored position of leading their fellow naval aviators from the decks of America's aircraft carriers. To be a CAG is to hold a special trust of responsibility bestowed upon an individual by the Navy's leaders (most, former CAGs themselves) after a careful and meticulously designed selection process involving years of preparation and training. Although the air groups are now air wings, the title CAG has a permanent position in naval lexicon.

To produce today's CAG, first begin with a starry-eyed youngster fresh out of college who is willing to undergo the rigors of the Navy's arduous preflight training program at NAS Pensacola, FL. If the youngster has what it takes to survive the harsh environment of rigid discipline enforced by Marine drill instructors, the likes of which never appeared in our candidate's worst nightmare, the individual then begins an intensive 18-month flight training syllabus to reach the goal of "Wings of Gold"—naval aviator.

The next 18–20 years of our fledgling aviator's life will be spent sacrificing permanence of residence and continuity of family life, while enduring a succession of two- or three-year sea and shore duty rotations. During these tours the aviator will serve as an assistant division officer, division officer, assistant department head, and then, the key tour, department head. If the individual excels above all contemporaries in each of these billets, all the while developing the requisite skills of a fleet aviator, the next step is a screening board to decide if the individual is ready and fit for the next big hurdle—a tour as executive officer, then commanding officer of a carrier squadron. Now at about the 15-year career point, the starry eyes have transformed to a look of eagles as our naval aviator joins the big league.

Each carrier air wing typically has eight or nine squadron commanding officers and only one CAG. Simple arithmetic shows the odds of selection to CAG from this group—all outstanding naval aviators. The number is further reduced, as CAGs are selected mainly from only VA, VF, or VFA COs. However, one or two do make the cut and soon assume the coveted title. Now at the 18-20-year career point, our prospective CAG is selected for a tour, initially as deputy air wing commander, and then CAG, and pins on the silver eagles of a U.S. Navy captain. For most of these individuals, becoming an air wing commander will be the ultimate point of their career. Some of them will continue to high-level staff or command positions, and some will be selected for flag rank. For most of them, however, CAG will be remembered as the *real* job; it never got better than that.

How it all started

Historical documentation of the genesis of the CAG program is sparse at best. The earliest reference to an overall commander of a group of aircraft squadrons appears in the Bureau of Aeronautics Newsletter dated 23 Feb 1927. During the period 3-16 February, LCDR William K. Harrill received orders to detach as Commander Observation Wing One.

"Whiskey" Harrill, born 3 May 1892, was a 1914 graduate of the Naval Academy and held a Navy Cross for his service during WWI. He became a naval aviator 30 Dec 1921 and, in January 1924, joined Aircraft Squadrons Battle Fleet (ASBF) to serve as Wing Commander, Observation Wing. Harrill concurrently was skipper of Observation Squadron One (VO-1), and next, VO-2. He ultimately rose to the rank of rear admiral before his 30 Jun 1951 retirement as an honorary vice admiral. Admiral Harrill died 11 May 1962 at the Naval Hospital in San Diego, Calif.

At the time of Harrill's command of the Observation Wing, VO squadrons served on board only battleships and cruisers. Relieved in September 1927, Harrill was succeeded by LCDR Calvin T. Durgin, who also became CO of VO-1. Durgin commanded the Observation Wing until 1929. The BuAer newsletter and fleet organization tables make no further mention of an observation wing until the late '30s. Patrol wings also came into being and remain active to the present time. Not carrier related, observation and patrol wings are not covered in this work.

The first man mentioned in available Navy documents as a leader of a group of fleet aviation units was LCDR William K. Harrill. "Whiskey" Harrill became Commander, Observation Wing with Aircraft Squadrons Battle Fleet in January 1924. At the same time, he was CO of VO-1. The observation wing squadrons served only in battleships and cruisers. USN photo, National Archives 80-G-466317

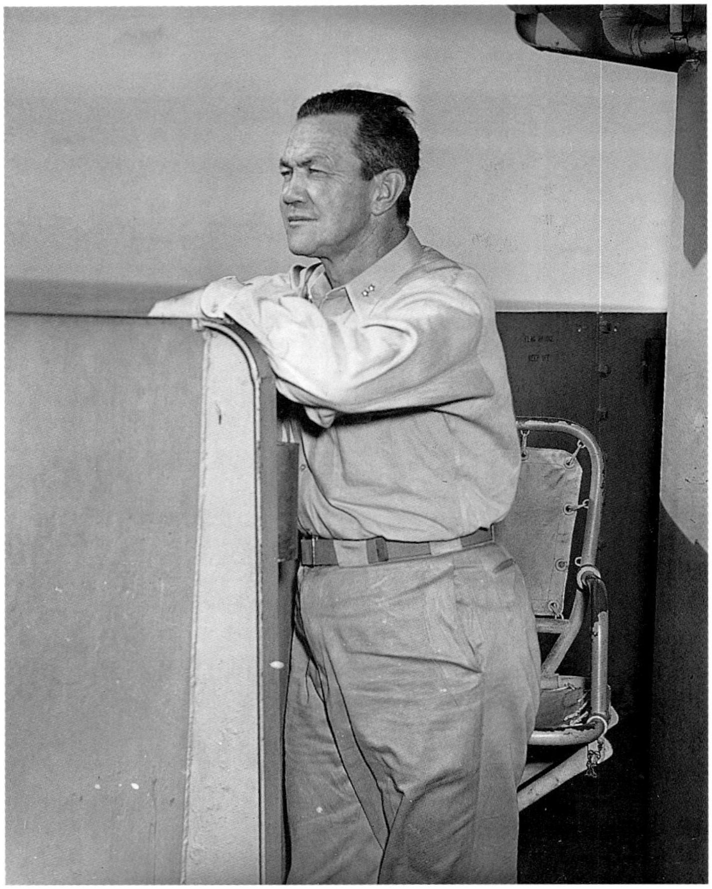

LCDR G.F. "Gerry" Bogan's name appears on the 1 Jan 1928 ASBF Fleet Organizational list as the first Fighting Wing Commander and CO of VF-1B. Seen here as a RADM on the flag bridge of his flagship, *Bunker Hill* (CV-17), Bogan became a leading carrier admiral of WWII. USN photo, National Archives 80-G-282801

By the time *Lexington* and *Saratoga* joined the fleet in 1927, and *Langley* had shed her experimental status, the Navy was operating four different types of squadrons from its carriers: fighting, bombing, torpedo, and scouting. The need for an individual to coordinate the training and develop tactics for these units, as well as oversee their administration, was recognized at the fleet command level. Consequently, as had been previously done with the observation and patrol squadrons, the carrier units were organized into wings and assigned a wing commander. The senior squadron commander of each mission type aircraft assumed these duties. During the late '20s, carrier wing organizations were established according to mission. The torpedo and bombing squadrons were combined under one wing, designated Torpedo Bombing Wing. Initially, the Navy also experimented with a Light Bombing Wing concept for the yet-undeveloped art of dive-bombing, but abandoned it in 1930. The exact dates of establishment of the wings are unknown. The first, Fighting Wing and Torpedo Bombing Wing, initially appear in the January 1928 ASBF organizational list. Biographical sketches of the men first selected to command these carrier wings follow.

Fighting Wing

LCDR Gerald F. "Gerry" Bogan, CO of VF-1B, appears on the 1 Jan 1928 ASBF Fleet Organization list with additional duty as Commander Fighting Wing in *Langley*. At the time, there were five fighting plane squadrons (changed to fighter squadron 1 Sep 1948): VF-1B, VF-2B, VF-3S, VF-5S, and VF-6B—three in the battle fleet and two in the scouting fleet, as indicated by their suffix letters.

Gerry Bogan was born 27 Jul 1894 in Michigan, and became a naval aviator 16 Mar 1925. His graduation from the Naval Academy in 1916 was followed by interim duty in battleships, cruisers, and destroyers. Bogan joined VF-1B as executive officer at Pearl Harbor in 1926, taking command in September. He served additional carrier tours and became CO of VF-3B in *Lexington*, then *Langley*, in June 1930. In 1932, he became a flight test officer at NAS Anacostia, D.C., at the time the Navy's flight test center.

Before America's entry into WWII, Bogan gained more carrier experience as *Lexington's* air officer, then navigator, and later XO of *Yorktown*. From August 1940 until October 1942 he commanded NAS Miami, FL, then assumed command of *Saratoga* until June 1943. He then became Commander 10th Fleet in the Atlantic until designated Commander Fleet Air Norfolk, 14 Oct 1943, and was promoted to rear admiral. He served there until becoming Commander, Carrier Division 25 on 18 Jan 1944. Throughout the remainder of the war, Bogan commanded several carrier divisions and task groups, as well as serving in the invasion of Southern France. He then went to the Pacific to become one of the Navy's leading carrier admirals while participating in the major carrier operations of 1944-45.

The pugnacious former Naval Academy boxer (his flat nose allegedly was a result of his boxing days) was one of the Navy's foremost Fast Carrier Task Force commanders. As Commander Task Group 38.2, he was awarded the Navy Cross for his heroism following two *kamikaze* hits on his flagship, *Intrepid,* off the Philippines 25 Nov 1944. He remained in the Pacific as a task group commander until the end of the war when he returned to the U.S. to assume command of Fleet Air Alameda in October 1945. He next became Commander Naval Air Force Atlantic 2 Feb 1946, donning the three stars of vice admiral. In July 1948, he became Commander, First Task Fleet, where he remained until his retirement. Always willing to fight for what he felt was right, Bogan adamantly opposed the unification plans for the armed forces being developed following the war. His outspoken manner resulted in his forced retirement from active service 1 Feb 1950. Gerry Bogan remained an active proponent for naval air from his La Jolla, CA, home until his death 8 Jun 1973.

Torpedo Bombing Wing

The Torpedo Bombing Wing vies with the Fighting Wing for the title of the first carrier air wing. It also appears for the first time on the ASBF organizational chart of 1 Jan 1928. The first Torpedo Bombing Wing Commander was LCDR Oscar W. Erickson, who was also CO of VJ-1B, a *Langley* utility squadron flying torpedo bombers.

Oscar Erickson was born 6 Feb 1896 in Oklahoma City, OK. He graduated from the Naval Academy with the class of 1916 and served in various surface ship commands before entering flight training in November 1920. He received his wings 11 Aug 1921. From November 1921 to September 1923 he was exec of VO-2B, followed by another XO tour at NAS Pearl Harbor until June 1924. He then returned to Annapolis as a postgraduate student in aeronautical engineering, studying at the Massachusetts Institute of Tech-

Also first appearing on the 1 Jan 1928 ASBF list was LCDR Oscar W. Erickson, CO of VJ-1B and Torpedo Bombing Wing Commander. Erickson was killed 5 Nov 1931 attempting to land aboard *Saratoga* (CV-3). USN photo, National Archives RG-80GX-465731

First commander of the short-lived Light Bombing Wing (1928-1930) was LCDR Arthur C. Davis, here with ACMM H.E. Wallace (left), after making the first landing aboard *Ranger* (CV-4) on 21 Jun 1934. Later, Davis earned a Navy Cross as CO of *Enterprise* (CV-6) and retired in 1955 as a vice admiral promoted to honorary four-star for his combat awards. USN photo, National Archives 80-G-931499, courtesy Bob Cressman

nology. Graduating in June 1926, Erickson next became inspector of aircraft at the Naval Aircraft Factory, Philadelphia, PA. On 11 Oct 1926 he reported to ASBF to become XO, then CO of VJ-1B, and subsequently Torpedo Bombing Wing commander. He was detached 25 Sep 1928 to command NRAB Rockaway, NY. On 27 Oct 1928, he became CO of NRAB Great Lakes, IL, remaining there until returning to NAF Philadelphia 6 Jun 1929. On 16 July, he successfully landed a Martin T3M without damage after the airplane's engine detached in flight and his rear seat crewman had parachuted from the plane.

Returning to sea, LCDR Erickson reported 30 Jun 1931 as CO of VO-4B, with additional duties as aide to Commander Battleship Division Four. He next assumed command of VS-2B in *Saratoga* 17 Aug 1931. He was killed 5 November during a landing attempt aboard Sara while she was operating off Point Loma, CA.

Light Bombing Wing

The Navy began experiments with dive bombing and developed a light bombing squadron concept during 1928–30, temporarily redesignating VF-6B to VB-2B and VF-5B to VB-1B. Although designated as bombing squadrons, the two units continued to fly fighter aircraft. On 1 Jul 1930, they reverted to fighting plane squadrons. Assigned to administer the Light Bombing Wing was LCDR Arthur C. Davis, who first appears on the 1 Jul 1928 ASBF organizational list as wing commander.

"Art" Davis was a Columbia, S.C., native, born 14 Mar 1893. He attended the University of Nebraska for two years before entering the Naval Academy, where he graduated with the Class of 1915. After graduation he was assigned to *Florida* (BB-30), where he received instruction in the new art of radios at Radio, VA. During WWI he served in ships as a radio officer at Queenstown, Ireland, and Brest, France. Following the war, he had further battleship tours and received instruction in ordnance engineering at the Postgraduate School, Annapolis; University of Chicago and Naval Proving Ground, Indianhead, MD. He began flight instruction at Pensacola with the 1 Jul 1923 class, and was designated naval aviator 11 Jan 1924. In April he was assigned to VO-1B and detached in June 1925 for duty at BuAer. He had further additional duty in connection with experiments in light bombing (dive-bombing), participating in the 1927 Fleet Maneuvers at Guantanamo Bay, Cuba. Davis is credited with being largely responsible for development of the Norden bombsight and for inaugurating the art of dive-bombing in the U.S. Navy.

LCDR Art Davis assumed command of VF-6B 26 Jun 1928, serving until May 1929, during which time he had additional duty as Commander, Light Bombing Wing. He next was an aide and then operations officer on the staff of ASBF (later named Commander Aircraft, Battle Force), where he remained until July 1931 when he returned to NAS Pensacola for one year to become Superintendent of Flight Training. During late 1932 until July 1933 he was on CNO's staff in Washington, D.C. Davis then joined *Ranger's* commissioning detail and became her air officer until transferring to *Langley* as XO in June 1936. He returned to Washington for three years as Head of Plans Division in BuAer. Davis then commanded *Langley* (June 1939–June 1940) with additional duties as Commander Aircraft, Asiatic Fleet.

From late 1940 until June 1942 Davis held the position of Fleet Aviation Officer on the staff of Commander in Chief, U.S. Fleet, serving there when the Japanese attacked Pearl Harbor. He is credited with helping plan the early carrier raids in the Central Pacific. On 30 June he assumed command of *Enterprise* in time to participate in the landing on Guadalcanal and the Battle of the Eastern Solomons. CAPT Davis was awarded a Navy Cross for his command of *Enterprise* during the latter battle when she was struck by enemy bombs.

Returning to the U.S., Davis assumed command in November 1942 of Carrier Replacement Squadrons, Atlantic Fleet at NAS Norfolk. In December he moved to become Commander Fleet Air Quonset Point, where he remained until February 1943, when he returned to CinC U.S. Fleet as assistant chief of staff. Detached in August 1944, he became chief of staff and aide to Commander, Fifth Fleet, remaining there until July 1945. For two months he commanded Carrier Division Five before returning to CNO's staff as a member of the Joint Strategic Survey Committee. Several more Washington staff jobs were followed by his assignment in March 1952 as deputy United States representative on the Standing Group, North Atlantic Treaty Organization in Washington.

By 1953, Davis was a vice admiral and continued his career in Washington, holding various positions within the Defense Department until he was transferred to the retired list 1 Apr 1955 as an honorary four-star admiral. However, he continued working in the office of Assistant Secretary of Defense until 1 October. During his Defense Department tour he was involved in the Geneva negotiations relating to the North Vietnamese siege against the French at Dien Bien Phu.

Admiral Davis was returned to active duty 10 Aug 1959 to serve as deputy for Military Matters to the Special Advisor to the Secretary of Defense for Disarmament Review. On 18 Jan 1960, he returned to retirement and lived in Chevy Chase, MD, until his death 10 Feb 1965.

Scouting Wing

The last of the early carrier air wings to be formed was the Scouting Wing, which first appears in ASBF's organization of October 1928. LCDR "Squash" Griffin, one of the original *Langley* pilots, became the first wing commander.

Virgil Childers Griffin, Jr., was born 18 Apr 1891 in Montgomery, AB. He attended the University of Alabama for three years and then entered the Naval Academy, graduating in 1912. From July 1912 until April 1916 he served in *South Carolina* (BB-26) and participated in the 1914 campaign against Veracruz, Mexico. He then went to Pensacola as a student and became naval aviator #41 on 19 Apr 1917.

Griffin remained at Pensacola as an instructor for one month before going to Europe to serve in WWI with the first U.S. Navy Aeronautical Detachment. He attended the French Naval Aviation School at San Rafael, France, and was designated (French) naval aviator No. 270. He then was assigned to Headquarters, Naval Aviation, Paris, from August 1917 until January 1918. After two months of instruction at the Great Yarmouth Royal Naval Air Station, England, he commanded the U.S. Naval Air Station St. Trojan, France, May 1918–April 1919.

Returning to the U.S., Griffin next was assigned to the Board of Inspection and Survey (BIS) in Washington, D.C., from August 1919 to October 1920. During spring 1920, he received instruction in landplanes at the Army Pursuit School, Carlstrom Field, Arcadia, FL. After his BIS tour, Griffin commanded the Ship-Plane Division, Atlantic Fleet, until he reported to *Langley* in September 1921. Following Godfrey Chevalier's death in November 1922, Griffin became her second air officer.

Griffin returned to Pensacola to act as Superintendent of Training May 1924–September 1926. He then went to *Lexington's* precommissioning detail and was her first air officer. In April 1928, he transferred to *Saratoga* for duty as Commander, Scouting Wing, in addition to being CO of VS-2B. During May 1929–May 1930 he was *Langley's* navigator before becoming exec of NAS Pearl Harbor for two years.

In June 1932 Griffin attended the senior course at the Naval War College and then returned to *Langley* as XO until June 1935. From August 1935 until June 1936 he attended the Army War College and then became OinC of NAS Anacostia, D.C., until February 1938. He next served in *Yorktown* until July, when he was assigned to the Staff of Commander, Carrier Division Two in CV-5. Following that tour he commanded Patrol Wing Five at Norfolk February 1939–June 1940.

The last of the early carrier air wings was the Scouting Wing, commanded by LCDR Virgil C. Griffin, Jr. "Squash" Griffin was one of the original *Langley* (CV-1) pilots. During WWII he commanded NATTC Norman, OK, and NAS Barbers Point, T.H., before his 1 Jan 1947 retirement as a CAPT. USN photo, courtesy Naval Aviation History Office

Having been promoted to captain 1 Nov 1939, Griffin next commanded NAS San Juan, PR, until spring 1943. He then was given command of the Navy's largest air technical training center, at Norman, OK, during April 1943–January 1944. Continuing his long list of commanding officer billets, Griffin took over NAS Barbers Point, T.H., in April 1945, where he remained until his retirement 1 Jan 1947. Squash Griffin died at the Balboa Naval Hospital, San Diego, CA, 27 Mar 1957.

Formation of the Air Groups

With *Lexington* and *Saratoga* joining *Langley* in the fleet during 1928, along with *Ranger* in 1934, and the near future promising more carriers, the Navy began to realize that one man would not be able to effectively oversee all the various squadrons of one type. In his book "*Carrier Admiral*," co-authored with Clark G. Reynolds (David McKay Co., Inc., 1967), RADM J.J. "Jocko" Clark mentions his being assigned as *Lexington's* "carrier representative":

> In June 1936, I was happy to get orders to *Lexington* as assistant air officer. But when I reported to the skipper, Captain Aubrey W. Fitch, an outstanding officer, he told me I was to be "carrier representative." This post had just been created for the senior active pilot on each carrier. Lieutenant Commander Matt Gardner had the job on *Saratoga*, Lieutenant Commander Hugh Sease on *Ranger*. This officer represented the captain in coordinating all training operations, especially when the planes were based ashore under the authority of Commander Aircraft Squadrons, Vice Admiral F.J. Horne. At first, the carrier representative was not expected to accompany the squadrons into battle, but in two years this billet was redesignated as Air Group Commander, and the officer assigned to it was given express command of the several squadrons of the carrier air group—aboard ship, on shore, and in battle.
>
> The idea of a senior aviator commanding the planes of the carrier at the scene of conflict was not new, but many officers had rejected it on the ground that it might infringe on the supreme authority of the ship's captain...Admiral "Billy Goat" Reeves had shunned the titles "air commodore" and "air group commander" because they suggested autonomy away from the authority of the captain. So "carrier representative" was used as a stopgap until the concept became standard doctrine. But an overall air commander was needed, and I was grateful to be one of the first when it became a reality.

In consonance with Admiral Clark's recollections, *Ranger's* roster of officers dated 1 Jun 1937 lists LCDR H.S. Seares (TorpBomb Wing Commander 1931-32) as air officer, with collateral duty as carrier representative FAD (Fleet Air Detachment). However, conflict with Clark's statement that the carrier representative billet evolved into that of the air group commander arises in fleet organizational lists, as well as historical documentation. Also,

Another of the early wing commanders was LCDR Harry R. "By Gosh" Bogusch, who commanded the Torpedo Bombing Wing during 1928-29. Bogusch graduated from the Naval Academy in 1911 and, as a LTJG, commanded submarines in the Pacific before undergoing flight training in 1927. His aviation commands were VT-2B, with collateral duties as Torpedo Bombing Wing Commander, and VP-9F (1929). He died suddenly at his home in Chevy Chase, MD, 7 Sep 1934 at the age of 47. USN photo, National Archives RG-80GX-465856

The second Fighting Wing Commander was LCDR Osborne B. Hardison, who relieved Gerry Bogan in 1928. Hardison was designated naval aviator 23 Jun 1923 and assumed command of VF-5S (later VF-5B) in August 1927. After commanding NAS Anacostia 1940-41, he became CO of En*terprise* (CV-6) from 8 Oct 1942 to 7 Apr 1943 and earned a Navy Cross for his performance at Santa Cruz, 26 Oct 1942. Following several high-level Pacific commands during WWII and later, Hardin retired 1 Jan 1955 with the honorary rank of VADM. USN photo 459136, from Steve Ewing

although one of WWII's great carrier admirals, Clark never became an air group commander.

The 1 Jul 1937 ComAirBatFor organization lists LCDR Wendell G. Switzer as CO VF-4 and senior squadron commander in *Ranger*. The list also identifies the following men as senior squadron commanders in the other carriers: *Lexington*, LCDR William Sinton, CO VT-2; *Saratoga*, LCDR Marion E. Crist, CO VT-3; *Yorktown*, LCDR Logan C. Ramsey, CO VT-5; and *Enterprise*, LCDR Thomas S. Combs, CO VS-6. Another conflict arises in the following fiscal quarter lists where the ASBF organization retains these officers as senior squadron commanders, but the U.S. Fleet organization (which should supersede the ASBF list) names the same men as air group commanders. The term "carrier representative" does not appear on either list. It is likely that the carrier representative was just what the name implies, a ship's company officer who acted as a liaison between the ship's captain and the air group, whereas the senior squadron commander was the true air group leader, taking direction from the carrier skipper through the carrier representative.

Interestingly, although the CAG billet was not formally established until 1 Jul 1938, the air group commander title began appearing in 1937 as related above. Recognition of the need for individual carrier air groups originated with the staff of Commander Aircraft, Battle Force, VADM Frederick J. Harve. First mention of air groups in official correspondence is contained in ComAirBatFor letter A3-1/11-Se/FF-2-3, dated 26 Aug 1937 and addressed to Commander, Battle Force. The subject was: "Organization, Aircraft, Battle Force, 1 October 1937." Paragraph two states:

O.B. Hardison was relieved as Fighting Wing Commander by LCDR Homer C. Wick, who was also skipper of VF-1B. Graduating from the Naval Academy in 1912, Wick served in many surface ships during WWI, before earning his wings 20 Dec 1921. He commanded NAS Anacostia 1925-26 and *Wright* (AV-1), before retiring in 1938 because of physical disability. Recalled to active duty in 1940, Wick served in various staff positions until again retiring 1 Jul 1947. He died 15 Jun 1951 at Bethesda, MD. USN photo, National Archives RG-80GX-28614

LCDR James B. Sykes was an interim Saratoga Air Group Commander during 1937-38 until relieved by Dixwell Ketcham 1 Jul 1938. An accelerated Naval Academy graduate (1918 with Class of 1919), Jim Sykes served in destroyers during WWI and submarines in the early '20s before his designation as a naval aviator 11 Sep 1925. He commanded VP-9F, then became air officer and air group commander in *Saratoga* (CV-3). After commanding Patrol Wing Five he became XO of *Ranger* (CV-4) and CO of *Card* (ACV-11), then *Bennington* (CV-20) from August 1944 until 30 Jun 1945. Sykes was awarded the Navy Cross for his service in *Bennington*. After the war he commanded NOTS Inyokern, CA, and retired 1 Jan 1948 as an honorary RADM. USN photo 305157, courtesy Robert J. Cressman

> In preparing the next Fleet organization it is recommended that the form in enclosure (c) (suggest 1 Oct 37 organization) be followed, i.e. that squadrons be listed under their respective carrier groups. This recommendation is made in order that the aircraft organization may be shown in the manner in which it is ordinarily operated. The wing organization is obsolete and should be abolished in favor of the carrier group organization.

This recommendation explains the conflict between the 1 Oct 1937 ASBF and U.S. Fleet organization lists. It is apparent the U.S. Fleet accepted ASBF's recommendation and implemented it on its 1 Oct 1937 list.

ComAirBatFor followed his air group recommendation with a formal request for establishment of air group commander billets. In ComAirBatFor confidential letter A3-1/11-Fd/FF-2-3(6a), dated 16 Nov 1937, he recommended the establishment of the CAG billet with the following duties:

> **Afloat**—Assistant to air officer; primarily concerned with air operations.
>
> **Ashore**— (1) Command squadrons of air group.
> (2) Exercise squadrons in group operations and tactics. 'His primary duties are those concerning the development and perfection of sound group tactical doctrine and through training and indoctrination of his air group toward the most effective employment in the time of war.'

(3) Coordinate and supervise administration and operating activities of his air group squadrons.

(4) Represent the C.O. of carrier and liaison between squadrons and ComAirBatFor.

(5) Senior C.O. of CAGs shall coordinate groups in the carrier division.

It was further recommended that the billets be established in Carrier Divisions One and Two prior to 1 Jul 1938. Also, CAG was to be considered as a department head in the carrier and that ComAirBatFor make changes in his duties as experience dictated.

ComAirBatFor's letter was routed through the usual chain of command and received the following endorsements, some with further recommendations:

1st: Commander, Battle Force: Suggested title be "Wing Commander"; Head of department status irrelevant; CAG should be senior officer in the air with approximately the same duties ashore as afloat; Concurred with establishment of CAG billets.

2nd: CinCUS: Forwarded to Commander, Air Scouting Force (VADM Ernest J. King) for comment, without action.

3rd: VADM King recommended title of Group Commander; Concurred in shore duties as set forth in original; Recommended ship-based duties be (a) Officer in Tactical Command (OTC) of group in the air. (b) Tactical Assistant to the carrier captain; (c) No carrier administrative duties.

4th: CinCUS concurred with title Group Commander.

5th: BuAer endorsement missing.

6th: CNO concurred and directed carrier squadrons be designated as group in the Aeronautical Organization for fiscal year 1939 that became effective 1 Jul 1938. He further directed that CAG duties be defined by ComAirBatFor for the present.

In Chief of Bureau of Navigation confidential letter FF-2-3(282)/Nav-313-SC, the chief of the bureau (who also headed personnel matters) directed: "Billet of Carrier Air Group Commander is hereby authorized effective 1 July 1938..." And thus, CAG was born. However, he would never be a mere mortal.

The basic CAG system as outlined above was implemented as directed. Air groups were formed in the late '30s for the new carriers that were building and scheduled for commissioning between 1936–41. They would be the last of the seven ship-named air groups (*Lexington* to *Hornet*). *Langley*, converted to a seaplane tender during 1936–37, was never assigned a formal air group.

As mentioned above, for a short time during 1937, although undocumented in correspondence, the senior squadron CO of each carrier appears on organization lists under that title, and there are no wing commanders assigned. It seems that for a few months the senior squadron skippers were acting as *de facto* CAGs while the bureaucracy was sorting out the paperwork. The title air group commander first appears on the 1 Oct 1937 U.S. Fleet list, but some personal biographies (not always reliable documents) claim an earlier individual assignment date. The October U.S. Fleet list itself contradicts the 1 Jul 1938 formal establishment date of CAG billets. It is probable that VADM Harve's recommendation was approved and the system was implemented in the fleet before the official establishment date.

Because of the confusion surrounding the date of origin of CAGs, profiles of only the first seven men to hold the title of air group commander on the 1 Jul 1938 official establishment date are presented here. Common practice of the time was to refer to these men using the full name of the air group, e.g., CLAG—Commander, Lexington Air Group; CEAG—Commander, Enterprise Air Group, etc.

Lexington Air Group

LCDR William Sinton was assigned to *Lexington* in 1937 as CO of VT-2, also becoming one of the interim "senior squadron command-

LCDR William Sinton was assigned to *Lexington* (CV-2) in 1937 as CO of VT-2. As the senior squadron commander in Lex, he became one of the early-style group commanders. Relieved in June 1937 as senior squadron commander, Sinton went on to become Lexington Air Group's first formal CAG 1 Jul 1938. He later commanded *Ticonderoga* (CV-14) during WWII. USN photo, courtesy Bob Cressman

ers." He was relieved of the latter in June by LCDR Valentine H. Schaeffer, who became Lexington Air Group's first unofficial CAG. Sinton in turn relieved Schaeffer to become the first formal CAG for CV-2 on 1 Jul 1938. Lexington Air Group was disbanded after the ship was sunk at Coral Sea 8 May 1942.

Born 9 May 1897 at Gurathmey, VA, Bill Sinton attended the University of Virginia before entering the Naval Academy with the Class of 1920. Accelerated by WWI requirements, he graduated 6 Jun 1919 to serve the next four years in the Mediterranean area in battleships and destroyers. He entered flight training at Pensacola with the class of 1 Jul 1923, receiving his wings 11 Jan 1924.

During 1924–29 Sinton had several tours in the Hawaiian area, including one with VT-2B. He then went to VS-1B in *Langley* until the following year when he commanded VT-6D14 at Pearl Harbor. One year later he returned to carriers in *Saratoga* as XO of VS-2, and then went back to battleships in 1934 as senior aviator in *Pennsylvania* (BB-38). Tours in the Hydrographic Office and Division of Air Navigation until 1937 in Washington, DC, were followed by his VT-2 and CAG tours in *Lexington*. He returned to the Naval Academy in 1939 as CO of the aviation detachment, where he remained until 1941. With the threat of war on the horizon, Sinton next commanded a joint Army-Navy-State Department Greenland Survey Expedition to investigate possible landing facilities and service sites in the Arctic area. In June 1941 he went to Norfolk to set up the Advanced Carrier Training Group Atlantic, and the following year he commanded *Chandeleur* (AV-10). He was CO of the seaplane tender during the early Pacific campaign until May 1943, when he joined the staff of Commander Air Force, Pacific as operations officer until April 1944.

Sinton next became Chief of Staff, Fleet Air West Coast, serving there until January 1945 when he took command of *Ticonderoga* (CV-14) in the Western Pacific after her CO, CAPT Dixie Kiefer, was severely wounded during a *kamikaze* attack. Under CAPT Sinton's command, Tico participated in the Philippine and Okinawa campaigns, as well as strikes against Formosa and the Japanese homeland. At the end of the war he brought Tico home and was ordered to become Commander Naval Air Training Bases, Pensacola, 19 Jun 1946. He served in that capacity until relieved of active duty and retired in the Pensacola area 30 Jun 1949 with the honorary rank of rear admiral based on his WWII combat. Bill Sinton died 25 Nov 1977 in Pensacola.

Saratoga Air Group

As with *Lexington, Saratoga's* air group also had a senior squadron commander assigned between the wing commander and CAG. LCDR Marion E. Crist, CO VT-3, held this title from early 1937 until relieved by LCDR James B. Sykes in June of that year. Sykes appears on the 1 Oct 1937 U.S. Fleet list as Saratoga Air Group Commander, which he retained until formalization of the billet and relieved by LCDR Dixwell Ketcham. Saratoga Air Group lives on today as Carrier Air Wing Three, the Navy's oldest air wing.

Dixwell Ketcham, also known as "Ketch" and "Clutch" by his Naval Academy classmates, was born 2 Dec 1889 in Buena Park, IL. Prior to his 1919 graduation (Class of 1920) he had WWI service as a midshipman in *New Jersey* (BB-16). He then served in cruisers and destroyers until June 1922 when he began flight training at Pensacola, receiving his wings in December. The following month he went to VO-1S in *Wright* (AZ-1), a unique ship fitted to tend lighter-than-air aircraft. He then assisted in fitting out *Richmond* (CL-9), where he served with the aviation unit from her commissioning 2 Jul 1923 until July 1924 when he rejoined *Wright*.

Following an October 1925-July 1928 tour at Pensacola, Ketcham joined VO-3S, the aviation unit in *Marblehead* (CL-12), remaining there until November 1929. After a one-year tour on the staff of Commander Aircraft Squadrons, Scouting Fleet, he transferred to the staff of ComCarDiv One, where he served until June 1931. Ketcham then returned to the Naval Academy as an instructor with VN-8D5 for three years. His first carrier squadron came as CO of VF-6B in *Saratoga* from June 1935 to June 1936. After a Washington, D.C., tour on CNO's staff, Ketcham returned to *Saratoga* as air group commander 1 Jul 1938 to May 1939, then remained on board as the carrier's air officer and operations officer for ComCarDiv One. In October 1940 he rejoined ComAirScout staff for one month before a June 1941-August 1943 stint in BuAer.

On 28 Aug 1943 CAPT Ketcham assumed command of *Chenango* (CVE-28). During the year of his command, the escort

As an interim measure between the early wing and air group commanders, the senior squadron CO of each carrier became a de facto CAG until the billet was officially established. LCDR Marion C. Crist (right), as CO of *Solomons* (CVE-67) in 1944, held Sara's senior squadron commander title during 1937. USN photo, National Archives 80-G-260610

LCDR Dixwell Ketcham was CO of VF-6B in 1936 before he became Saratoga Air Group's first official CAG. He relieved LCDR James B. Sykes, who was listed as Sara's CAG in 1937, before the billet became official. Ketcham commanded the escort carrier *Chenango* (CVE-28) during WWII and died in 1993 at the age of 103. USN photo NH 95719, courtesy Bob Cressman

Ranger Air Group's first CAG was LCDR Morton G. Seligman, former CO of VN8-D5 at Annapolis and holder of the Navy Cross for his WWI actions as CO of a sub-chaser. Although he would earn another Navy Cross as XO of *Lexington* (CV-2) at Coral Sea, his career was overshadowed by a suspected security violation on his part. USN photo 46608, courtesy Bob Cressman

carrier participated in combat operations in the Gilberts, Marshalls, Palau, Yap, Ulithi, Woleai, Western New Guinea, and the Marianas. In September 1944 he became Commander Fleet Air Wing One with the rank of commodore, where he directed patrol plane operations against the Japanese throughout the late-war campaigns. With the rank of rear admiral, Ketcham assumed duty as ComCarDiv 27, serving in further combat operations until the end of the war. He assumed additional duties in November 1945 as Commander Escort Carrier Force, Pacific Fleet, remaining in the dual capacity until January 1946 when he became ComCarDiv 17.

Relieved of his sea duty command in January 1947, RADM Ketcham reported in April as Commander Fleet Air Alameda, where he remained until returning to sea 19 Nov 1948 as ComCarDiv Five. On 14 Sep 1949 he returned to Alameda in a twilight status until his 1 November retirement date as an honorary vice admiral. VADM Ketcham resided in California until his death 7 Jul 1993 in San Diego at the age of 103.

Ranger's (CV-4) squadrons were initially led by senior squadron commander LCDR Wendell G. Switzer, who also commanded VF-4 in 1937. Switzer commanded two CVEs and *Wasp* (CV-18) during WWII and rose to the rank of VADM before retiring in 1959. Photo courtesy Mrs. W.G. Switzer and Nancy Schwartz

Ranger Air Group

Senior squadron commander in *Ranger*, LCDR Wendell G. Switzer, was relieved 1 Oct 1937 by LCDR Morton Tinslar "Mort" Seligman, who, after becoming one of the informal CAGs, retained the official title after 1 Jul 1938. Seligman remained as *Ranger* CAG until relieved in June 1940 by LCDR Oscar A. Weller. Ranger Air Group also survives, now as CVW-1.

Mort Seligman hailed from Salt Lake City, UT, where he was born 1 Jul 1895. Another WWI-accelerated graduate of the Naval Academy, he was commissioned in June 1918 with the Class of 1919. For the next 20 months he had surface ship duty at Base 18, Inverness, Scotland. He also commanded the sub-chaser SC-272 and received a Navy Cross for his actions as her CO during WWI mine-sweeping operations.

Seligman next had cruiser duty before undergoing flight training at Pensacola, which he completed in June 1922. He served in squadrons attached to *Oklahoma* (BB-37), *Aroostook* (CM-3), and *Langley* until May 1925 when he went to NAS Pearl Harbor. Joining VF-6B in *Saratoga*, he remained with that unit until a tour of duty at BuAer from July 1929 to June 1932 when he assumed command of VF-1B in Sara. In 1934 he took command of VN-8D5 at the Naval Academy, followed one year later by two years back in Washington at BuAer.

In March 1937 he reported to *Ranger*, first as assistant air officer and then as CAG. Leaving CV-4 in June 1939, he went to NAS San Diego before joining *Lexington* in January 1942 as executive officer. Seligman earned a second Navy Cross for directing damage control and fire-fighting parties after the ship was hit during the Battle of Coral Sea 8 May 1942. CDR Seligman suffered a broken back during the battle.

Following the sinking of *Lexington,* Seligman was suspected of being involved in a breach of security involving a newspaper reporter who subsequently published an article revealing the U.S. had broken the Japanese Navy code. Although never charged, the situation cast a shadow over his career.

Mort Seligman's post-*Lexington* assignments included command of NARB Peru, IN, and ComFair Alameda staff, but his injuries received at Coral Sea and the unfortunate breach of security incident forced an early retirement 1 Nov 1944. Based on his combat citations, he retired with the honorary rank of captain. Seligman died 9 Jul 1967 at San Diego as a result of a stroke.

Yorktown Air Group

LCDR Miles R. Browning succeeded senior squadron commander LCDR Logan C. Ramsey to become Yorktown Air Group's first CAG 1 Jul 1938. Yorktown Air Group disbanded following the carrier's sinking at the Battle of Midway 7 Jun 1942.

A profile of Miles Browning reveals a highly-intelligent, heavy-smoking, high-strung, complex individual. He was born 10 Apr 1897 in Perth Amboy, NJ, and entered the Naval Academy in 1914. At the academy, where he was known as "Skeeter," he excelled in swimming and was a talented artist. He graduated in 1917 with the Class of 1918 and went to *Oklahoma* (BB-40). On 18 June he joined the French steamship *Lutitia* and was senior American Naval Officer aboard while she operated with Cruiser Force, U.S. Fleet, for several months.

Returning to the U.S. in January 1919, Browning served for the next two years in a variety of surface ships until entering flight training at Pensacola 31 Jan 1924. Following his 29 September designation as naval aviator, he joined *Langley* until she arrived in San Diego, where he was then assigned to VO-2S, serving there from January 1925 to May 1927. He next went to NAS Hampton Roads as operations officer. Reporting 20 Jul 1929 to Aircraft Squadrons, Scouting Fleet, Browning took command of VS-5S until June 1931. On 14 July he reported to BuAer, where he served until June 1934.

LCDR Browning next had command of VF-3B and the Fighting Wing until June 1936 when he went to the Naval War College as a student. He then went to the Air Corps Tactical School, Maxwell Field, AB, 1 Jun 1937, as a naval instructor. When detached in June of the next year, he joined *Yorktown*, serving as CAG until 26 Jun 1939. He then organized and for two years commanded the Fleet Aircraft Tactical Unit at NAS San Diego.

From July 1941 until July 1943, Browning served as operations and war plans officer, then later, as chief of staff and aide to VADM William F. Halsey, Jr. Halsey was Commander Aircraft, Battle Force, and later Commander South Pacific Area, as well as

CAPT Logan C. Ramsey reads his orders to command *Block Island* (CVE-21) on 8 Mar 1943, during her commissioning ceremonies. Ramsey led *Yorktown's* aviators as senior squadron commander during 1937-38 . USN photo, courtesy CDR Charles C. Luce, USNR(Ret)

LCDR Miles R. Browning relieved Logan Ramsey 1 Jul 1938 to become Yorktown Air Group's first CAG. His career was marred by poor leadership and came to an inglorious end when he was relieved for cause as CO of *Hornet* (CV-12) in 1944. USN photo, courtesy Bob Cressman

Enterprise's (CV-6) squadrons were first led by senior squadron commander LCDR Thomas S. Combs, CO of VS-6. Combs commanded *Yorktown* (CV-10) during WWII and became Chief of the Bureau of Aeronautics 1 May 1951. He later commanded successively the Second and Sixth Fleets. He became Deputy chief of Naval Operations (Air) 1 Aug 1956 and retired as a VADM 1 Apr 1960. USN photo NH 1015391, courtesy Chuck Haberlein

South Pacific Force. During this tour Browning impressed Halsey and was decorated for his staff actions at Midway under VADM Raymond A. Spruance, who had succeeded the ailing Halsey just prior to the battle. However, Browning's performance at Midway is disputed by historians who feel he was given credit for only following Spruance's orders. Spruance later described Browning: "A lean hawk-like man, he was emotionally unstable and evil-tempered, becoming angry, excited, and irrational with little provocation. He drank too much, too often, and had a capacity for insulting behavior, especially when drunk."

Although unpopular with Spruance, as well as his subordinates, Browning, after Halsey's return, was given command of the new *Hornet* (CV-12), heading her precommissioning detail in September 1943. He commanded CV-12 from her commissioning 29 Nov 1943 until 29 May 1944 when he was relieved for cause. During his command, she fought at Palau, Yap, Ulithi, Woleai, Truk, Satawan, and Ponape, in the Hollandia operations.

Browning's vicious attitude toward juniors—he was infamous for dressing down personnel in public—as well as his erratic behavior, finally caused his downfall. After several confrontations with RADM Jocko Clark, who flew his flag in *Hornet*, and an incident on the carrier's hangar deck that resulted in the death of a crew member, Browning was relieved of command. He was banished to the staff of the Command and General Staff School, Fort Leavenworth, KS, where he remained until his 1 Jan 1947 retirement. His wartime combat service allowed him honorary promotion on the retired list as a rear admiral. These promotions, awarded to naval officers who had received combat citations, were discontinued in the late '50s. They are commonly called "Tombstone Promotions," and resulted in only a change of title. Miles Rutherford Browning died 29 Sep 1954 at the Naval Hospital, Chelsea, MD.

Enterprise Air Group

Nearly one year before her commissioning, *Enterprise's* air group began forming at Norfolk. LCDR Thomas S. Combs, CO of VS-6, was assigned as senior squadron commander and was relieved 1 Jul 1938 by LCDR Giles E. Short, her first formal CAG. Enterprise Air Group ceased to exist after September 1942.

Giles Ezra Short was born 25 Mar 1895 in Lohrville, Iowa. He graduated with the Naval Academy Class of 1919 in June 1918 and went to destroyers, operating from Queenstown, Ireland. After the armistice he was transferred to U.S. Headquarters, Brest, France, and in February 1919 was assigned to the ex-German ship *Grad Waldersee*. He next helped commission and served in *Thornton* (DD-270) as her senior engineer until December 1920 when he began flight training at Pensacola.

Short's initial duty as an aviator was at NAS Anacostia, D.C., where he went in June 1922, remaining until December. That month he began service with VO-1S attached to *Wright* (AZ-1) until detached in August 1924 to go to NAS Coco Solo, Canal Zone. Departing Panama in June 1927, Short returned to *Wright*, serving in her squadrons until April 1932 when he went to Norfolk and VS-1B, flying from *Langley*. As was common practice at the time, LCDR Short switched temporarily to patrol planes, taking command of VP-5F in June 1934 at Coco Solo. He returned to Norfolk to form and command VB-7B (soon changed to VB-5) in *Yorktown* from January 1937 to June 1938 when he became *Enterprise's* CAG.

Detached from CV-6 in June 1939, Short commissioned and commanded *Patoka* (AO-9) until June 1940. Following graduation from the Naval War College, he remained there as a staff member until June 1942 when he went to Tacoma, WA, to commission and command *Bogue* (CVE-9). Taking his escort carrier to the Atlantic, he also commanded an ASW task group operating against German submarines. Temporarily promoted to commodore in July 1943, he was next assigned as CO of Midway Islands until 16 Sep 1944, when he reverted to captain and transferred to Commander Naval Air Force Pacific staff for temporary duty. After two months he briefly commanded *Ticonderoga* (CV-14) until taking command of *Intrepid* (CV-11), which he retained until 24 Dec 1945.

After a brief assignment to CNO's staff, Short became Commander Alameda Group, Pacific Reserve Fleet from 2 Mar 1946 until 20 Apr 1948. CAPT Short returned to the Navy Department until his retirement 30 Jun 1949 in the honorary rank of rear admiral. RADM Short died 1 Jul 1973 while living in Laguna Hills, CA.

Wasp Air Group

Although *Wasp* was not commissioned until 25 Apr 1940, her air group began forming in 1939, and on 1 July, LCDR Grover B.H. Hall became her CAG. The air group's final selected CAG, LCDR Courtney Shands, never got a chance to lead the group as the ship was sunk the day after his selection, before he could take command. Wasp Air Group was disbanded after the carrier's loss at Guadalcanal 15 Sep 1942.

Seen here as a LTJG, Giles Ezra Short became Commander, Enterprise Air Group 1 Jul 1938 after commanding VP-5F and VB-5. During WWII he commanded *Bogue* (CVE-9) and distinguished himself as an ASW task group commander in the Atlantic. In the Pacific, he commanded *Ticonderoga* (CV-14) and *Intrepid* (CV-11). USN photo, courtesy Bob Cressman

Wasp Air Group began forming in 1939 under CAG LCDR Grover B.H. Hall. He previously had commanded VS-42 in *Ranger* (CV-4) and was CO of *Charger* (ACV-30) at the beginning of WWII. Hall earned a Navy Cross as CO of *Enterprise* (CV-6) when she was struck by Japanese dive bombers off Kyushu, Japan. He retired in 1961 as a RADM. USN photo, courtesy Mrs. Marie Hall

Born 28 Jan 1900 in Franklinville, NY, Grover Bud Hartley Hall II graduated from the Naval Academy with the Class of 1920. He served in battleships and destroyers until August 1924 when he entered flight training at Pensacola. Designated naval aviator in March 1925, for the next three years he flew observation planes from *Maryland* (BB-46) and *Arizona* (BB-39). From July 1928 until June 1930 he served at Coco Solo, after which he joined *Nevada* (BB-36) for duty with V0-3B.

Leaving battleships, Hall attended postgraduate school at Annapolis, focusing on electrical engineering there, and at the University of California, Berkeley, receiving a Master of Science degree in June 1933. He then returned to sea duty on the staff of Commander Aircraft, Base Force, attached to *Wright* (AV-1) until June 1935. Hall then went to NAS Anacostia as radio test officer.

In June 1938 LCDR Hall took command of VS-42 in *Ranger*. The following year he was transferred to Patrol Wing Five at Norfolk. On 1 Jul 1939 he became *Wasp's* first CAG, remaining until April 1941. He then assumed command of *Charger* (ACV-30), an Atlantic Fleet escort training carrier.

Hall returned to Washington in November 1943 with orders to the Naval Research Laboratory at Anacostia, where he worked on Identification, Friend or Foe (IFF) and other electronic projects before commanding the Combined Research Group. In December 1944 he returned to sea to command *Enterprise*. During his command of CV-6, the carrier participated in the attacks on Formosa, Luzon, and the China Coast; landings at Lingayen Gulf; Fifth Fleet raids against Honshu and Nansei Shoto; occupation of Iwo Jima; along with the Fifth and Third Fleet raids in support of Okinawa Gunto. CAPT Hall was awarded the Navy Cross for his heroism in action off Kyushu, Japan, after *Enterprise* was hit by dive bombers. He relinquished command of CV-6 in September 1945.

Following a second BuAer tour from September 1945 to January 1948, he took command of Carrier Division 14 on 15 February. Promoted to rear admiral 18 Jun 1948, Hall next became, in September 1949, Director of Guided Missiles Division on CNO's staff. In March 1952 he assumed command of Fleet Air Wings, U.S. Atlantic Fleet, then in July of the following year, command of Carrier Division Six, flying his flag in *Coral Sea* with the Sixth Fleet in the Mediterranean.

On 20 Apr 1954 RADM Hall reported as Commander, Caribbean Sea Frontier, as well as Commandant, Tenth Naval District, headquartered at San Juan, Puerto Rico. He was detached 18 Jan 1957 for duty as Commander Naval Air Bases, Eleventh and Twelfth Naval Districts on the West Coast. On 14 Aug 1959 he assumed the triad command of Commander Barrier Pacific, Commander Fleet Air Hawaii, and Commander Airborne Early Warning Wing Pacific, at NAS Barbers Point, T.H. He also took on additional duties of Commander Naval Air Bases, Fourteenth Naval District. RADM Hall remained in Hawaii until his retirement from active duty 1 Oct 1961. He died 13 Mar 1987 in San Diego, CA.

Hornet Air Group

The last of the prewar ship-named air groups, Hornet Air Group had only two commanders. It was disbanded shortly after the June 1942 Battle of Midway. LCDR Stanhope C. Ring was the first CAG, followed 30 June by LCDR Walter F. Rodee.

A profile of Stanhope Cotton Ring reveals one of the few unfortunate examples of failed leadership in the CAG system. Ring rivals Browning for the title of least loved or respected of the prewar air group commanders. Born 13 Oct 1902 in Norfolk, VA, the son of a naval officer, Ring graduated from the Naval Academy in the Class of 1923. He first helped commission *Colorado* (BB-45), remaining in her ship's company until January 1925. The following month he reported for flight training at Pensacola, but did not receive his wings until 13 Dec 1926. The reason for this delay is unknown. He then went to VF-5S in *Wright* and then *Lexington*. Detached in June 1929, Ring went to BuAer for three years, with additional duty as White House aide, 1930–32.

In June 1932 Ring joined VF-3S in *Langley*, and in December became a member of her ship's company. In June 1933 he rejoined Fighting Three, and in August 1934 became assistant operations officer on the staff of ComAirBatFor in *Saratoga*. Detached in June of the following year, Ring then went to NAS San Diego for one year and then to VP-17. He commanded that patrol squadron from October 1938 until May 1939. He next returned to BuAer until March 1941, when he reported as Assistant Naval Attaché for Air at the American Embassy in London.

Another of the rare examples of failure of the CAG system was LCDR Stanhope C. Ring, first commander of Hornet Air Group. A martinet, Ring's poor leadership qualities undoubtedly were contributing factors to the marginal performance of his air group at the Battle of Midway. USN photo NH 02394, courtesy Chuck Haberlein

Returning to the U.S. in June 1941, Ring, after a six-year hiatus from carrier aviation, began forming Hornet Air Group, which was established 20 October. He became CAG two weeks earlier. After the 18 Apr 1942 Doolittle raid launched from *Hornet* against Tokyo, the carrier and air group returned to Pearl Harbor before proceeding to the vicinity of Midway Islands.

Ring, by this time, had established himself with his air group personnel as a vindictive martinet and "sundowner," who instead of leading his men was given to driving them. (He was unaffectionately called "Sea Hag" behind his back.) Just prior to the Battle of Midway, he reportedly placed several pilots in hack (restricted to their rooms) for not coming to attention when he entered their ready room. With less than a two-month training period for his air group before it was rushed into combat, Ring apparently felt protocol of his pilots more important than their training. The performance of Hornet Air Group at Midway bore the bitter fruit of Ring's style of leadership. (See, "*The First Team*" John B. Lundstrom, Naval Institute Press, 1984.) Flying an SBD-3 on the 4 and 5 June missions, Ring returned to *Hornet* with his bomb still aboard his aircraft. On the evening of the 5th, he asked his ensign wingman what the procedure was for dropping the bomb.

On the third day of battle, the air group did perform well in an attack led by Ring against the Japanese cruiser *Mikuma*, for which he received a Navy Cross. With his torpedo squadron annihilated and most of his fighter squadron's aircraft gone (mainly from ditching from fuel starvation after never engaging the enemy the first day of battle), some of Ring's survivors had dispersed to other carriers and Midway. Hornet Air Group disbanded in October. Incredibly, of the three air group commanders at Midway, Ring was the only one to later be selected for active-duty flag rank.

Relieved 30 June, Ring reported 1 July as operations officer for Patrol Wing Two. In March 1943 he became chief of staff for Commander Fleet Air Noumea, then from 1 April to 25 July, he was operations officer for Commander Aircraft Solomons. He returned to the States and became Director of Aviation Training for Deputy Chief of Naval Operations (Air) in Washington.

VT-8 TBD-1 BuNo 0308, flown by LTJG Jeff Woodson and radioman-gunner ARM2/c Otway Creasy, Jr., prepares to launch from *Hornet* (CV-8) for the Battle of Midway. VT-8 was annihilated during the battle, partly as a result of a lack of pre-war training of Hornet Air Group. USN photo, courtesy Mark Horan

Stanhope Ring next commissioned and commanded *Siboney* (CVE-112) from 14 May 1945 until 2 Feb 1946, when he took command of *Saratoga*, serving until August when he attended the National War College. He assumed command of *Boxer* (CV-21), commanding her from August 1947 until July 1948, before returning to the War College as a staff member. On 1 Aug 1950 he assumed command of the Naval Operating Base and NAS at Kwajalein, remaining in these positions until October when he was detached for duty as deputy chief of staff to Commander in Chief, Allied Forces, Southern Europe.

On 21 Jan 1954 RADM Ring became ComCarDiv One, commanding naval aviation forces supporting the evacuation of the Tachen Islands. On 16 May 1955 he was assigned as Vice Commander Military Air Transport Service. In August he was hospitalized, where he remained until his retirement, effective 1 November. Promoted upon retirement to the honorary rank of vice admiral, Stanhope Ring died 25 May 1963 in San Diego.

Hornet's 20 Oct 1941 commissioning and the establishment of her air group brought to a close the prewar era of carrier aviation, often referred to as the "Golden Age" of naval aviation. America was gearing up for war, and the Navy's air groups were being readied. However, they were in reality still land-based units that occasionally operated from the carriers. World events soon changed this condition. The Navy's air groups were going to war, and their CAGs would lead them into battle.

The duties of the air group commander remained fairly consistent throughout WWII. His billet in the ship's organization was effectively that of a department head. CAG was responsible for the administration of his squadrons, directed through his squadron COs. He oversaw combined air group training and was further responsible for the performance of his air group. CAG served at the pleasure and under the direction of the carrier skipper, usually without much interference or close supervision. The individual personalities of CAG and CV CO had great bearing on their relationship and how much direction CAG received. Overall, it was a good system, and it served the Navy well.

Normally, in combat, CAG desired and was expected to lead his men, and in most circumstances, this was the case. However, there have been CAGs who, for whatever reason, opted to remain on board during combat and direct the group from the carrier. In other cases, higher authority (carrier CO or task group commander) directed CAG to remain on board, or not actively engage the enemy during strikes. While researching his sequel "*The First Team and the Guadalcanal Campaign*" (Naval Institute Press 1994), John Lundstrom discovered a case in point in a 10 Jul 1942 directive issued by RADM Aubrey W. Fitch, Acting Commander Carriers Pacific: "Air Group Commanders, Duties of." This order directed CAGs not to lead strikes. Instead, the senior squadron commander on each mission was to be strike leader. CAG was to fly a specially-modified aircraft with no payload and carrying extra fuel to serve as a strike evaluator, often with a photographer on board. It was almost as if CAG was to be a passive observer rather than a combat leader.

Lundstrom further discovered Fitch's directive did not reach the carriers off Guadalcanal until late August or early September at the earliest. CAGs during the Guadalcanal landings served as target coordinators, and on the 24–25 August missions, CAGs Harry D. Felt (*Saratoga*) and Wally Beakley (*Wasp*) both led their strikes in armed aircraft.

The first carrier to adopt the new policy was *Hornet*, which left Pearl 17 August. She was experimenting with deck-load strikes, a new tactic designed to allow continuous manageable operations. Approximately half of the air group was launched, about the greatest number that the carrier could effectively handle. While the first group was out a second group was readied for launch just prior to the first group's recovery.

Hornet's new CAG, Walt Rodee, in a modified TBF, usually led the second strike with a pair of F4F Wildcat escorts. On 26 October he broke off with the escorts to photograph the ships under attack; he did not select targets or order any attacks. Unable to recover aboard the sinking *Hornet*, Rodee diverted south to the island of Espiritu Santo.

Dick Gaines, CAG 10 from *Enterprise*, also flew a modified TBF during the 25–26 October battle. He did not lead the strikes.

Most pilots, especially the CAGs, resented the new policy, feeling it restricted aggressiveness. They felt CAG had to show leadership and command the strikes. It is unknown when, or if, Fitch's directive was ever modified or rescinded. Throughout the war, and thereafter, CAGs flew a variety of missions, on some occasions aggressively engaging the enemy in combat, and at other times acting as a strike coordinator, standing off and directing the attack.

A classic example of WWII CAG selection and how they handled their jobs was LCDR Edgar E. "Steb" Stebbins, CO of VB-5, who was an emergency replacement for LCDR Charles L. Crommelin as CAG 5 in *Yorktown* (CV-10). The Crommelin/Stebbins epic is related in Clark G. Reynolds' previously mentioned "*Carrier Admiral*," and "*The Fighting Lady: The New Yorktown in the Pacific War*" (Pictorial Histories, Missoula, Mont., 1986). In "*Carrier Admiral*," quoting Jocko Clark, then-CO of CV-10:

> ...Crommelin's fighters shot up several bombers on the ground at Mili. When this strike returned, Crommelin begged me to let him go back and finish the job. I told him Mili was not worth the effort, but he persuaded me to let him go. Crommelin, adventurous and courageous, seemed to want to conduct a one-man war of his own...

Reynolds continues the saga in the *Yorktown* book:

> ...Coming in at 350 feet, Charlie took a 20mm shell burst only inches in front of his cockpit. Shrapnel 'frosted' his plexiglas greenhouse and windshield and peppered his body with over 200 wounds. Instinctively pulling back on the stick, he leveled off at 300 feet, but was hurt badly—right wrist fractured; a bolt imbedded in one leg; one finger nipped off; several teeth chipped; penis, right shoulder and mouth bloodied; and eyes hit, leaving him only partial vision in one eye.
>
> But his mind was alert, and he broke his pencil in half and wedged part of it under the choke to keep gas feeding the engine. ...he couldn't see forward, so he looked out the side of his cockpit to stay alongside wingman Tim Tyler. Tyler coached Crommelin by radio and guided him the full 120 miles back to the ship. Remarkable though this feat was, Charlie still faced the challenge of landing.
>
> ...LSO Dick Tripp faced the unusual phenomenon of two planes coming up the groove together, Crommelin heading for the deck, Tyler exactly on his wing to one side. At the precise moment, Tripp gave the 'cut' and Tim yelled the order through his radio to Charlie. Charlie obeyed, dropped to the deck and caught the wire. He taxied out of it past the barrier and tried to climb out but had to be lifted. Trying to walk away, he finally collapsed on the flight deck.
>
> Dr. Gard managed to save Crommelin's eyes, but, for the second time, Air Group Five had lost the services of its skipper. Jocko Clark could only think of one possible replacement. 'Lieutenant Commander Stebbins, report to the bridge,' beckoned the speaker. The bombing squadron leader soon confronted the captain.
>
> 'Stebbins, can you fly a Hellcat?'
>
> 'Negative.'
>
> 'O.K., you take off in one this afternoon.'
>
> Ed Owen immediately checked out Steb in an F6F, whereupon he took off, circled the ship a few times and made a perfect landing to gain the confidence of everyone. It was an unorthodox move. Normally, a more senior aviator should have been assigned by the Navy, but Jocko preferred *Yorktowners* he knew and trusted. Flatley and Crommelin were tough acts to follow, but Steb laid down his guidelines that night. He would coordinate all air strikes and avoid dogfighting, while the fighter pilots must protect the bombing planes before dogfighting and making their strafing runs.

Stebbins, using "Cairo 99" as his radio callsign, successfully led CVG-5 from 21 Nov 43 until relieved by CDR Edwin B. Parker, 25 Jun 1944. Steb, however, did have one failing—his radio discipline was a little weak, causing him to be known as "Yap, Yap Cairo." Crommelin recovered from his injuries to become CAG 12 in June, but was subsequently killed during a recce mission off Okinawa in March 1945.

The Marine Carrier Air Groups

Documentation is scarce regarding the Marine Carrier Air Groups (MCVG) that were formed during late-WWII. By mid-1944 some Marine Corps aviation leaders had recognized the need for Marine aviators to fly close-air support missions from offshore carriers for the mud-Marines during amphibious landings. As a result of high-level meetings between the Marines and the Navy, in October, 18 MCVGs were programmed for activation to be assigned to the new *Commencement Bay* CVEs. How many groups that were ultimately formed is unclear. What is known is that six, and possibly a seventh, were definitely operational before the end of the war and assigned to CVEs. Five of them made it to the Western Pacific combat area. The 7 Sep 1945 Location of U.S. Naval Aircraft list shows 17 MCVGs, all with squadrons assigned, one VMF (18 aircraft) and one VMTB (12 aircraft). However, several groups have notations such as "rdy Oct 45," "rdy Feb 46," etc. No Marine Corps records have been found to confirm if all 17 of these groups were actually operational with an air group commander, or not. The end of the war resulted in an apparent neglect of record-keeping and paperwork within the Navy and Marine Corps for several months. Although the Marines continued limited CVE operations after the war and into Korea, the MCVGs were deactivated by mid-1946.

Postwar and Korea—The Air Task Groups

A variation of the CAG theme occurred during the Korean War. When the war began 25 Jun 1950, the Navy had three CVB *Midway*-class carriers, none of which served in Korea. Additionally, there were four *Essex*-class CVs in commission and nine air groups to man the carriers. By the end of 1952 there were eight *Essexes* in commission with 16 active air groups. In July 1951 two new air groups had been authorized by Congress, CVG-101 and -102, both composed of activated Reserve squadrons. (On 3 Feb 1953 they became CVG-14 and -12, respectively.) However, to reach 16 air groups, the Navy was forced to engage in a bit of legerdemain.

Operating levels off Korea for the carriers were such that the ships could remain in the combat zone with no threat of attack for extended periods of time, broken by brief periods of rest and upkeep in nearby Japan. The same was not true for pilots and aircrew. Although they did get the occasional trips to Japan, when the carrier was on the line, the aircrews were facing combat on a daily basis for much longer periods than had been experienced in WWII. It soon became obvious that the air groups could not maintain the high-level combat situation for the same amount of time the carriers were capable of remaining on station. Additionally, carriers were able to accomplish much shorter turnaround times between their deployments to Korea. Limited by Congress in the number of air groups authorized, Navy planners came up with the innovative solution of air task groups.

The standard post-WWII attack carrier air group composition was set 20 Jul 1948 at three fighter and two attack squadrons, augmented by specialized detachments of airborne early-warning, night attack, and photo reconnaissance aircraft, with one helicopter for rescue and logistic missions. By January 1950 some air groups had

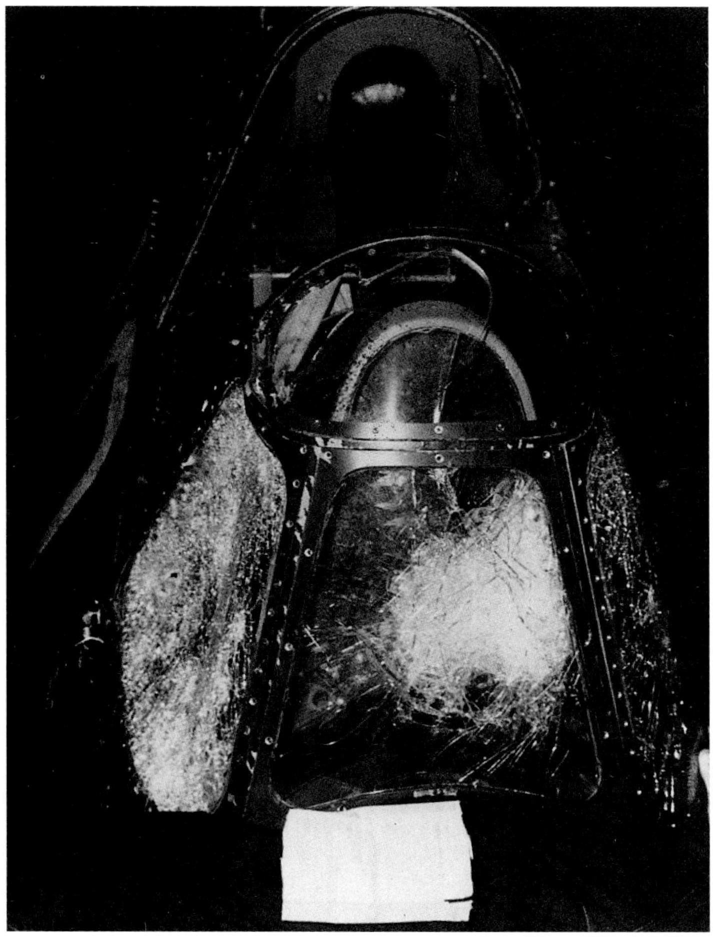

At left, Charlie Crommelin drags on a stogie in *Yorktown's* (CV-10) fighter ready room following a mission. LT Woodie McVay is at his left. As CAG 5, Crommelin took a 20mm AA shell in the windscreen of his Hellcat (right) that produced more than 200 wounds and nearly blinded him. Crommelin's warrior instincts and his wingman brought him back to land safely aboard *Yorktown*. He recovered to become CAG 12 and was later killed at Okinawa in March 1945. USN photos,courtesy Steve Ewing

four VF and one VA squadron assigned. Two VF squadrons flew Vought F4U-4B Corsair prop-driven fighter bombers, backing up the two all-jet primary VF squadrons. The attack guys were flying Douglas AD Skyraiders. Each squadron was authorized 18 aircraft, for a total of 90 planes. The addition of special-mission aircraft brought the total to an excess of 100, as related previously—more aircraft than the carrier could effectively operate at one time.

To solve the problem of its shortage of air groups, the Navy "stole" one squadron from some existing air groups and formed air task groups, thereby circumventing congressional limitations. The air task groups were never formally established and do not appear in either Naval Aeronautical Organization or Location and Allowance of Aircraft lists. Instead, the individual squadrons continue to appear with their parent air group for the entire time of their air task group service. The shortage of spare attack squadrons compared to the abundance of fighter squadrons brought about the unique situation of VF-54 and VF-194 flying Skyraiders, functioning as attack squadrons while retaining their fighter designations.

The initial remedy to the combat exhaustion problem in Korea was the formation of CVG-19X, which replaced CVG-19 in *Princeton* (CV-37) in June 1951 during mid-cruise. Fresh personnel from Stateside squadrons were sent out to relieve CVG-19's personnel. The original aircraft remained on board. On paper, CVG-19 was still deployed, but in reality, all-new squadrons were on board *Princeton*. CVG-19X was only a temporary fix, and soon ATG-1 and -2 were born, respectively, under CDRs Fritz Crabill and Jim Daniels. Profiles of these men appear on following pages. Although technically not air group commanders on paper, the ATG commanders functioned in exactly the same manner as such and are recognized in this work as CAGs.

CAPT Jim Daniels IL, USN(Ret), tells how he became commander of ATG-2:

> An air task group was formed by a type or force commander using his *existing* (legal) units as a temporary operational unit for a specific mission.
>
> An air group was a permanent commissioned unit (legal), composed normally of four or five squadrons led by a designated officer known as air group commander (CAG).
>
> I reported for duty to ComNavAirPac as a prospective VF CO or CO of an air group on 30 Nov 1951... I was told that day I was Commander Air Task Group Two and that in due course designated squadrons and splinter groups would be assigned for a routine deployment to Korea (my special assignment), all verbal.
>
> I assumed command by reporting to ComNavAirPac by letter and reporting as Commander Air Task Group Two for temporary additional duty on the staff of Air Pac. Listed in the roster only! No written orders!

As leaders of unofficially established units, the air task group commanders were forced to operate somewhat differently than official CAGs. ATG staff personnel had to be borrowed from within the individual squadrons. CAG did not have non-judicial punishment authority, nor even an official address. All correspondence and mail were routed through the ship. However, there was never any question who was in charge and the ATG organizations functioned well.

Between 1951 and 1959, when the last of the already nearly invisible air task groups disappeared, a total of eight groups were formed, four on each coast.

Reserve Air Groups

In late 1945 a resurgence of the Naval Air Reserve program was initiated that would grow to vast proportions in the 1950s. Included in this new program were plans for 30 carrier air groups to begin forming in 1946. A wealth of surplus aircraft was available from the Navy inventory and, at least for awhile, the Reserves would fly first-line aircraft, such as the F6F-5, Goodyear FG-1D Corsair, TBM-3E, and Curtiss SB2C-4E Helldiver. By September 1946, the Reserves had 15 CVGs, five CVLGs, and 35 CVEGs in operation. The CVG and CVEGs were designed much the same as their regular Navy counterparts, with one each VF, VBF, VB, and VT squadron for the CVGs. The CVEGs had one each VF and VT squadron. Unaccountably, the CVLGs had only a VF squadron assigned. Prior to WWII, Reserve units did not operate from carriers.

The Reserves had no carriers in the new program, but instead relied on regular Navy decks to maintain their proficiency. The carrier requalification started slowly in August 1948 when detachments from NAS Dallas' CVG-55 and -83 went aboard *Wright* (CVL-49) off Pensacola. Ten air groups were scheduled for requal in 1949, and 18 the following year. The Reserve carrier qualification program was fortuitous, as it was a vital asset in 1950 with the outbreak of the Korean War and the recall to active duty of many Reserve aviators. By 1953, four complete new active duty Reserve air groups were formed with recalled squadrons from all over the country.

In July 1949 the Reserve air groups had been re-formed. The air group commander billet and organization of the groups were changed to an air wing concept. The new air wings would include all types of squadrons, such as VP and VR, in addition to the carrier units. The existing 55 air groups were consolidated into 27, one at each major Reserve facility. They were flying 2,051 aircraft—988 fighters alone.

The quality of the Reserve program was clearly demonstrated during the Korean War, as individual aviators and squadrons were recalled to active duty. At one time, 50 percent of *Princeton's,* 90 percent of *Boxer's*, and 100 percent of *Bon Homme Richard's* pilots were Reserves. However, following the war, the Reserve program rapidly fell into decline and sadly became little more than a weekend flying club. On 27 Jan 1968, six carrier squadrons were recalled to active duty for the *Pueblo* (AGR-2) crisis. All but one of these units were found to be unsatisfactory in training readiness and manpower qualifications. On 1 July a reorganization of the Reserves was ordered, and the present Naval Reserve Force concept was initiated. Two Reserve carrier air wings were established: CVWR-20 at NAS Cecil Field, FL, and CVWR-30 at NAS Miramar, CA. The CVWRs were set up nearly identical to regular Navy CVWs, with the exception of not having a dedicated helicopter squadron assigned. Additionally, two CVSGR air groups were formed in 1969 for the ASW task, but were phased out in 1975 and '76 with the demise of the regular Navy CVSGs.

Under the new Reserve program, the CVWRs flourished and became top-notch outfits with a wealth of talent and first-line aircraft. Selected squadrons began returning to the carriers for requalification in 1970-71 and, in November 1976, CVWR-30 demonstrated its capabilities by successfully passing on Operational Readiness Evaluation aboard *Ranger*—a first-time evolution for the Reserves. CVWR-30's aircrews accomplished an 88.75 percent boarding rate—a rate equal to most regular Navy air wings. However, the end of the Cold War during the early '90s resulted in drastic cutbacks in the Reserve program. In 1994, CVWR-30 was disestablished, leaving only CVWR-20.

Presently, the Reserve mission is twofold: peacetime support to the fleet and wartime mobilization readiness as an integrated USN/USMC air wing. CAPT Robert L. McLane, former CAG 20, states:

> Basically, during peacetime operations the wing supports the Navy's active component while maintaining a training posture from which it can deploy as a fully-functioning carrier air wing following mobilization. To strengthen this concept, *John F. Kennedy* (CV-67) was recently designated as the Navy's operational reserve carrier for training, contributary support, and mobilization operations. Besides being aligned with CV-67 for support and training, CVWR-20 may deploy on relatively short notice for crisis response. The air wing mirrors active air wings, with the exception of an S-3 squadron. VMFA-142 has been designated as the USMC component of CVWR-20, supplementing the existing two VFA squadrons.
>
> Presently, CVWR-20 is composed of: VFA-201, VFA-203 and -204 (12 F/A-18A each), VAQ-209 (4 EA-6B), and VAW-78 (4 E-2C).

Additionally, HS-75 augments the air wing as required for training and mobilization purposes with its SH-3Hs, along with VMFA-142's F/A-18As as the Marine component. For administrative purposes only, VFC-12 and -13 are assigned to CVWR-20. A new VAW squadron (VAW-77) was established in FY 1996, and, although assigned to CVWR-20, is funded by and works strictly under the control of the Drug Enforcement Agency (DEA) drug interdiction program. VAW-77 relieved fleet VAW squadrons and VAW-78 of their DEA commitments.

The RAG Concept

Prior to WWII, following flight training carrier aviators were assigned directly to a fleet squadron and received advanced training there in that type squadron's aircraft, as well as mission. Concurrent with the Navy's expansion to a wartime footing, advanced carrier training groups were set up on each coast to train pilots and crews, as well as carrier ship's company. Early in the war, the Naval Air Operational Training Command was established for pilot and aircrew type training. During 1944, CVG-97 on the East Coast and CVG-98 on the West Coast were formed as initial replacement training air groups, while CVG-99 in the Marianas and CVG-100 at Barbers Point were formed as advanced combat training groups. All but CVG-98 were disestablished in 1945 and '46. CVG-98 (CVG-21 after 15 Nov 1946) remained until 5 Aug 1947, at which time specific type training in individual aircraft types reverted to the squadron level.

With the advent of jet aviation and higher-performance aircraft entering the fleet, the Navy recognized that handing a new pilot fresh out of flight school an aircraft handbook and "showing him how to start it" was no longer going to work. Squadron COs had their own training methods, and the accident rate usually varied accordingly. Finding new pilots wasn't the problem—airplanes were becoming too expensive.

Lack of fleet training in type became critical in the early '50s. The swept-wing Grumman F9F-6 Cougar had a high accident rate, and new aircraft, such as the North American FJ-3 Fury, Vought F7U-3 Cutlass, and Douglas F4D-1 Skyray were either entering, or about to enter, fleet service. ComNavAirPac innovatively set up a special training unit in VC-3, the West Coast night fighter squadron, to introduce new aircraft and train personnel, much like the WWII replacement air groups had done. This unit was called Transitional Training Unit Pacific and based at NAS Moffett Field, CA, under LCDR "Bud" Sickel.

CAPT Jerry O'Rourke missed getting a carrier CAG tour, but his two assignments as a replacement training squadron CO during the fleet introduction of the F-4 Phantom were major contributions to the fleet CAGs. USN photo, courtesy CAPT G.G. O'Rourke, USN(Ret)

Eventually, the Navy designed two outstanding programs, Naval Air Training and Operating Procedures Standardization (NATOPS), and a new Replacement Air Group (RAG) concept. The late CAPT Gerry O'Rourke, USN(Ret), told about the latter. His story also shows some of the complexities of bringing a new aircraft—the McDonnell F-4 Phantom II—into the fleet. CAPT O'Rourke died 30 Dec 1997.

After a long series of pilot training schemes, each of which fell short in one way or another, the "RAG concept" was begun in the late '50s and is still going strong today. In 1958 shorebound Readiness Air Groups (immediately dubbed "RAGs") were formed under direct control of AirLant and AirPac, with individual "RAG squadrons" each majoring in a single operational aircraft type (VF, VA, VAM, VAH, VAW, VS, and later VAQ). Each RAG squadron was conveniently co-located at a major coastal naval air station where all the fleet squadrons of that type that were not deployed were concentrated.

The single noteworthy concentration disadvantage was that the fleet air group squadrons, while shore-based between deployments, were widely scattered, making it tough on the CAGs and their needs for joint tactical training. By far the major benefit introduced by the RAG concept, however, was to free the fleet squadrons from the onerous, time-consuming and oft-frustrating task of getting both the nugget pilot from the training command and the second- or third-tour pilot coming off shore duty up to speed on tactics and carrier operations. All RAG graduates were expected to be fully combat ready on arrival at their squadron, a goal that previous systems had never really achieved since WWII.

The early years of the RAG were made particularly demanding by the widespread replacement of props by jets, the new need for *all* carrier pilots to fly in all-weather conditions, and the newer breed of multi-place carrier aircraft. Requirement for "team smarts" included historic pilot skills and overall proficiency in using new and highly-sophisticated sensors and weapons, as well as mastering the different tactics of hi-tech warfare. An entirely new breed of tailhooker, the naval flight officer (NFO), to man and operate the systems of the A-6s, F-4s, A-3s, A-5s, and E-2s was born. Integration of the pilot-NFO team generally presented the most demanding training challenge of the RAG.

The "CAG of the RAG" on each coast was a junior captain. By necessity, he traveled a hel-uv-a-lot, had to take any heat from the fleet and OpNav bigwigs, and couldn't devote much individual attention to the intricacies of all the new systems or the highly-integrated team training. He could and did, however, manage to snag a few traps on almost every qual cruise of his far-flung RAG squadrons and was, therefore, the envy of all the rest of his colleagues whose fourth stripe had irrevocably promoted them out of an operational cockpit.

On the other hand, the CO of the individual RAG squadron was a commander fresh from his tour as a fleet squadron skipper and therefore fully up to speed on equipment and tactics. Nevertheless, he had a multitude of challenges to face almost every day and night. The sheer size of these squadrons, usually outnumbering that of a full fleet air group, was bad enough, but with more than half of its personnel "in transition" while en route to their designated fleet squadron, there were literally mountains of routine, basic organizational chores. Aircraft utilization needs were extremely high, so much so that seven-day, around-the-clock scheduling was sometimes the case. While this routine was considered normal for a fleet squadron during pre-deployment workup periods, the RAG squadron, serving as many as nine sea-going "customers," rarely had any respite. Keeping up the morale of the maintenance crews was always a worry, but not often a problem, since most of the senior enlisted men were, like the CO and instructor pilots, also fresh from a fleet squadron tour and well understood the origin of the needs. The training of the second- and third-tour pilots, at least in actually flying the aircraft and getting aboard the boat, was relatively easy—getting them to truly believe in the fancy new systems, the need for all-weather skills and, most importantly, accepting the NFO as a full team member was often a gut-wrenching task.

The training of the nugget pilots, who had already gotten to know and appreciate NFOs in flight training, had to place prime emphasis on flying and tactics, often leaving inadequate time for developing the technical skills for fully exploiting the sophisticated weapon systems. The new NFOs had to be trained on an almost individual basis and, in the absence of experienced NFOs, almost totally by the RAG squadron IPs (Instructor Pilots). At least a few of these IPs initially approached this "newness" with all the enthusiasm of an old dog. In short, large-scale, but highly-individualized "attitude changing" was far and away the most demanding challenge to the early RAG COs. My solution to it was the historic one, i.e., throw enough flight time at the problem and it will go away!

Development of tactics and improvement of operational procedures, however, was the foremost of these "type responsibilities." As any good new idea surfaced, ashore or afloat, it quickly became an item for intensive discussion, trials, testing, possible inclusion in training syllabi and, most importantly, recommendations for further changes involving other elements or assets of the fleet or even the entire Navy. In my day, one of these programs was the opening of full career opportunities to the new breed of NFO. That fight was extremely tough, starting in the 1959-60 era, but not really laid to rest, as least on emotional terms, until VADM Dick Dunleavy, an ex-A-5 NFO, became the top boss of all naval aviation in the early '90s. Throughout all the heated discussions, studies, and surveys, as well as supplying much of the documentary and practical arguments for such an earth-shaking change were the F-4, A-6, and A-5 RAG squadron COs from both fleets.

I was something of an oddity—I had two tours as the boss of RAG outfits, sandwiched around a fleet squadron command. My first RAG tour, as a LCDR, was as OinC of VF-101 Det A during the initial introduction of the F-4. Det A was set up specifically for this task at NAS Oceana, VA, while its parent squadron, VF-101, remained at NAS Key West, FL, finishing off the F4D-1 training pipeline. "101 Alpha" started off in early 1960 with a handful of people, some embryo radar intercept officers (then called "observers") fresh from a hurry-up course at James Connally AFB in Texas. A few of the pilots, like myself, came out of the VF-101 F4D pipeline; others entered directly from shore duty tours. At a short "Fleet Introduction Program" at VF-121, the PacFlt F3H and new F4H RAG squadron, we got our very first rides in the brand new, "Phabulous Phantom," then called 128 series F4H-1s. Our short stay at VF-121 was not a particularly happy one for me, since I felt (in a very outspoken manner) that F4 training there was not being accorded the priority it deserved. Needless to say, I was not a popular fellow at Miramar. Just as soon as we could escape that scene, we brought about a half-dozen F4Hs back to Oceana to join the growing assemblage of "Druts," F3D-2T2 *Skyknights*, scrounged out of Navy and Marine inventories and hastily fitted out with, essentially, F3H radars. We were also given a few F9F-8Ts, as well. The sleek -8Ts, by the way, were so naturally stealthy that we had to install "spinners" under their wings to enhance their radar reflectivity. Oddly enough, at that time, our most critical assets were these ancient Druts, simply because they were all we had for giving basic radar training to the pilots and basic intercept training to the nugget RIOs.

Our biggest "competition" at 101 Alpha was the PacFlt RAG, VF-121. We fought like tigers with them for new production birds, kept close tabs on how well they were doing, how they were doing it, etc. Almost all of our many essential interchanges with them tended to be rather cool, largely as the aftermath of our unhappiness during the FIP. We also fostered a healthy, spirited competition with the single-seat F8U Crusader community, often referring to them as the "uneducated" segment of the fighter world, who didn't know how to spell radar, much less use one.

By the time I left 101 Alpha in early '62, it had become a very big outfit, with some 300 or more RAG people alone, plus a full transitioning fleet squadron, plus the beginnings of replacement pipelines. VF-74, VF-102, and VF-41 had been transitioned into the Phantom. Eventually, VF-101 moved from Key West to Oceana, and the training reverted to the normal RAG style of training individuals rather than entire squadrons.

After leaving 101 Alpha I commanded VF-102 aboard the brand-new *Enterprise* (CVA(N)-65). We had two Med cruises, several Soviet overflights, and the Cuban Missile Crisis to contend with, but no combat. Low points for me were a cold cat shot which my fine young RIO, Dave Philo, did not survive, and a selection interview in Washington by which I became one of the many Rickover Rejects.

When rotation time came around for me at VF-102, I had a firm eye on a CAG billet, hopefully in PacFlt. Instead, I was shocked to be told that I was to take over VF-121, the F-4 RAG at Miramar. The only good news was that CAPT Larry Flint, CAG of RCVG-12 out there, was a prime instigator of the orders, so that come what may, I could count on solid support from my immediate boss. And that was exactly the way it worked out.

The organization and function of the air group commander and his staff were well established by the end of WWII and further refined during Korea. Two wars had proved the efficiency of the system, but in 11 short years after Korea it would face its greatest test during the 10-year long Vietnam War.

The Super CAGs

By the time of the 1973 U.S. withdrawal from the Vietnam War, the complexity of carrier warfare was increasing dramatically. New "smart" weapons were being developed, while aircraft were getting bigger and becoming more sophisticated. By the early '80s, each carrier squadron mission had become an intense warfare specialty of its own, to the point that it was felt by some naval leaders that it was no longer possible for one man to be an expert in all these fields while keeping himself prepared to fly and fight. CAG was no longer expected to be only a leader and warrior; the "new world order" coming over the horizon dictated that he additionally be a political scientist as well as a sea and land warfare tactician. He was also expected to have a firm grasp of such areas as elec-

Secretary of the Navy, John F. Lehman, Jr., instituted the only major change in the CAG system that had been in effect for more than 50 years. The current senior air wing commander ("Super CAG") program was initiated in 1985 under Lehman's close direction. USN photo, PH2 Don Koralewski

tronic warfare and high-tech munitions, along with being an intelligence specialist, to name a few of his expanding roles.

Tragically, it became apparent that the lessons of the Vietnam War had not been well learned. On the early morning of 4 Dec 1983, a two-carrier raid was launched from the Med against Syrian positions in Lebanon. The raid was primarily planned, and strictly controlled from Washington, as well as other higher authority, hauntingly similar to Vietnam strikes. Little damage of military consequence was suffered by the Syrians, resulting in great embarrassment to the Navy. The CAGs, who had litle control over the raid, were unfairly criticized by the Secretary of the Navy, John F. Lehman, Jr., for the failure of the attack, which had resulted in the loss of an A-6, and an A-7 flown by CAG 6, CDR Edward K. "Hunyak" Andrews. The A-6 pilot was killed and his B/N captured (later returned), but Andrews was rescued. Lehman used the raid as a fulcrum to institute new far-ranging carrier warfare policies. At the time of the Lebanon raid, air wing tactics were esentially developed by the individual wings during their turnaround training period.

The Lebanon raid, coupled with the new warfare and political complexities, caused a questioning of the CAG selection process—was the Navy getting the "best and the brightest"? Another problem recognized was that carrier aviators were being denied the same opportunity to reach flag rank that their fellow naval officers were in other fields. The lengthy system carrier aviators had to pass through and the limited number of qualifying billets to make flag rank were the culprits. The system additionally made them older than their contemporaries from other communities, thereby shortening the aviator's flag careers.

Lehman, one of the youngest men (often referred to as "John-Boy") to hold the SecNav job, was forceful, bright, and headstrong. He was also a Reserve A-6 B/N, although he had never seen active naval service, having transferred from the Air Force Reserve. An extreme "hands on" secretary, Lehman began issuing edicts addressing many facets of the Navy. His directives concerning naval aviation tactics and CAGs were controversial, and some were met with considerable resistance from Navy traditionalists. Resistance from his admirals didn't bother Lehman much, who often had little regard for their counsel. He was noted for conducting impromptu, informal late-evening "conferences" with junior officers at the NAS Oceana officer's club bar following his Reserve weekend drills. How much those evenings influenced his decision making process has been widely debated among naval aviators. In 1984 Lehman ordered a study of the existing CAG program that eventually resulted in the senior air wing commander concept, or "Super CAG," as it was commonly known.

In December 1984, then-COMO Jeremy D. "Bear" Taylor, a special assistant on CNO's staff, was tasked by the Vice Chief of Naval Operations, (VCNO) ADM Ronald J. Hays, to draft a proposal for Lehman's Super CAG program. Bear had already sent a letter to 50 flag officers to canvass their opinions on the existing CAG situation and solicit their proposal for change. One exchange between Bear and the VCNO is enlightening:

> ...I resisted by asking the VCNO to get a delay until the improvements to the old CAG program suggested by 50 admirals had been tried. Admiral Hays was persistent and adamant. I kept trying for the delay, and in one parry suggested that somebody needed to tell the emperor that he wasn't wearing any clothes. At that, VCNO admonished me with: "...that is disrespectful, Bear. Now go to your desk and draft a proposal that I can take forward." The die had been cast, and even the leaders of naval aviation didn't know it.

VADM Robert F. Dunn, USN(Ret), was Commander Naval Air Forces Atlantic during the time of Lehman's directive. A former CAG 7, Bob Dunn's comments on the situation are revealing:

> ...None of the senior aviators at the time liked the idea, but when Lehman pushed something there was no turning back. He perceived most aviation admirals to be mossbacks and drags on the wheels of progress, anyhow.
>
> Ostensibly, the reason for Super CAG was so that aviators did not have to spend valuable time learning all that "Blackshoe stuff." Statements were made, I am told, like, "Anyone can drive a steamboat." I feel there was also a gross misunderstanding of the relationship between the carrier CO and the CAG. Part of that was brought on by Lehman's penchant for adopting untested ideas heard at Happy Hour in Oceana, and part was brought on by eager battle group commanders who had been neither CAGs nor carrier COs and dealt directly with the embarked CAG. It was easy to conclude that the ship CO was an unnecessary layer between the admiral and the wing.
>
> Another reason was the long time it took for an aviator to go through all the training, including deep draft, before going to carrier command. This was a special problem with the nuclear ships, and with more and more of them coming into the inventory it was seen as an ever-increasing problem. Several of us proposed alternatives to solve the training pipeline problem and preserve the CAG system that had served the Navy so well over the years, but to no avail. Lehman's mind was made up.
>
> His point man on the issue was Bear Taylor. Bear was merely trying to do what SecNav wanted, and I'm not sure how much he ever really believed in the concept. At any rate, the Super CAG was established and guaranteed successful right off the bat, with assignment of the first two front runners destined for flag, anyhow.
>
> After Lehman left the Pentagon, CNO ADM Carl Trost asked me (I was Op-05 [Deputy Chief of Naval Operations (Air)] by then) whether we ought to retain the Super CAG or go back to the earlier system. My reply was that we had to keep it for a while because too many outstanding people were now caught up in the pipeline, and we'd sacrifice them with a precipitous change. Instead, along with Jim Service (ComNavAirPac) and Dick Dunleavy (ComNavAirLant), we planned to use the screening process to lower quietly, gradually, and subtly the screen point for CAG (Super CAG) back to

what it had been pre-Lehman. We were making progress on that when the plan became unglued due to inattention when the principal players changed.

Gerry O'Rourke also gives his views on Super CAG:

> Initially, I was philosophically opposed to the Super CAG concept as unnecessary condescension to overweening bureaucracy, as well as a violation of the one-ship, one-captain tradition. However, after a few hours of conversation with then-Commodore Fred Lewis on the subject, I became a convert. Modern air combat is just too complex and involves too much integration with other naval and military elements to be properly handled by a commander CAG with part-time supervision by a non-flying ship CO captain. The stakes are too high for any miscue, and the intricacies of the command relations too demanding. Still, the move does introduce more bureaucracy. I hope to hell that the four-striper CAGs chosen can retain the spirit, emotion, and dedication to people and ideals of a younger commander who is closer to the kids in the cockpits and on the deck—the truly great operators of naval aviation who have always had to do the best they could with what they had on hand, but who always seem to get the job done.

And so Super CAG was born—to be even more immortal than before.

The new system was implemented in 1985 and was described by Lehman in an interview published in the Fall issue of "*Wings of Gold*" magazine: "The Super CAG recognizes the air wing commander as the 'on scene' expert in carrier battle group strike warfare. It underlines that fact by formalizing the tie between the CAG and the battle group commander, filling the position with a more experienced 0-6 (captain) aviator with more extensive intelligence grounding and tactical training, and recognizing this as a major command."

One of the major objections to the new concept was that it effectively took the carrier CO out of the loop and connected CAG directly to the battle group commander. Under the previous system, CAG had worked directly for the carrier CO. Lehman addressed the question of who was to fight the ship, CAG, or the skipper?: "...Super CAG does not change the role of the ship's CO fighting the ship. Projection of power at sea or ashore is not defined as 'fighting the ship'—the strategy and execution of power projection and strike warfare is the responsibility of CAG and the battle group commander." This statement did, however, vastly alter the position of the carrier CO. Although nearly always under the direction of a flag officer in battle, the carrier skipper previously had been the ultimate authority of the ship, with ultimate responsibility for her performance and safety.

Two existing air wings were converted to Super CAG status on a trial basis for evaluation: CVW-8 on the East Coast at Virginia Beach and CVW-15 at Miramar. They were, however, to be different. CAG 8 was to have one deputy commander as an assistant for the daily administration of the air wing. The wing intelligence and maintenance officer billets were to be upgraded in rank to the more experienced commander level. CAG 15 was to be supported by two "group commander" assistants, air strike (offensive) warfare and air sea (defensive) warfare. The air strike commander would be responsible for the training and operational readiness of the medium attack (VAM), light attack (VAL), and electronic warfare (VAQ) squadrons. The air sea warfare commander would be responsible for the fighter (VF), anti-submarine warfare (VS) (HS), and early warning (VAW) squadrons. The group commanders were to take over the responsibilities of the old CAGs. Squadron COs would no longer report directly to CAG, but to the group commander instead.

To qualify an officer for Super CAG, an extensive training pipeline was designed; this, along with the normal progression of billets to be held before selection to CAG. The prospective CAG would undergo exhaustive training that included an expanded Strike Leaders Training Syllabus (SLATS) at the Naval Strike Warfare Center, the Commanders Tactical Training Course, and a special training syllabus at the Navy Fighter Weapons School (Topgun). He would receive additional training in Tomahawk cruise missile employment,

CAPT Fredrick L. "Bad Fred" Lewis (RADM USN) held the CAG 8 job in 1981 and was heading for command of a carrier when he was selected as one of two men to initiate the senior air wing commander program. Bad Fred became the first Super CAG, 8 Jan 1986. Bob Lawson

platform weapon systems, and at the Marine Aviation Warfare Training Squadron (MAWTS) full Weapons Tactic Instructor (WTI) course. Then there was intelligence training and briefings at NWSC, Fleet Intelligence Centers, and the various intelligence agencies in Washington, followed by policy briefings by CNO's staff and participation in a major fleet exercise. No, CAG was no longer a mere leader and warrior.

The men selected to initiate the new program were CAPTs Frederick L. "Bad Fred" Lewis (CAG 8) and Ronald J. "Zap" Zlatoper (CAG 15). Both men were exemplary naval officers well established on the path to flag rank. Bad Fred had already held the CAG 8 job once before, and Zap was a former CAG 1.

Fred Lewis was born 16 Dec 1940 in Los Angeles, CA. He graduated from the Naval Academy in the Class of 1962 and was designated naval aviator in November of the following year. After a one-year instructor tour in VT-2 at NAS Whiting Field, FL, and F-4 pilot training, he reported to VF-74 where he participated in the 1965–66 Mediterranean and 1967 Tonkin Gulf deployments of *Forrestal.* Next assigned to VF-33, Lewis deployed to WestPac during 1968 in *America* (CVA-66) where he flew combat missions in Vietnam. He was then selected for test pilot training at Pax River, graduating in October 1969.

After serving two years as a project test pilot with Service Test at NATC, LT Lewis went to VF-14 in July 1972 after F-4 refresher training in VF-101. Subsequent to two Med deployments in *Kennedy* (CVA-67), Bad Fred transitioned to the F-14. He then reported to VF-101 as OinC of the Atlantic Fleet's Tomcat replacement training squadron.

In November 1976, Bad Fred was selected for command and reported to VF-142 as XO until assuming command in January 1978, which he held until March 1979. He then served as chief of staff for Commander, Fighter Wing One at Oceana. Following that tour, he commanded CVW-8 in *Nimitz* from June 1980 until September 1981, leading the wing during highly-successful Gulf of Sidra operations. Upon completion of his next assignment as air wing training officer for ComNavAirLant, Fred commanded Fighter Wing One from April 1982 until October 1983. Next came a CNO staff tour accompanied by participation in the Program for Senior Defense Managers at Harvard University before he became Senior CAG 8 from January 1986 until July 1987. Selected for flag rank, Fred assumed duties the following month as Director, Strike and Amphibious Warfare on the CNO staff. In October 1988, RADM Lewis assumed command of Tactical Wings Atlantic at Oceana, with additional duties of Commander, Naval Safety Center, in Norfolk, relinquishing the latter in May 1989.

Bad Fred Lewis was selected in March 1993 to head the Navy's new Naval Doctrine Command at Norfolk, the position he held until his retirement in September 1995. Fred gives his observations of the Senior CAG program:

> In 1984–1985 I had been tracking through the pipeline training en route to *Iwo Jima* (LPH-2), including completion of SOSMRC (Senior Officers Material Readiness Course) in Idaho Falls and PCO (Prospective Commanding Officer) training at Newport, RI. (I guess that makes me the *only* Senior CAG who had any ship's engineering training before or since.) In May 1985, before my pipeline training was completed, while embarked on board *Guadalcanal* (LPH-7) with Tom Treanor, I received a call from VADM Ed Martin informing me that I had been selected along with Ron Zlatoper to initiate the Super CAG program. To say that I was surprised, shocked, and upset is putting it mildly, as I had my sights set on deep draft and ultimately CV command. Zlatoper and I met with some senior aviation officers to discuss the program to try to understand the vision as to where the new program should go. Zlatoper's disappointment in not going to his deep draft command was much more severe than mine. I had to continuously encourage him that what we were about to do was exciting and in the best interests of the Navy. It turned out, though, that we basically had a blank check to design our own training track, and we had access to any intelligence and combat command or staff in the U.S. and Europe. The next six months were spent carrying out this plan.
>
> Originally, the Super CAG concept included having two deputies—one for strike and one for ASW. I thought this was overkill in the extreme and got the concept changed so that only one deputy would be assigned.
>
> As we looked at the CV deployments at the time, *Nimitz* (East Coast) and *Carl Vinson* (West Coast) would be the ships in which this concept would be applied. I was senior to my partner, so I naturally had first pick and chose *Nimitz.* I targeted January 1986 for my change of command in CVW-8. I relieved Dan Rainey then and assumed command as CAG 8 for the second time, but this time with Fox Fallon in the personage of my deputy. Zlatoper's change of command followed later. Thus, I was the *first* Super CAG.
>
> The Super CAG program has evolved over the years to the point that it is institutionalized and universally accepted. As you can imagine when we first started, there was great resistance to the profound change that the Super CAG program brought to naval aviation.
>
> The Deputy CAG program has changed several times during the interim. Op-05 was never in favor of the Deputy CAG idea, and just prior to Tailhook '90, with the concurrence of other senior naval aviation leadership, including Zlatoper, disestablished the program. Needless to say, this caused great turbulence in the system, especially to those in pipeline training, and I'm afraid some careers were hurt. Joe Prueher and I fought against the disestablishment, and Desert Storm proved us right. Air wings having deputies during Desert Storm were well served, and those who did not, were not. Since Desert Storm, the Deputy CAG program has been reinstated and has become a "fleet-up" program for succession of command.
>
> In terms of warfighting enhancement, the Super CAG program clearly succeeded in the areas of SW/ASUW (strike warfare/anti-ship underwater warfare). The ASUW role was a new one for naval aviation, as now we had control of surface assets and submarine-launched TLAM missiles (Tomahawk land attack missiles). Basically, the Senior CAG program gave the battle group commander an expert in strike warfare who could

recommend to the embarked flag whether or not it was a good day for 16-inch guns, TLAM, snake eaters (special operations), or all the weight of firepower a carrier air wing could bring to bear.

Admiral Lewis now serves as executive director of the National Training Systems Association, as well as on the board of directors for the Tailhook Association. He lives with his "beautiful Aussie wife" Allison in Alexandria, VA. At this writing, their son, 1st Lt. Lance Lewis, USMC, was on his way to flying F/A-18s, and their daughter, Ashley, was a language instructor in France.

The West Coast goes Super

Ron Zlatoper headed for the West Coast and CVW-15. Zap was a Cleveland, Ohio, native and had received his commission in 1963 through the Naval ROTC program at Rensselaer Polytechnic Institute. After flight training he went to VA-65 to fly A-6 *Intruders*. During 1965-68 he flew combat missions over North Vietnam from *Constellation* and *Forrestal*. In subsequent assignments, he flew the A-6 Intruder with VA-42 and VA-34, while deploying with the latter in *Kennedy*. In 1978–80 he served as XO, then CO, of VA-85 in *Forrestal*.

Ashore, Zap was Air-Surface Guided Missile Program Coordinator and F/A-18 Program Coordinator in the office of CNO from 1975–77. He also was Executive Assistant to the Director, Office of Program Appraisal, for SecNav during 1980–81. He then served as Military Assistant to the Secretary of Defense 1983–85.

Also headed for carrier command before being sidetracked to the Senior CAG program was CAPT Ronald J. "Zap" Zlatoper (ADM USN). Zap became the senior air wing commander of CVW-15 on 16 Feb 1986. USN photo

In senior operations assignments, Zlatoper commanded CVW-1 in *America* 1982–83, was Senior CAG 15 in *Vinson* and chief of staff for ComSeventhFlt in *Blue Ridge* (LCC-19) 1987–88. After selection to rear admiral in 1988, Zap went to the staff of Chief of Naval Personnel.

Returning to sea in July 1990, RADM Zlatoper took command of Carrier Group Seven. Five months later, commanding the eight ships of Ranger Battle Group, he deployed to the Arabian Gulf for Operations Desert Shield and Desert Storm. Under his leadership, the battle group executed the first cruise missile and aircraft attacks on Iraqi forces. As the anti-surface warfare commander in the Arabian Gulf, he was responsible for the destruction of the Iraqi Navy.

Promoted to vice admiral in 1991, Zap took on the dual role as Chief of Naval Personnel and Deputy Chief of Naval Operations for Manpower and Personnel. On 6 Aug 1994 he became Commander in Chief, U.S. Pacific Fleet, the position he held until his retirement 1 Jan 1997. Zap's Super CAG observations follow:

> The assignment as one of the first two Senior CAGs came to me as a bolt out of the blue. I was within two days of checking out as the Military Assistant to the Secretary of Defense and reporting to Idaho Falls for my SOSMRC school when I received a call from then Op-05 VADM Ed Martin, who said that I was to become a Senior CAG and be assigned to Air Wing 15 in Miramar. Fred Lewis, the other initial Senior CAG, and I spent a week in the Op-15 spaces in the Pentagon drafting a rough charter to explain what the concept was. In essence, the only real differences I sensed, at least on paper, were that the command was now designated "major," and the air wing commander would now report directly to the battle group commander with additional duty to the CO of the carrier, vice reporting directly to the CO of the carrier and additional duty to the battle group commander.
>
> When I showed up in California, I was somewhat of an anomaly in that no one was quite sure what I was supposed to do or what type of "backing" I had. The initial tour was somewhat complicated by the fact that within a week of showing up on the West Coast we began air wing/carrier workups. Hence, the press of keeping aviators safely airborne and recovering them back aboard overcame the more leisurely process of establishing an area of "turf," if you will. From my personal observations during the period that I was CAG 15 (Feb 1986–May 1987) there was more a period of "feeling out" relationships than anything else between the carrier CO and myself. The then serving AirPac was not overly supportive of the concept. The battle group commander considered it a *fait accompli,* and the CO of the carrier was of the impression that it was a diminution of his past areas of authority and therefore was not totally supportive of the concept.
>
> During the intervening years, I think the program has been well structured and now serves the Navy well. Specifically, in a downsizing Navy as we go from 15 carriers down to 12 and

perhaps fewer, major command becomes a more scarce opportunity while remaining an important career attainment for anyone who desires to assume positions of higher authority in the Navy. As we downsize, submarine major command selection remains very high, and surface major command opportunities stay about where they have been historically—but absent the designation of air wings as major commands, aviation command opportunities would shrink drastically. Based on any criteria that one desires to use (people, mission, resources, importance), major command designation is definitely warranted.

Fred Lewis' Deputy CAG, CAPT Bill "Fox" Fallon, went on to become one of the new Senior CAGs, taking over CVW-8 in 1990. Fox presents his view of the program:

> After Navy Secretary John Lehman directed a trial of the new Super CAG concept, two trial air wings were set up to operate somewhat differently. I became CAPT "Bad Fred" Lewis' Deputy CAG in CVW-8, while CAPT "Zap" Zlatoper was to have two deputies, CDRs Lyle "Ho Chi" Bien and "Packo" Campbell in CVW-15. All of us began a series of *ad hoc* training tracks of 3–6 months, during which time both coasts were directed to go to the single DCAG route (Ho Chi Bien became the West Coast DCAG). Also about this time (late '85) Lehman directed that the program be implemented Navy-wide without waiting for the trial results.
>
> The program lurched forward during the next few years, hindered by continued resistance from some senior admirals and an ill-defined CAG role in the battle group. Highly successful with increased combat readiness in its initial deployments, Senior CAG's responsibilities and roles varied widely until the early '90s when the position stabilized into the principal advisor for strike warfare to the battle group commander, along with being ASUW and strike warfare commander in the battle group command structure. Directly subordinate to the battle group commander, CAG was assigned additionally to the CV CO for ship related matters.

The current CAG program has evolved to initial screen (in first year after selection for 0-6) as DCAG for a nominal 18-month tour. Training track is 4–6 months, with carrier qualification in two types of aircraft and familiarization/beach NATOPS qualification in others, plus ground schools. CVW staffs have 8–9 other staff officers, as well as 12–15 enlisted members.

Wing to Wing and Leadership

From wing commander to air group commander to air wing commander; from air group to air wing; the titles of the men and the units they have led for the past 70-plus years have changed. The ships and aircraft have evolved with an incredible increase in technology never dreamed of when it all started in 1922. However, although much more sophisticated, and with vastly-greater responsibility, the fundamental requirement of the Navy's CAGs remains the same—leadership. Saburo Sakai, the Imperial Japanese Navy ace and former enemy of U.S. Navy carrier aviators, says it well: "What I was most enthusiastic about was to be a truly good leader (who) could produce more good pilots, to take care of them...I feel much more proud of myself when I think that I never lost a wingman in combat, than when I think about the number of my victories."

3

The Men

"Long live the fighting hearts. And when we pass on—no problem. We will be back. Our fighting hearts soar like eagles. Long live the fighting hearts."

RADM Jeremy D. Taylor, USN, former CAG 3
Tailhook Symposium, 1990

The Men

For more than seventy years the Navy has had selected aviators assigned the task of leading its aviation groups from aircraft carriers. During that time the United States has been involved in four wars and almost inumerable confrontations around the world that have required the services of carrier aviators. The following profiles provide a spectrum of carrier aviation leaders and carrier operations from 1941 until 1994. No special criteria was used in the selection of these men to appear in this book. Some were selected because of their reputations as outstanding CAGs. Others were selected at random, or because of their involvement in noteworthy "peacetime" carrier operations. Wide latitude was given to each CAG to tell his story in the manner in which he saw fit, and to discuss areas of his tour of duty that he felt were memorable.

CDR MALCOLM T. WORDELL, USN (CAPT Ret)
CAG 44 2 Feb 44-11 Jun 45
USS *Langley* (CVL-27)

"Mac" Wordell enlisted in the Navy as a 17-year-old recruit 1 February 1929; 15 years later to the day, he assumed command of Light Carrier Air Group Forty-Four as a commander. He already had one Navy Cross and was on his way to earning a second. Serving two years in the fleet as an enlisted man, Mac won an appointment to the Naval Academy and graduated with the class of '35. After his mandatory two-year stint of sea duty, he entered flight training and was awarded his wings in December 1939.

LTJG Wordell's first squadron tour was with the "Red Rippers" of VF-4 flying F3F-1s from USS Ranger *(CV-4). Transitioning*

"Mac" Wordell, Navy ace with 7.5 victories and winner of two Navy Crosses, one in the Atlantic theater and one in the Pacific. Photo courtesy CAPT M.T. Wordell, USN(Ret)

to the F4F Wildcat, The Rippers (now VF-41) and Ranger *supported the invasion of North Africa in Operation Torch during 8-11 November 1942. In a four-day battle against the Vichy French, Ranger Air Group and aircraft from the escort carriers* Sangamon *(ACV-26),* Suwannee *(ACV-27), and* Santee *(ACV-29) covered the landings of Army troops in Morocco. VF-41 skipper Mac Wordell led a flight of 16 Wildcats in an attack against French ships exiting Casablanca harbor, where he was shot down by anti-aircraft fire. This action earned him his first Navy Cross. Mac wrote of his experiences in "*Wildcats over Casablanca,*" first published in 1943 by Little-Brown.*

During 1943 Mac became executive officer of a test squadron at Patuxent River, MD, then took over the newly-formed CVGL-44 as its first commander 1 Feb 1944. CAG Wordell led the air group through its combat tour in Langley *(CVL-27). While participating in several air engagements in the Pacific, he became an ace by downing 7.5 Japanese aircraft. He earned his second Navy Cross 25 Oct 1944 at the Battle of the Philippine Sea for his participation in strikes against the Japanese carriers* Zuikako *and* Zuiho, *both of which were sunk during the battle. After leading his division on strafing attacks against the carriers and heavily armed cruiser escort, Mac assumed the duty of target coordinator and remained overhead the battle for five hours, effectively directing sea and air attacks against enemy forces.*

Leaving CVLG-44 in spring 1945 after its combat tour, Mac next assumed command of an airborne early warning squadron. He completed his naval career in 1946 with a series of peacetime command and staff assignments. Retiring in La Mesa, CA, Mac Wordell did airport consulting work in Nevada and Utah and remained active in civic affairs until his death in 1996 at La Mesa, CA. He describes his CAG tour:

"I had Air Group 44 from February 1st 1944 until the spring of 1945, while deployed in *Langley* to the South Pacific. The light carriers operated a little differently from the bigger ones in those days; CVL COs were in charge of the air groups unless we were on a mission conducted by the task group commander. Air group commanders had authority when airborne under operation orders.

We never used '00' on our aircraft to identify CAG's plane. My callsign was either 'Bronco 1,' or '99 Bronco.'

Fifty years later the memory is pretty poor, but I do recall some of the details of the 25 Oct 1944 battle (Battle for Leyte Gulf), on which I was the target coordinator over the Japanese fleet. My duty was to direct the incoming strikes to the best targets remaining afloat. Often my suggested targets brought responses like, 'Are you kidding?', etc. With all the humor, though, we got the job done. On the way back to the *Langley* that night, I had the pleasure of steering some of our surface ships to rendezvous with the 'cripples' that were forced to ditch while returning to the task force. Upon returning to the carrier, I gave Admiral Mitscher an up-to-date report on what was left of the Japanese fleet, and that's when he told me I had a second Navy Cross."

CDR DAVID McCAMPBELL, USN (CAPT Ret)
CAG 15 9 Feb 44-10 Jan 45
USS *Hornet* (CV-12); USS *Essex* (CV-9)

Dave McCampbell was a 1933 graduate of the U.S. Naval Academy and became a naval aviator in 1938. He first did a squadron tour with VF-4 in USS Ranger *(CV-4) and then an early-war stint as landing signal officer in* Wasp *(CV-7). McCampbell became commanding officer of VF-15 upon its establishment in September 1943 as part of Air Group 15, assigned to the newly-commissioned* Hornet *(CV-12). One week after CVG-15's return from* Hornet's *brief shakedown (13 Jan–2 Feb 1944), McCampbell relieved CDR William M. Drane as air group commander.*

Accompanying Hornet *to Hawaii in late February, CVG-15 switched to* Essex *(CV-9), commanded by CAPT Ralph A. Oftsie, flying aboard 3 May to commence one of the most successful combat tours of any WWII air group. McCambell led CVG-15 for its entire tour (3 May–17 Nov 44), during which time he became the U.S. Navy's leading ace.*

Essex *and CVG-15 first conducted raids against Marcus and Wake islands. They next participated in the June Marianas Campaign, strikes against Chichi Jima and Iwo Jima in the Bonins, as well as the Battle of the Philippine Sea. After a brief rest and rearmament period, the Fast Carrier/air group team returned to combat at Palau and Morotai, then moved to the Philippines. An October respite was followed by strikes against Okinawa and Formosa before departing the combat area 17 November in* Bunker Hill *(CV-17), for a triumphant return to the United States.*

CDR Dave McCampbell, CAG 15, in F6F-5 MINSI III adorned with 21 victory flags. McCampbell became the U.S. Navy's most successful fighter pilot with a record 34 aerial victories during his six-month combat tour in *Essex* (CV-9). USN Photo

Dave McCampbell was awarded the Medal of Honor, the only Fast Carrier Task Force pilot to achieve this status. The U.S. Navy's most successful fighter pilot lived a quiet life of retirement and disdained the role of hero, feeling that he merely "did his job." He recalled some of his experiences as CAG 15 prior to his death in June 1996:

"My relations with *Essex's* skippers (Oftsie and CAPT C.W. Wieber after 7 Aug 1944) were quite good; they allowed me to run my own show." However, RADM Fredrick C. Sherman, Commander Carrier Division One, in *Essex*, could sometimes be another matter. Sherman wasn't happy with his CAG operating in the fighter pilot mode, preferring instead McCampbell take a more supervisory role. However, CAG still managed to add five enemy aircraft to his score in one sortie during the "Marianas Turkey Shoot," 19 June 1944, along with two more on his second sortie of the day.

On 24 October, during the Battle of Leyte Gulf, Dave nearly missed going at all. "The air officer passed the word for all fighter pilots to man their planes. I asked if that included me, since I was a fighter pilot." After some discussion at the flag bridge level, permission was granted for McCampbell to go. His F6F-5 Hellcat "Minsi III" was located on the hangar deck, where it had to be quickly brought to the flight deck and spotted on the catapult for launch. A hurried refueling resulted in a full drop tank with the main fuselage tanks only partially full; not an ideal situation to participate in a full-blown air battle. During the ensuing long melee, McCampbell, possibly taking the aggressive fighter pilot syndrome to the extreme, and with only his wingman, took on a formation of 60 enemy aircraft. He downed nine Japanese aircraft and scored two probables, bringing his total to 30. He was the only U.S. pilot to ever achieve nine victories in one mission. Out of ammunition and low on fuel, CAG 15 diverted to *Langley* (CVL-27) for recovery, where his engine quit from fuel starvation as he taxied out of the landing area. McCampbell's reward for this feat? "I got my ass chewed out by the admiral." Presumably, RADM Sherman was able to forgive his CAG's enthusiasm for combat after telling him "not to ever let it happen again," and later endorsing a recommendation for the Medal of Honor.

As CAG 15, McCampbell ended his six-month combat tour and the war with a total of 34 confirmed aerial victories, the highest scoring Navy pilot and third highest U.S. pilot, after Army Air Force MAJs Dick Bong and Thomas S. McGuire, both of whom were later killed. McCampbell was the only pilot to achieve such a record in one combat tour.

Dave McCampbell's postwar career included command of NATTC Jacksonville, FL; CO *Bon Homme Richard* (CVA-31); as well as tours with the Joint Chiefs of Staff and North American Defense Command before his 1964 retirement in Lake Worth, FL.

CDR JACKSON D. ARNOLD, USN (ADM Ret)
CAG 2 1 Jun 44 – 1 Dec 44
USS *Hornet* (CV-12)

A 1934 graduate of the Naval Academy, Jack Arnold won his wings of gold at Pensacola in 1937. First assigned to Torpedo Squadron Six in USS Enterprise *(CV-6), he next served in the cruiser* Savannah *(CL-42) as senior aviator during 1938-40. He was a test pilot and inspector at the Assembly and Repair (A&R) unit at NAS Ford Island, Pearl Harbor, during the 7 Dec 1941 Japanese attack.*

Arnold's next assignment was CO of VT-2 beginning with its June 1943 formation, leading the squadron into combat from Hornet *until he became CAG 2 on 1 Jun 1944. CAG Arnold led his air group through the Marianas, Bonin Islands, and Philippines campaigns, as well as others, until CVG-2 was ordered to return to the U.S. in November 1944. Admiral Arnold is now retired in Rancho Santa Fe, CA.*

"I was assigned to Air Group Two formed at Quonset Pt., RI, in June 1943. At that time, CDR Roy L. Johnson had the group, and I was CO of Torpedo Two.

We were assigned to *Hornet* in February 1944 in Honolulu. Our first strike was in March to mine the harbor of Palau, then in April we struck New Guinea and Truk. During May and early June we worked over Saipan, Tinian, and Guam. On 1 June I relieved Roy Johnson as air group commander.

Action against the Bonins continued with the first strike on Iwo Jima, my first as CAG 2. Strikes continued against Guam, and

CDR Jackson D. Arnold, CAG 2 in *Hornet* (CV-12) June–Dec 1944, during some of the heaviest fighting of the war in the Pacific. A torpedo bomber pilot, Arnold normally led his air group in a TBF Avenger. USN photo, courtesy ADM J.D. Arnold, USN(Ret)

on 19 June, the Japanese fleet launched their major attack against our forces. Their intent was to strike and rearm, then refuel on Guam to return to their fleet. The major air battle of the war took place—later called the 'Marianas Turkey Shoot.' Our planes shot down 428 Japanese aircraft that day. *(Postwar analysis placed Japanese losses at 275–350 aircraft from all causes.)* That day RADM 'Jocko' Clark, CTG 58.1 aboard *Hornet,* insisted I fly a torpedo plane to coordinate an attack on the airfield at Agana, Guam, as the Japanese attempted to land. (Communications were better in a TBF than the F6F.)

Next Day, 20 June, I led several hundred planes in pursuit of the Japanese Fleet (later called 'First Battle of the Philippine Sea'). We were launched in spite of the fact we all knew it was beyond our range to strike and return. We returned after dark, and in the melee that followed, about 80 planes made night water landings; most of the pilots and crewmen were later picked up. Several Japanese carriers were sunk, including *Shokaku,* which was damaged and later sunk by a submarine.

Strikes in July continued against Guam and the Bonin and Volcano Islands. During 6–21 July, we supported Army and Marine landings on Guam, and 21 September I led the first strike on Manila. We sank a number of ships, then dropped cigars and candy in the prison (Bilibid in Manila) where some of our troops were prisoners. During September we continued working on the Visayan Islands and New Guinea landings.

These were our last real missions before we returned to the States in November 1944.

As far as communication between CAG 2 and the CO of *Hornet* was concerned, we had three different skippers. The first was CAPT Miles Browning—a difficult one to work with, and relieved by Admiral Clark 24 May 1944. CAPT W.D. Sample replaced him, and communications were greatly improved—a fine CO. In August, CAPT Sample was relieved by CAPT 'Arty' Doyle. He was like a football coach, and a real pleasure to work with. With Jocko Clark aboard as TG commander, he had to be briefed by CAG after each strike, but his interest was appreciated, and his enthusiasm stemmed from his early VF experiences.

One of my key officers was the flight surgeon, LCDR Victor Stratton. He was close to all the pilots, and would advise me if he detected apprehension or combat fatigue in a pilot so that I could suspend a flight or two, or do a little discreet counseling.

As for the original CAG use of '00,' I am not certain. When I received my wings in 1937 and joined the formation of Enterprise Air Group as a torpedo squadron pilot, our air group commander, LCDR D.P. Johnson, did not use the 00 designation. When I assumed command of the six-plane unit on the *Savannah* (CL-42) in 1938–40, I don't believe 00 had been used in carrier air groups. I was test flying aircraft for the Assembly & Repair unit on Ford Island, Pearl Harbor, when the Japanese struck on 7 Dec 1941, and I don't recall 00 had yet been born. As CAG 2, my radio callsign was '99 Ripper.'"

CDR ROBERT E. RIERA, USN (RADM Ret)
CAG 11 13 Oct 44-9 Apr 45
USS *Hornet* (CV–12)
CAG 1 (CVG-1) May 47-2 Aug 48
USS *Midway* (CVB-41)

Robert Emmett "Rebel" Riera was probably destined from birth to be a naval aviator. Born in Pensacola, FL, 30 Nov 1912, he entered the Naval Academy in 1931, graduating with the class of '35. He initially served in USS New York *(BB-34) as a junior officer in the gunnery department. He reported to NAS Pensacola for flight training in January 1938 and was designated naval aviator one year later. His first squadron assignment was with Bombing Two in January 1939, embarked in* Lexington *(CV-2).*

Emmett returned to Annapolis in September 1941 for postgraduate training, then reported to Scouting Squadron 16 in July 1942. He participated extensively in the Pacific war, flying from Enterprise *(CV-6),* Lexington *(CV-16), and* Hornet *(CV-12). His numerous combat decorations include the Navy Cross "For extraordinary heroism as Commanding Officer of Bombing Squadron TWENTY in action against large enemy Japanese Fleet Units during the Battle For Leyte Gulf, October 24 and 24, 1944...."*

Riera's first command was VB-20 in Enterprise *before he assumed the leadership of CVG-11 in* Hornet *following the 13 Oct 1944 combat death over Formosa of CDR Fred Schrader, his Annapolis classmate. CAG Riera led his air group against targets in the Ryukyus, Formosa, and Philippine campaigns until it departed* Hornet *in February 1945.*

RADM Emmett Riera led CVG-11 from *Hornet* (CV-12) during the Pacific war where he earned a Navy Cross. During 1947-48 as CAG 1, he took CVBG-1 to the Med embarked in *Midway* (CVB-41) for her first operational deployment. Credit: USN photo, L.W. Wilson.

From March 1945 to March 1947, Emmett served as assistant head of the attack aircraft branch of the Bureau of Aeronautics in Washington, D.C. During the latter month he assumed command of CVBG-1 to lead one of the Navy's largest air groups (120 planes) from its largest aircraft carrier type, the new 55,000-ton Midway. *In July 1948 CDR Riera relinquished command of CVBG-1 and headed for NAS Corpus Christi, TX, for a tour as single-engine training officer. He next was a student at the Naval War College, Newport, RI, from July 1950 to June 1951, then served until November 1952 as operations officer on the staff of Commander Carrier Division Four.*

After a year's service as executive officer of Coral Sea *(CVA-43), Emmett joined the staff of Commander Air Force, Atlantic Fleet in November 1953 as training officer. Returning to Pensacola in May 1955, he reported to Chief of Naval Air Training staff, and in September became chief of staff. Two years later he commanded* Greenwich Bay *(AVP-41), and in March 1960 assumed command of the navy's first supercarrier* Forrestal *(CVA-59).*

Returning to Washington in June 1961, CAPT Riera was a member of the staff of the Chief of Naval Operations and was selected for rear admiral the following month. In February 1963 he assumed command of Carrier Division 14, followed in March 1964 by a tour as Commander Alaskan Sea Frontier, with additional duties as Commandant 17th Naval District/Commander Fleet Air Alaska.

RADM Riera departed the frigid north in March 1966 for the more moderate climate of the Mediterranean Sea when he became Commander Fleet Air Mediterranean and ASW commander for the Sixth Fleet. In June 1968 it was back to CNO's staff until February 1970 when he became Commander Fleet Air, Western Pacific. His last duty station was Commandant of the Eighth Naval District in New Orleans, LA, the post he held from May 1972 until his 1 July 1974 retirement. Admiral Riera has returned to his roots and currently resides in Pensacola. His brief comments about his CAG tours describe his command relationships and the origination of the use of "00" for CAG aircraft.

"During WWII the CAG worked in conjunction with the air officer of the carrier, who was generally a former CAG himself. Both the air officer and CAG worked for and with the carrier CO. The CAGs also worked closely with the task force staff and were often called to the TF flagship for conferences.

As CAG 1 in *Midway* the chain of command was much the same. Since the carrier division staff was normally embarked in *Midway*, the CAG and flag staff worked together very closely. My tours as CAG were both great.

Regarding CAG markings on the aircraft, I believe the '00' came about after WWII started. When I was in VB-2 on board old Lex, as I recall, CAG flew one of the Utility Unit J2Fs, which carried no special CAG markings.

For our combat operations in WWII CAG flew with one of the squadrons, normally in a fighter. (CAG Schrader was killed while strafing in an F6F.) The radio callsign of the plane was the aircraft number plus the ship's callsign. For example, the callsign for *Enterprise* was, 'Texas,' *Hornet* was 'Ginger,' etc. Thus, the radio call for the number one fighter in the *Hornet* VF squadron was 'Ginger One.' This arrangement left no callsign or side number for CAG, and so he was given '00.' This, to the best of my recollection, is how 'double-nuts' got started."

LCOL WILLIAM A. MILLINGTON, USMC (BGEN Ret.)
CAG 4 15 Jan 45-Feb 45
USS *Essex* (CV-9)

BGEN William A. Millington, USMC(Ret) was the first of only two Marine aviators to command a U.S. Navy air group. As the commanding officer of VMF-124 assigned to Air Group Four in Essex, *and senior squadron commander of the air group, Bill assumed the duties of CAG 4 upon the combat loss of CDR Otto Klinsmann, 15 Jan 1945, during a raid against Takao, Formosa. (Hit by flak while leading the strike that day, CAG Klinsmann managed to return to the task force screen and ditch alongside a destroyer. However, he was unable to escape from the cockpit and went down with his aircraft.)*

Graduating from the University of Washington in 1936, Bill soon entered basic flight training at NAS Sand Point, WA, and then went on to Pensacola in 1937 as a Marine Aviation Cadet. His first operational assignment was VMF-2, Marine Aircraft Group Two, at North Island, NAS San Diego, CA. In 1940 he returned to Pensacola as a flight instructor, where he stayed until the following year when he went to NAS Floyd Bennett Field, NY, again as a flight instructor. In 1942 he reported to American Airlines' multi-

***Right:* LCOL Bill Millington was the first of only two Marines to ever command a Navy air group when he became CAG 4 in *Essex* 15 Jan 1945. USMC photo, courtesy BGEN Wm. A. Millington, USMC(Ret).**

engine training at Forth Worth, Texas. However, the Marines needed fighter pilots, and Bill was ordered as executive officer to VMF-123, MAG-12, at Camp Kearny Mesa (later NAS Miramar), CA, and then on to Guadalcanal. On 14 July 1943, he became commanding officer of VMF-124 on Guadalcanal. Bill scored his first aerial victory 8 August when he downed a Japanese Pete seaplane. He brought the squadron back to MCAS Mojave, CA, in October to re-form. VMF-124 departed San Diego in late October in Ticonderoga *(CV-14) for transportation to Pearl Harbor. The squadron trained at MCAS Ewa and conducted carrier qualifications on board* Bataan *(CVL-29) before embarking in* Essex *two months later at Ulithi with VMF-213 as units of CVG-4. Assignment of Marine squadrons to the carriers was to increase the number of fighters on the carriers as defense against Japanese* kamikaze *suicide plane attacks, which began during the Philippines campaign.*

While assigned to CVG-4 the two Marine F4U-1D Corsair squadrons operated as a single unit under the command of LCOL Millington. As they began combat operations, most of the Marine pilots each had only about 12 carrier landings, those gained in Hawaii during the previous months. Bill relates some of his Essex *experiences:*

"We (VMF-124 and VMF-213) reported aboard *Essex* at Ulithi 28 Dec 1944 (54 pilots, four ground officers, and 120 enlisted). The rest of our squadron personnel remained at MCAS Ewa. We had a total of 36 Corsairs aboard. Our complement was the first of several Marine fighter squadrons sent to the Pacific Fleet to relieve a shortage of Navy pilots, and to introduce the Corsair to carrier combat duty.

On returning from a strike against the enemy on Formosa, I learned of CDR Otto Klinsmann's death. I was summoned to the bridge, where our task group commander, RADM Frederick Sherman, informed me I was now CAG 4. CDR Klinsmann was a fine officer and very able air group commander. His death was a tragic loss. CVG-4 consisted of one Navy fighter, one bomber, and one torpedo squadron, along with the two Marine fighter squadrons. Fortunately, we quickly learned to work and operate as a team. We were readily accepted wholeheartedly by the entire ship's company and air group personnel. Liaison with the ship's captain and staff, as well as Admiral Sherman and staff, was very smooth. The landing signal officer, LT Billy Sunday, was outstanding; he had no previous experience with Corsairs, but quickly adjusted in fine fashion. All our pilots had a high regard for him.

As to our young Marine pilots (average age 22-23), I was proud of their performance. They had only 400 hours flight time and 12 carrier landings each, and operated in some *extremely* adverse weather, but they did well. We had more than our share of operational losses, but it's a wonder we didn't have more, considering their limited experience. Our early operational losses were mostly a result of insufficient instrument flight experience and overwater navigation under fleet conditions. Our carrier operation losses again can only be attributed to our lack of fleet experience. Most of these were simply stalling out on launch or during landing approach.

While serving as CAG I had no particular plane assigned to me—I flew whatever Corsair was available, and so my plane had no special markings."

As CO of the two Marine fighter squadrons, Bill first led his pilots into combat 3 Jan 1945 escorting VT-4's TBM Avengers on a strike against Kogi Airfield, Formosa. On this mission, he became the first carrier-based Marine to score an aerial victory when he downed a Japanese Frances night fighter. He also led attacks against the enemy in the Philippines, followed by missions over French Indochina, before becoming CAG 4 in mid-month. Under Millington CVG-4 continued its combat operations against Hainan, Tokyo, Iwo Jima, and Okinawa. On 25 February CAG Millington shot down a Tony fighter over Kumagaya airfield to end the war with three victories before Essex *returned to Ulithi in early March.*

Departing Essex *and CVG-4 that month, Bill took his Marines back to the States and assumed his next assignment as MAG-44's XO at Ewa.*

In 1946 he attended Command and Staff School, after which he reported to Headquarters Marine Corps in Washington until 1948. During 1948-50 he was Attaché at Oslo, Norway, then attended Senior Amphibious Warfare School in 1951. In 1952 Bill became CO of the Marine Air Detachment at NAS Seattle, WA, before his next command as CO of MAG-24's all-weather jet fighters at MCAS Cherry Pt., NC. Returning to Washington, D.C., he was Operations Officer, Division of Aviation, Marine Corps Headquarters 1954-56.

BGEN Millington retired from the Corps in 1956 to join a private business firm in Seattle. He now resides in El Cajon, CA.

CDR LOUIS H. BAUER, USN (CAPT Ret)
CAG 3 2 Aug 45-30 Jul 47
USS *Kearsarge* (CV-33)/USS *Randolph* (CV-15)

Lou Bauer graduated from the U.S. Naval Academy with the class of 1935. His first fleet assignment before entering flight training in January 1938 was as a junior officer under training in USS Ranger *(CV-4). Upon winning his wings in 1939 at Pensacola, Lou went immediately into fighters with VF-2, the famed "Flying Chiefs" squadron, composed of all enlisted Naval Aviation Pilots (NAPs) with officers in key billets. As officer-in-charge of a VF-6 detachment from* Enterprise *(CV-6), temporarily land-based at Guadalcanal, Lou distinguished himself by leading his F4F Wildcat-flying squadronmates through combat that included the confirmed downing of 28 Japanese aircraft in one day.*

After subsequent commanding officer tours of VF-6 and VF-3 flying combat operations in the Pacific from Enterprise, Saratoga *(CV-3), and HMS* Victorious, *Lou became CAG 3 in August 1945 to prepare for the invasion of Japan. Instead, the end of WWII provided CAG Bauer with the opportunity to transition his group to a peacetime mode.*

His later tours before retiring 1 Jul 1965 included assignments as executive officer, Bon Homme Richard *(CVA-31) during the Korean War and several staff tours. He was commanding officer of the seaplane tender* Orca *(AVP-49) and* Yorktown *(CVA-10), before his final tour as CO, NAS Miramar.*

Following retirement, Lou became a patent agent and worked for General Atomic Co., then Salk Institute before again retiring in 1982 at La Jolla, CA, where he now enjoys his home above the Pacific Ocean. Lou reminisces:

"During my tour as CAG 3 at NAAS Oceana, VA, the air group was composed as follows: VF-3 (F6Fs), LCDR Ed Bayers; VBF-3 (F6Fs), LCDR Frank Lawler; VB-3 (SB2Cs), LCDR "Badge" Badger; and VT-3 (TBMs), LCDR Roy Belcher. The four squadrons were formed at various locations, such as Atlantic City and Quonset Point, and assembled at Oceana as an air group, at which time I reported as air group commander.

Other than routine squadron and air group training (group gropes), we enjoyed such assignments as group formation flights over New York for the Navy Day celebration and participation in the August 1946 Cleveland Air Races by flying 16 of Ed Bayers' F6Fs in formation and tail-chasing over the airfield. We attended the Miami All American Air Maneuvers in January 1947 and participated in an air show at Danville, VA, as well as performing various demonstrations in the Norfolk and Oceana areas. (For which I received a few letters complaining about the low-altitude noise.)

Both the VF and VBF squadrons (and of course CAG) transitioned to F8Fs during October 1946.

We made the usual training cruise to Guantanamo Bay, Cuba, and during June and July 1947, the group was deployed in both *Kearsarge* (CV-33) and *Randolph* for the summer Midshipman Cruise. The midshipmen all had flights in the multi-seaters, and of course, the fighters flew with them. We visited the Med and made a courtesy call at Edinburgh, Scotland. The Flag was aboard *Randolph*, and I was advised to keep pajamas and a toothbrush aboard both ships. Frequently, while airborne, I was directed to land aboard *Randolph* for meetings with the admiral and staff.

I first used the 'CAG' letters designation on the F6F, and also on the F8F. I later changed to the '00' designation, but I don't remember why.

We initially were getting ready to go back out to the Pacific, but with the end of the war, we had a pretty routine tour. There were no problems with demobilization, and I had a stable group with no losses to civilian life. With the cutbacks in active carriers and virtually no extended deployments, we pretty much enjoyed ourselves."

CDR Lou Bauer, CAG 3 in his F8F-1 Bearcat on board USS *Kearsarge* (CV-33) during his 1945-47 CAG tour. Lou initially identified his personal aircraft by use of "CAG," but later switched to "00." USN photo, courtesy CAPT L.H. Bauer, USN(Ret)

CDR ROBERT G. DOSÉ, USN (CAPT Ret)
CAG 14 14 Dec 45-March 46
CAG 81/13 13 Mar 46-May 47
USS *Princeton* (CV-37)
CAG 4 (Replacement Air Group)
3 Jul 58-22 Jan 60

From Great Lakes TG-1 biplane torpedo bombers to supersonic F-8 Crusaders, *Bob Dosé brought them all aboard a carrier. Entering the Navy during September 1936 as an Aviation Cadet, Bob completed flight training one year later and reported to Torpedo Squadron Three in USS* Saratoga (CV-3) *as a commissioned Aviation Cadet.*

His next duty was as a flight instructor at Pensacola and then Corpus Christi before becoming XO of VC-13 in the escort carrier Core *(ACV-13).*

Bob joined VF-12 as XO under LCDR "Jumpin' Joe" Clifton, before fleeting up to CO in November 1943. He led VF-12 in more than 200 hours of combat, flying F6F-3s from Saratoga, *and is credited with shooting down two Japanese fighters while sharing a third.*

Combat was followed by another instructor tour, this time in fighters at NAAS Green Cove Springs, FL. Bob also was operations officer in Gilbert Islands *(CVE-107) before his two fleet CAG assignments, 1945–47. Demobilization forced CAG Dosé to lose nearly half of his experienced CVAG-13 pilots and to ready 40 neophyte aviators direct from the training command—most of whom had not flown a fleet-type aircraft—for an impending deployment in* Princeton *(CV-37). However, demobilization also allowed Bob two air group commands, albeit the first, CVG-14, being a short tour as a result of the air group's disestablishment in the drastic postwar cutbacks.*

One of the Navy's early jet pilots, Bob checked out in the TV-1 Shooting Star and F2H Banshee at NATC Patuxent River in 1948. He held two staff assignments before becoming XO of Kearsarge *(CVA-33) in 1954, then took command of Air Development Squadron Three (VX-3) during 1955-57. During this tour Bob established several naval aviation "firsts." He was the first squadron pilot to fire a Sidewinder air-air missile, and he made the first U.S. Navy mirror carrier landing, on board* Bennington *(CVA-20). On 6 June 1957, he and his wingman, LCDR Paul Miller, established a transcontinental speed record while flying F8U-1* Crusaders *from* Bon Homme Richard *(CVA-31) 50 miles west of San Diego, CA, across the U.S. to land aboard* Saratoga *(CVA-60), 50 miles east of Jacksonville, FL, in 3 hr. and 28 min.*

A dubious "first" also held by Bob Dosé is that of becoming the first pilot, and a captain at that, to eject from the Crusader *using the fully-automatic Martin-Baker ejection system. After attending the Naval War College, Bob was tasked with establishing the Navy's first modern replacement air group (RAG), a concept abandoned by the Navy after WWII. He then commanded* Mispillion *(AO-105),* Midway *(CVA-41), and the Military Assistance Advisory Group, the Netherlands, before retiring 1 Jul 1968.*

Bob Dosé flew more than 7,000 hours in 86 separate aircraft types during his Navy career and lived in La Jolla, CA, at the time of his death in October 1998.

"My first CAG job was in December 1945 as CAG 14 in Alameda. Of note were the COs assigned to that group. They were a cut above most: Fighting Fourteen, 'Blackie' Weinel; Bombing Fourteen, 'Snapper' Knapp; and Torpedo Fourteen, 'Stud' House.

I remember changing the type of formation flying radically for the bombers and torpeckers. We scrapped the three-plane parade formation and went into free-cruise (finger) formation of four planes as the basic units. They loved the flexibility of that type of formation for the maneuverability it gave them. Actually, one month after we formed, we flew our first group grope. However, we were ordered to stand down CVG-14 in June 1946, and I was ordered to Air Group 81.

It was CVG-81 when I became CAG and was redesignated CVAG-13 in November 1946. I received despatch orders to CVG-81 and took the 'Red-eye' that night to Norfolk. I flew one period of bounce in the F4U-4 which I had been flying, and took the group out to our ship, the *Princeton,* the next morning. I remember watching in horror as the group made a mock attack on the task force. I had never in my life seen such a disorganized attack. They flew like a swarm of bees. I remember one string of planes pulling out of their dives into a cloud from the east as another string was entering the same cloud from the west. They didn't know anything about carrier discipline. For instance, they took a whole hour to come aboard. When they all finally got back aboard ship I called the COs into the fighter ready room. I was angry! My opening remark to them was, 'Gentlemen, amateur hour is over!' That was for starters. They later worked themselves into a wonderfully functional, experienced, and happy air group.

My next CAG tour, Air Group Four, was the first replacement air group. With wonderful staff assistance from ComAirLant, VADM William Rees, and people on his staff, such as Chief of Staff Tommy Booth (who had been exec of the *Princeton* when I was aboard), and Sam Brown as operations, we set up the group in 1958.

This new concept meant a lot to all of us who worked with it, to the point of sentimentality. We realized that it would save many young lives. I remember one evening in Florida when we came back to our dock in my little cabin cruiser, I was having a final drink with my wife Betty and Sam Brown. In all sincerity, I told Sam that I thought he, personally, with the AirLant staff and others behind him, had done more good for naval aviation during their shepherding of the group than anyone or anything else in the Navy. When I looked at him, he had tears running down his cheeks! Those tears were not from the drinks, either. The RAGs have reduced the accident rate dramatically, and have produced much more professionally trained air groups in the Navy.

This was, I think, the most productive job of my Navy career. I feel I did more good for the Navy in this job than in any other.

To begin with, I was in the Naval War College in Newport, RI, when I learned that I was to be given this new air group with its new mission of establishing Level Readiness. I was told in the middle of my scholastic year in War College that I was to put this new concept into practice.

I chose as my big thesis 'The Level Readiness Concept.' I planned what squadrons I wanted and where I wanted them. AirLant staff and BuPers backed us completely. I had complete control of who would be in command of the squadrons. I flew many visits to AirLant staff and to BuPers. Have you ever heard of any person in any armed service who has ever been blessed with such a break?

Note that our syllabus called for four and one-half months of training and 110 hours in type. We graduated our first student in four and one-half months. He was Bob McDonough, a Crusader pilot, who went on to make a good name for himself with the Blues and in other important duties.

I don't know much about the present or future CAG duties, since I've been retired now for 25 years. I do consider the CAG job to be vital to the smooth running and flying of the group. I always felt more like a lion tamer than a commanding officer. With all these young, ambitious, and aggressive flyers in the group (including the COs), I think the air group commander will always have a vital part to play in command and coordination."

Bob Dosé also commanded Replacement Air Group Four at NAS Cecil Field, FL. CAG 4's F9F-8 is aboard *Saratoga's* deck edge elevator 24 Aug 56 while under command of CDR Wm. E. Fly. USN photo, courtesy Dick Starinchak

Holder of three CAG jobs, Bob Dosé flew everything from Great Lakes TG-1 biplanes as a commissioned aviation cadet to supersonic F8U Crusaders as a captain. USN photo, courtesy CAPT R.G. Dosé, USN(Ret)

CDR RALPH WEYMOUTH, USN (VADM Ret)
CAG 11 19 Aug 50-22 May 51
USS *Philippine Sea* (CV-47)

Ralph Weymouth was born 26 May 1917 and entered the Naval Academy in 1934, graduating with the class of '38. He served his obligatory two-year blackshoe stint as an ensign in surface ships before reporting for flight training in September 1940 at NAS Pensacola and NRAB Opa Locka, FL.

Winning his naval aviator wings in March 1941, LTJG Weymouth began his flying career in SBD Dauntlesses with Scouting Squadron Three embarked in Saratoga *(CV-3), where he participated in Pacific combat operations, including being shorebased at Guadalcanal during September to October 1942.*

Remaining with the faithful SBD, Ralph next went to Bombing 16 in Lexington *(CV-16) where he flew more combat, including operations in the Marshalls, Kwajalien, Hollandia, Peleliu, Truk, Saipan, Battle of the Philippine Sea, and Guam. He was executive officer of VB-16 January–April 1943, then CO until July 1944 when he returned Stateside to become Dive Bombing Training Officer of the Naval Air Operational Training staff at NAS Jacksonville, FL, flying SBDs and SB2Cs until June 1946.*

Weymouth attended the Naval Postgraduate School until June 1948 where he studied aeronautical engineering. He then continued his studies at Massachusetts Institute of Technology where he earned a masters degree in aeronautical engineering in September 1949. He then became Navigator and Inactivation Officer in Kearsarge *(CV-33) until May 1950.*

Returning to the cockpit, Ralph became CO of VF-112 (May-July 1950) and flew combat operations in Korea from Philippine Sea. *On 19 August he assumed command of Air Group 11. From June 1951 to July 1956 he held staff jobs in Washington, D.C., and Honolulu, and also attended the Armed Forces Staff College. In August 1956, as a captain, he took over as Head, Aviation Department, U.S. Naval Academy.*

CAPT Weymouth's deep draft command was Duxbury Bay *(AVP-38) during July 1959–July 1960. He then commanded* Lake Champlain *(CVS-39) July 1960–July 1961 before returning to Washington and studies at the National War College. After Graduation in June 1962, Weymouth served several more Pentagon staff tours before becoming Commander Iceland Defense Force in January 1965 as a rear admiral. In May 1968 he became Commander ASW Group Three in the Tonkin Gulf, then in March 1970, Commander Fleet Air Wings Atlantic.*

VADM Weymouth's final duties were as Director Navy Program Planning, then Director Research, Development Test and Evaluation on CNO's staff. He retired from active duty 1 Jan 1973 and currently resides in Wonalancet, NH, where he is active in several organizations, such as Veterans for Peace and the National Peace Foundation.

Having served in three wars Admiral Weymouth understands martial combat and the importance of peace. One of his training grounds was the initial stages of the Korean War, where he flew from the second U.S. carrier to enter that war in 1950.

CDR Ralph Weymouth in early June 1950 as skipper of VF-112. In just a few days he would be heading for war in *Philippine Sea* and become CAG 11 upon the death of CDR Sully Vogel. Photo by Mrs. Laurel Weymouth

"I was CAG 11 during a combat tour off Korea from about early August 1950 until my transfer in May 1951 when Air Group 11 returned to the States. I had commanded an SBD squadron during WWII and felt the need to move into the jet era, and so had gotten command of VF-112 flying F9Fs just before the Korean action started. We were embarked in *Philippine Sea*, which departed San Diego just after the war began.

I took over the group when CDR Sully Vogel was killed over Seoul during the Inchon landings. Sully's F4U was hit by flak during an attack; his Corsair caught fire and he bailed out, but his parachute also caught fire and failed to open properly. We later found Sully's shallow grave and returned him home.

The position of CAG had certain anomalies. Shorebased, CAG had full command authority given to all commanding officers. But aboard ship the air group became an integral part of the ship's company. While CAG continued to run the group in just about every way, he lost his command status, including disciplinary authority, to the commanding officer of the ship and, thereby, was quite downgraded in the eyes of the CO and executive officer of the ship. This situation created an uncomfortable problem for the carrier CO, XO, CAG, and squadron skippers. Generally common sense and good leadership fostered good relations despite this peculiar arrangement. Occasionally, a carrier exec and, less often, a carrier skipper would view the situation as an usurpation of responsibility and authority and would make trouble for the air group. The common sense split was that the ship would accept full handling of the housekeeping and deck handling while deferring to the air group to handle all personnel management, aircraft maintenance, tactics in the air with the OTC (Officer in Tactical Command), etc. The carrier skipper, being a fellow of possibly the maximum experience (though often the pace of technology growth left him poorly equipped), and well placed to observe the whole picture, put his oar into affairs, but most often he did so ably and through a personal touch with CAG.

In my own CAG experience I had two different carrier executive officers. Under the first one all went quite smoothly; problems got worked out between us on a face to face basis, and none of the problems went deep into the group. The second exec, who came on scene in the middle of a long tour, refused to accept the inherent friction as something to be solved, but rather something for him to personally manage. Yet he was without much sense of how men are constituted, how they behave under stress, or how they are to be swayed in appropriate directions. Under the anxiety of his own personal views and the difficulties of getting things done in his often injudicious way, he became more and more difficult to deal with and more and more of a martinet who generated antipathy. This fellow might have caused serious problems no matter what the organizational arrangement, but the anomalies of the then existing relationships aggravated the problems enormously.

Let me outline just one small issue this exec posed for the group. In *Philippine Sea*, the designed access for pilots going between ready rooms and flight deck was up a ladder that went directly past the executive officer's cabin. The exec was upset with the noise that the ladder passage would make at odd hours when he might be at work at his desk or trying to sleep. We and ship's company attempted to be thoughtfully quiet on the ladder and to silence its fastenings, but never successfully enough to satisfy the exec. So he had the hatch bolted shut, and the route became very long and inconvenient for pilots loaded with their gear trying to reach their cockpits on deck. The issue became a sore point and exploded when, while I was away in Korea, someone (surely a somewhat tight pilot returning from the once-a-month two-to-four day liberty in Sasebo) defecated on the exec's carpet in his cabin. Never before or since had I known of so open a revolt in any command with which I was associated. Of course, this incident only aggravated our tensions. The whole affair went on for several months and got me an unsatisfactory draft fitness report, which the carrier skipper refused to sign.

My predecessor as CAG flew all the group's aircraft types: F4U, AD, and F9F. Since I had relieved at sea and in mid-war, my opportunities to check out were limited; though I did get all but CarQual'd in the other types, most of my work was done in the F9F. On several occasions I did some close support coordination work over the Pusan perimeter from ADs, but the majority of my coordination of distant attacks was done in F9s. A principal example was an attack on the Yalu bridges by three air groups. With the F9 I was on scene early and encouraging the troops to keep on coming, since it was CAFB ('clear as a f...... bell', the F-86 pilot's expression of that period) over the bridge, and the three air groups were pressing on across the peninsula in a tough snowstorm. I was able to stay on station to provide coordination until the attack was completed, despite the lesser endurance of the jet. But, that situation did not ameliorate the complaints from the group that I should be there in a Corsair or Skyraider, which remained always with the attacking aircraft. From my view I had a superior vantage in getting there early and being able to see better and have more freedom from Chinese MiGs. The fellows did have a real point in that CAG did not share all the experience of an attack group...in the cited case it was a rough storm, and from my seat above I had not let them return before making the attack. I had crossed the mountains in the clear at 35,000 feet while they were forced to transit the storm.

CVG 11 gained much experience early. We were the second carrier on line in July 1950 and outlasted some nine carriers that came and went before we were relieved. Part of this extension of tour came because our losses were low. Though one out of four of our pilots had a serious incident like ditching, land crash, parachuting, etc., we lost only four pilots during our nine-month deployment due to good and fortuitous rescue. Deployment extensions (we went past some four scheduled return dates) were hard to handle, and there were not any goodies that could be bestowed to make up for it.

During the early Pusan perimeter days when things got desperate for the Army, we went in to give some intensive close support of a negative nature (one usually associates close support with an offensive stance, but, in this case, it was defensive—we even had a request to 'shoot anything that moves within five miles of the front.') At the outset we couldn't correlate targets, and so I went into Pusan headquarters to learn the trouble. There were four sets of map coordinates in use by Army, Air Force, Marines, and Navy! Thereafter, I learned to carry all four systems with me in the back of the AD to be able to interpret between whatever was requested from the ground over to the charts naval aviators were using.

Out of this experience I got a call in November 1950, in Yokosuka, when we all thought the war was winding down and were ready to head home, to report to the Seventh Fleet flagship where ComSeventh briefed me on the Chinese invasion and the need to improve coordination with the several forces ashore in Korea. Coordination had been terrible; we had seemed to have completely forgotten what had been learned in WWII and so well demonstrated in Okinawa and elsewhere. This was the time of the Air Force touchiness and 'the revolt of the admirals,' so that each service was defending its independence even on the battlefield! At the time, the Air Force had a joint operation manual—I've forgotten its title now—but the Army had not yet subscribed to it, nor the Navy, but it was being followed somewhat in that JOC (Joint Operations Center) had been established and was then operating under Eighth Army. The Navy was there with three naval officers, one an aviator, with the title of Navy Liaison Group. Their workings at the time had been limited to communication with the Far East Command in Tokyo—there was no contact with Seventh Fleet and its carriers. Seventh Fleet had sketchy contact directly with Fifth Air Force, and it was that which had governed the Navy through area divisions and occasional target requests (like the Yalu bridges mentioned above). Altogether a very messy affair stemming from interservice rivalry. At the time there was even a further rivalry in that MacArthur had established a division within the Army by creating an east coast command (under General Almond, I believe), including the Marines, totally independent from the Eighth Army—it did not even have anything but mail communication with Seoul and the Eighty Army.

With the Chinese coming down on Almond and the Marines, as well as the Eighth Army, Admiral Struble recognized the carriers were not being used to full potential, so he put me in the plane for Seoul with orders to 'go over there for a week and find out how we can talk to them.' I armed myself with two frequencies and the promise that the carriers would monitor them until I could get some talk going. In Seoul the Air Force and the Army each gave me a communication team to monitor these two frequencies, and within a few days ComSeventh and the carriers could talk to me. It was up to me to find a way to coordinate carrier effort with Fifth Air Force and Eighth Army. I read the manual and saw that it afforded entry to the JOC by the Navy, even though Navy was as obstinate as the others and would never sign that manual. So, I usurped the Navy Liaison Group and called ourselves an equal in the JOC. Since the manual was there, the others had to accept (and because I was offering a way to get badly needed air more effectively into the fray). So the job became one of getting the carriers to give me what they would for the western three-quarters of the peninsula (they did have a line finally to Almond's Marines and worked hard to help them out coming down from Chosin Resevoir), and getting targeting info for this offered air support. By Christmas the JOC and all Americans were chased out of Seoul to reestablish in Taegu and the Almond command was folded, so the JOC became the focus for all Seventh Fleet support ashore. It was not so ideal as it sounds. The timing required to get offers from the fleet, or requests from Eighth Army, and the difficulties of getting weapons requests or assignments across, left a lot to be desired. In addition, there was a real problem of enough aircraft radio frequencies to handle all those aircraft in so small an area. The AF wanted to use a system that would only just manage to handle their fighters but which would not work for all unless one brought the aircraft down low enough to use mountain shielding. It was a mess, but use of the JOC method, even though the Navy would have none of it officially, did get more effective use of the carriers. And it did give us the chance to educate that generation of Air Force and Army people in Korea about how to get Navy help. I spent two months ashore in Korea doing this work, and finally got relieved by a team of four men.

That absence from my group, worthwhile as it was from the total fleet view, was not in my best interest as CAG. My squadrons had kept up their work well, no losses while I was gone, but the continuation of the war and setback by the Chinese had fostered some perspectives that were not good battle behavior. One of my squadrons had taken up the 'shoot anything that moves' attitude, and without adequate attention to Geneva Conventions or good sense. It took me some difficult sessions to slow down the blowing away of farmers' oxcarts in the back country. Certainly, some carts near the front lines were carrying war supplies, but the majority well away from action were just survival farmers, and their destruction could have only negative results. I don't think I ever did eliminate this trend which was later so exaggerated in Vietnam. Being absent those two months did not help the infusion of my desires on the conduct of the air war.

CDR WILLIAM H. LEONARD, USN (RADM Ret)
CAG 17 2 Nov 50-Feb 51
USS *Coral Sea* (CVB-43)

Bill Leonard was born in the U.S. Army at Douglas, AZ, in January 1916. He graduated with the U.S. Naval Academy Class of '38. After two years in the cruiser and battleship navy he completed flight training in early 1941. His first aviation tour was VF-41 (later, VF-42), soon to be based on board USS Yorktown *(CV-5) for the Atlantic Neutrality Patrol. As VF-42, his squadron went to the Pacific Fleet in December 1941 with* Yorktown *and fought in various early carrier actions, including Coral Sea and Midway. In the latter, he was in air combat when* Yorktown *was disabled and managed to save his fighter to carry on the battle flying from* Enterprise *and* Hornet *(CV-8).*

After Midway he reported to VF-11 and was in action at Guadalcanal April to July 1943. In November 1944 he was assigned to the staff of Commander Task Force 38, VADM J.S. McCain, as assistant to CAPT J.S. Thach, the operations officer. This assignment lasted to war's end and entailed flying from Hancock *(CV-19),* Wasp *(CV-18), and* Shangri-la (CV-38). *He is credited with six aerial victories.*

Bill served as a test pilot at the Naval Air Test Center from late 1945 to mid-1948, when he took command of the Navy's first carrier-based jet squadron, VF-171 flying FH-1 Phantoms. He served under CAPT Arleigh A. Burke, August-October 1949 in OpNav during the B-36 controversy. Bill's commands include Air Group 17, VX-5, Salamonie *(AO-26), and* Ranger *(CVA-61). As a flag officer, Commander Carrier Division 14 he was on scene commander for three Gemini mission astronaut recoveries.*

As a rear admiral in an office under the Secretary of Defense, he helped expose critical deficiencies in the F-111B, such that the Navy version of this airplane was canceled and the way cleared for the Navy to develop and procure in its place the F-14. In an earlier assignment in OpNav, he helped originate the specific requirement and pursue the actions that resulted in the Navy's invaluable A-6 all-weather carrier attack airplane.

Leonard flew and tested more than 59 types of military aircraft, including two versions of the Japanese Zero. He was carrier qualified in 15 Navy aircraft types on 17 axial deck carriers and one angled deck.

RADM Leonard retired from the Navy in mid-1971 from his last command, the Naval Safety Center. Bill enjoys a quiet family life on his one-acre spread in the Tidewater, VA, area. He remains active in professional affairs, appearing in symposiums, and has loaned vast support to serious historical works over the years.

RADM Bill Leonard's extensive WWII fighter pilot experiences earned him ace status and a successful postwar tour as CAG 17. USN photo

"I had been CO of VF-171 flying FH-1s, F2H-1s, and F2H-2s since August 1948. In September 1950 we deployed to the Med in *Coral Sea* at the outbreak of the Korean War. It was thought we might be needed to cover the European side of the global communist adventure. *Coral Sea* was a splendid example of the well-appointed carrier, but the XO, CDR De Long, had a run-in with CO, CAPT F.M. 'Trap' Trapnell, and was sent home in October 1950. He was relieved by Dick Kibbe, who went from ops to XO, and Bruce Weber went from CAG 17 to ops CVB-43. In Late October I turned VF-171 over to R.I. Myers, my most able exec, and fleeted up to CAG 17. I had the job for the duration of the deployment, but was snagged for duty ashore (OpNav). I turned CVG-17 over to George 'Spanky' Carter in late February 1951. Mark Russell had been CAG when I reported to VF-17A (later designated VF-171) in August 1948.

As CAG I was carrier qualified in all CVG-17 planes prior to deployment, so enjoyed a mix of F2H-2, F4U-5, and AD-4 flying during my tenure. CAG's assigned '00' plane was AD-4Q 124056. Squadron COs—all crackerjacks—were: VF-171, R.I. Myers; VF-172, C.A. Blouin; VF-173, I.G. Peters; VF-174, R.E. Elliott; and VA-175, J.E. Kennedy. The group had two squadrons of F2H-2s, two squadrons of F4U-5s, and one AD-4 squadron. The jets had two exchange pilots, each from the U.S. Air Force, steady and good people. We had a USMC photo detachment flying F4U-4Ps. CVG-17 was a potent outfit. We did cloak-and-dagger photo work over the Balkans from Trieste on the west to Salonika on the east. ComSixth Fleet was an up-and-at-'em, air-minded type, VADM John Ballentine, and his number one boy was no less than CNO-to be, George Anderson.

CVG-17 began operating with CVB-43 during summer 1950, working up for a routine Med deployment. There were the usual get-acquainted encounters, CVG-vs.-ship, but no notable clashes. I had served with Trap in TF 38 combat when he was Admiral

Radford's chief of staff, and during December 1945 to August 1948. I had previously worked to please him (not easy) as a test pilot at Patuxent where he was Coordinator of Test. As CO CVB-43 he was as I had expected, exacting and somewhat cool; the DeLong episode emphasized this. But once we started operating seriously (and you must be serious in Banshees), on an axial deck with no bingo fields anywhere overseas, we settled into what was a great ship/CVG relationship. Kibbe, a seasoned and fine XO having come up through combat, knew his stuff. Weber was likewise and, as a strong partisan for the air group, combined as part of the magic. Trap, above it all and, as time went on, obviously enjoying his responsibility, was keen to try out new ways to handle the ship and the planes. We used to have pleasant sessions on the bridge as we worked out various ploys. The flag on board, CarDiv Two RADM W.G. Switzer, and company were not inclined to meddle with the air show and left us alone. In all, it was a happy arrangement.

Our safety record was quite good, but we had a fatal F2H-2 crash when the aileron boost went wild, going full-throw at 300 feet in the landing pattern. This accident cost us one of VF-171's Air Force pilots. 'Red Horse' Myers had a similar failure at higher altitude, saving the bacon and bringing back the evidence, for which they yielded a cure, a factory fix. Oh, yes, we had the 'mother of all hangar crashes.' Number one elevator went down out of control with an F2H-2 teetering one wheel on–one wheel off. The elevator went all the way down, and then the Banshee majestically rolled into the pit, with the plane captain jumping free just in time. The ship accepted the blame for the strike damage to the aircraft.

Since this was a happy cruise it generated very little in the way of recommended changes, a tribute to the idea that when you have the right people, things go well, and when you don't...I have seen the 'don't,' too. Both *Enterprise* and *Hornet* seemed to have a sour atmosphere when I served on each during the Battle of Midway. This disparity was in strong contrast to the happy relationship we enjoyed in *Yorktown*, as long as she lasted. In late 1944, CVG-14 in *Wasp* was not a cheerful association. CVG-7 in *Hancock*, in early 1945, was fairly pleasant. CVG-85 in *Shangri-la* May–August 1945 was very unhappy. In each of these exposures I was air ops for CTF 38 and got the group's version of the troubles. The common denominator seemed to me to be personalities. When the captain was a confident and able airman, things went well; when not, the XO or air boss would set the tone, and it might be unpleasant.

Going back to CV-5, the captain was a 'Johnny-come-lately,' so not a respected airman. However, his XOs, surly 'Jocko' Clark and later sunny 'Dixie' Kiefer, along with the strong man of the air department, Murr Arnold, set the tone. Though strict, it was fair, made good air sense, and Yorktown Air Group was confident it was in good hands even though an enemy was making things difficult. CAPT Buckmaster deserves great credit for handling his ship and people in the way he did, cultivating the capabilities to best effect. In the year I served in *Yorktown's* fighter squadron, we had two CAGs. Curt Smiley left in February 1942, having done one combat mission on the first of the month over the Marshalls and Gilberts. He struck me as being an administrator keen on flying little and resting much. He barely got in his flight time.

The next *Yorktown* CAG was Pete Pederson, who fleeted up from command of the VF squadron. He was a good and able squadron commander, but carrier qualified only in the F4F—no platform for a proper CAG. Furthermore, Buckmaster and the Flag preferred to use him as the ship's fighter direction officer. In those primitive days of that function, he was certainly the best one in the ship for that work. He made one foray as CAG when we attacked Salamaua and Lae at New Guinea 10 March 1942. He led the VF sweep with no air threat to CV-2 or CV-5. That flight was his last as CAG. He did well as FDO in *Coral Sea* and *Midway*, but that's not CAG work.

In the WWII Task Force Instructions, the air section was written largely by Jimmy Flatley, Mitscher's Ops Officer. Flatley was wildly indignant at the failure of attack coordination at Coral Sea and Midway, and 'wrote the book' to develop proper employment of CAGs, senior pilots, and squadron commanders to understand and exercise the role of strike coordinator. He also made it a point that op orders and directives that generated attack operations should designate strike coordinators even to the point of indicating who should succeed to the job when necessary. I never got wartime CAG on-the-job training, but worked almost a year for Flatley, so I knew the subject and what he did about it."

CDR CHARLES H. "FRITZ" CRABILL, JR., USN(CAPT Ret)
CAG ATG-1 28 Sep 51-3 July 52
USS *Valley Forge* (CV-45)

"Fritz" Crabill entered the Navy during May 1937 as a seaman second-class through the V-5 Aviation Cadet program. Earning his wings in June 1938, his early squadron assignments include VT-2, USS Lexington *(CV-2), and flight instructor at NAS Miami, FL, before he was sent to Coronado, CA, for training with Commander Amphibious Training Group Pacific. In July 1944 he reported to Commander Support Aircraft Pacific Fleet, under whom he participated in seven amphibious operations in WestPac as OinC of Tactical Air Support Unit 12.*

Fritz states: "During the period July 1944 to May 1946, Tactical Air Support Unit 12 (now called tactical squadron) was like a floating crap game, except that we stayed put while a stream of rear admirals arrived in SoPac, took command for an operation, collected their Legion of Merit, and went back to the Pentagon, or wherever. I worked for nine different admirals and managed to get along well with each in turn. I did not realize how well until 1957 when BuPers finally permitted us to see our fitness reports for that period. Had I known what was in those reports, I might not have fought so hard to get back to my first love—flying!"

CDR "Fritz" Crabill, CAG ATG-1, arrives back on board USS *Valley Forge* (CV-47) after his ditching off of Chongjin, North Korea, 17 April 1952. While flying a strike over Korea in an AD-3, Fritz was shot down by enemy ground fire and rescued by a minesweeper. USN photo

General Line School and a flying refresher course at Jacksonville, FL, was followed by command of VA-55, then more schooling and staff work before Fritz became Chief Staff Officer, ComTacAir Control Group One. He commanded ATG-1 during September 1951-July 1952.

Another amphib tour came before he was operations officer in Ticonderoga *(CVA-14) and then CO of VAW-12. Fritz completed his Navy career with staff assignments and as Inspector General for 11th Naval District before retiring 30 Jun 1963. Since his retirement Frtiz Crabill says: "You asked what I did after retirement. The answer is, essentially, nothing except manage my own affairs for fun and profit. Well, maybe it was more fun than profit, but I have never regretted the decision."*

When Fritz received orders to report to ComAirPac for duty as a prospective air group commander, what he didn't know was that he was getting an air group, but then again, he wasn't—he was to become the first CAG of Air Task Group One, a group of squadrons collected from other air groups to form ATG-1. The air task groups were a successful ploy by the Navy to circumvent congressional limitations on the number of air groups the Navy was allowed. As such, they were never formally established (or "commissioned," a term often erroneously used; only Navy ships are commissioned, all other units are established and disestablished).

ATG-1 fought hard in the first of America's no-win wars and paid the price. During ATG-1's six-month Korean combat tour, the group dropped more bomb tonnage than all carrier air groups combined during the most intense six-month period of WWII in the Pacific. Most of this effort by ATG-1 was in support of Operation Strangle, a WWII Air Force design to interdict enemy supply lines. Many Navy pilots saw Operation Strangle as a cruel hoax and failed policy. Twenty-five ATG-1 aviators failed to return to Valley Forge, *with many others shot down and rescued. One of the latter was Fritz Crabill, who to this day feels "we were never assigned a target that was worth the life of one of my pilots." He continues:*

"By 1950, I was convinced that I was being type-cast with the amphib forces. All of my contemporaries wore ribbons that went with their wings—I didn't even have an Air Medal. So when I heard that I was being ordered to ComTacGruOne, I went, hat in hand, to the BuPers detail officer. I wanted an air group, preferably one consigned to close air support. Our Marines had been chased by hordes of Chinese south from Chosin Reservoir. I pointed out that I had worked with Marines before, knew how their forward air controllers worked with air, artillery, and infantry. That argument might have been counterproductive; in fact, I might have convinced him that I was just the fellow to help ComTacGruOne train and deploy two new TacRons. However, I was promised that, once the TacRons were operational, I would be 'given priority consideration for command of an air group.' But!! When the second TacRon was ready to go in September 1951, BuPers wanted me to take command of it. Note that my little WWII air support control unit was now a command, and as CO, I was being offered a job in a captain's billet. Mind you, I had not seen my fitness reports from WWII. Even if I had I would probably have been blinded by my desperation to get back into flying and would have wondered what the hell good was a fitness report from a blackshoe admiral to a naval aviator. At any rate, I saw the whole plot as more type-casting—'*deja vous* all over again.' So I flew to Washington screaming, 'YOU PROMISED, YOU PROMISED.' At that point, I guess BuPers thought, 'Okay, you bastard, we will give you a bastard outfit and see if you can make it work.' That is how I got ATG-1.

I saw my squadron commanders all together in one room for the first time a week before we sailed aboard *Valley Forge*. Less than half the pilots were currently carrier qualified. We spent a month in Hawaii hurrying that job through. I lost a squadron commander in the rush. By the time we got to Korea, it was too late to provide close air support to anybody. The State Department and JCS had decided to settle for a stalemate. Nevertheless, I had two squadrons of Reserves from the mid-west who were all charged up with the idea that if this war was important enough to take them away from their families and businesses, it was worth their best shot. I lost 25 pilots over Korea. That works out to be one out of every four who sailed with me in *Valley Forge* that did not come home. Ever since, I have wondered what I could (or would) have done to change that fact. Getting shot down was one of my most memorable moments.

I was leader of a flight of ADs and Corsairs making rail cuts along the east coast of North Korea opposite Mayang Do island. We had been making runs in the area for about one and one-half hours and were ready to return to base. I felt a thump while pulling out of a strafing run at about 800 feet. I thought I had been hit, but since the plane continued to function normally, I did not think it was serious and went in for another pass. On pulling out, one of the planes behind me called and said I appeared to be smoking or losing hydraulic fluid. My wingman then pulled in close to inspect my plane and said the whole belly was covered with oil. I continued to climb out over the water and looked for a minesweeper which I had previously noted as being in the area.

At 7,000 feet I was almost directly over the minesweeper when I picked him up about seven miles south of Mayang Do. At about the same time, my oil pressure started to drop off rapidly, and I knew I would have to ditch.

I chopped off throttle and started to let down as rapidly as possible. I wanted to get down before my engine froze, so I could make a power landing. I got out of my parachute, unsnapped the retainer strap across the pararaft, tightened my belt and shoulder straps as tight as I could get them, and checked to insure that the pararaft lanyard was snapped to my Mae West. Then I pulled down my goggles, opened the canopy and headed for a spot about a hundred yards upwind and ahead of the minesweeper, wheels up, flaps down.

The sea was fairly rough with whitecaps whipped up by a wind that was gusting from 25 to 30 knots. At about 100 feet I advanced throttle to make a power landing, but all I got was a clatter of noise and violent vibration—no power. I immediately chopped the throttle again and tried to pick the crest of a wave to set down on. I saw a big one coming, pushed over slightly, and hauled back to catch its top. I caught it about right, although my speed was a little high—80 to 85 knots. The first jar was up against the belly of the plane, then it plunged down the back side of the wave and came to rest just as the next big wave broke over me, filling the cockpit. The landing was abrupt, but not at all bad considering the state of the sea.

I unsnapped my safety belt and climbed up until I was straddling the cockpit and sitting on the canopy. From this position I yanked on my pararaft lanyard until it came out of the pocket in the seat pack. Still sitting on the canopy, I unsnapped the pararaft and pulled the CO^2 toggle. While it was inflating, I jumped down on the wing, but my feet slipped and I fell off the leading edge into the water. I was wearing an Mk III exposure suit and floated very comfortably with water up to my armpits. When the raft was fully inflated (about 30 seconds) I climbed in and started to paddle away from the plane. In the meantime, the pararaft cover and survival pack had floated across the wing, and the lanyard had apparently fouled between the aileron and flap. I made several attempts to free it, but failed and, fearing that the plane would sink and drag the pararaft down, I took out my sheath knife and cut the lanyard.

Once away from the plane, I looked around and saw the minesweeper bearing down on me about 25 yards away. It hove to about 30 feet away when I was thrown a line and quickly hauled aboard. I was in the water less than five minutes.

Aboard the *Ptarmigan* (AM-376) I was treated like a king. Theirs is a monotonous job of patrolling and sweeping, and you would have thought it was I doing them the favor. Later that evening, I was transferred by highline to *Buck* (DD-761), and the following day by helicopter to *Valley Forge*."

CDR JAMES G. DANIELS III, USN (CAPT Ret)
CAG ATG-2 30 Nov 51-6 Feb 53
USS *Essex* (CVA-9)

James Ganson Daniels III, a name as impressive as the man who owns it. Jim Daniels' Navy career spanned an era that brought him into three wars, in which he acquitted himself much more than adequately, or even admirably. Enlisting in the V-5 Aviation Cadet program during May 1938 as an apprentice seaman, Jim completed flight training and was commissioned an ensign in September 1939; his first assignment was VF-6 in USS Enterprise *(CV-6). His WWII combat experiences covered a great deal of the war in the Pacific, including flying the first wartime Combat Air Patrol (CAP) mission (VF-6 from CV-6, 7 Dec 1941) off Oahu, T.H., and the final CAP mission (CO VBF-2 from* Attu *(CVE-102)) 1945 over* Missouri *(BB-63) off Tokyo in the Sea of Japan.*

Jim's second mission on the Day of Infamy was a late afternoon search. His flight was ordered to land at Ford Island that night after failing to locate the Japanese carriers. Five of the six F4F-3 VF-6 Wildcats were shot down, with three pilots killed by friendly fire upon arrival at Pearl. Jim was the only aviator to escape with his life and his airplane while landing at Ford Island.

During the February 1942 carrier attacks on the Marshall Islands, Jim was credited with shooting down one enemy aircraft and sharing another. During a Japanese attack at the Battle of Santa Cruz 26 October 1942, Enterprise's *senior LSO, Robin Lindsey, and his assistant, Jim Daniels, dashed from the LSO platform and manned the rear seat .30 cal. guns of two SBDs, whereupon they shot down an attacking torpedo plane. Taking a breather from combat, Jim next served as an LSO with the Advanced Carrier Training Group Pacific before becoming CO of VBF-2 in January 1945 and returning to the war in the Western Pacific.*

Postwar tours included several school and staff assignments before he assumed command of ATG-2 during fall 1951. More staff duty and schools followed ATG-2 until Daniels became CO of Navasota *(AO-106) 1961–62. His next command was* Ticonderoga *(CVA-14) 1962–63, and Jim finished his career with tours as Carrier Division Five Chief of Staff, Vietnam Operations Officer for the Joint Chiefs of Staff, and, finally, Commander Fleet Air Hawaii, before his 1 July 1970 retirement.*

Jim Daniels was another CAG of one of the Navy's most unusual carrier air groups, Air Task Group Two, formed to fight the Korean War.

"Everyone thought I was nuts when, one day in March 1951, I talked to my CO at Annapolis and asked permission to visit my detailer at BuPers. My next assignment was due in about three months, and I wanted a combat air group in the Pacific, hopefully to get to Korea. The detailer also thought I was crazy, having offered me a number of jobs in the Atlantic. However, he came around and said I could have any one of three air groups deploying to WestPac during the summer of '52. I also requested all the refresher training available, as I hadn't flown a fighter since my Corsair squadron in November '45. He agreed, so after detachment in late August, it was a month in F8Fs at Corpus Christi, with lots of gunnery, then all-weather flight training for another month; just flying, no administrative duties. I then reported to ComNavAirPac as prospective CAG of ATG-2. Next came more schools, close air support (CAS) at the Amphibious Base at Coronado, CA, and special weapons (nuclear) training at Sandia Base, Albuquerque, NM. Later I checked out in jets for Korean combat and requaled in the F4U.

NAS San Diego was homeport for ATG-2 and, later, *Essex*, our deployment carrier. The four squadrons assigned to make up the group were VF-23 (F9F-2s), LCDR C.C. Aikins, CO; VF-821 (F9F-2s), CDR D.W. Cooper, CO; VF-871 (F4U-4s), LCDR F.O. Harrell, CO; and VA-55 (AD-4s), CDR Lewis Chick, CO; plus detachments of night-attack, night-fighters, special weapons attack, photos, and a helicopter or two.

The air group concentrated on instrument flying because of the lousy weather in Korea; field carrier landings, day and night, in order to CarQual; air-to-ground gunnery; rockets; and dive-bombing. Six months of intense work with the squadrons paid dividends,

Jim Daniels served his country during three wars. Flying into Pearl Harbor the night of 7 Dec 1941, Jim was the only pilot of six VF-6 Wildcats to survive friendly fire and land his F4F at Ford Island. Three of the pilots were killed, and two others were forced to bail out. USN photo, courtesy CAPT J.G. Daniels III, USN(Ret)

as CarQuals aboard *Bon Homme Richard* (CVA-31) in April and, in June, aboard *Essex*, resulted in no accidents. More training and more training in Hawaiian waters followed.

In mid-June, Air Task Group Two deployed for Korean waters with 130 officers, 622 enlisted men, and 80 aircraft aboard *Essex*, CAPT P.D. Stroop, commanding. 'Hutch' Cooper knew that when operating with VF-23 all VF aircraft would be operating interchangeably for ease in scheduling and handling. Cooper, strongly religious, had a St. Christopher medal in the cockpit of each of his aircraft and wanted a medal in each of VF-23's F9s. All VF aircraft were so equipped—a good idea. Incidentally, Cooper's squadron didn't lose a pilot!

We reached the operating area after a show of force in the Formosa Strait in late July and, on 1 Aug 1952, we began regular operations over North Korea, averaging some 90 sorties per day, strikes, recco photos, etc. Normal operations consisted of three days of strikes plus one day of replenishment. Ordnance requirements were extreme. The first period on the line was 18 operating days. Major strikes against Pyongyang (*Essex*), Namyang-ni (*Essex*), and the Aoji oil refinery (*Princeton* (CVA-37) and *Essex*) were executed with no losses.

The Namyang-ni strike on 20 August 1952 was the first of five major strikes against North Korea involving ATG-2. As strike coordinator flying an F9F, I led a 105-plane strike to the west coast of Korea with no damage to any aircraft—going, coming, or during. U.S. Air Force intelligence later estimated 70 to 80 percent of the targets destroyed. On our climb-out after the attack, USAF F-86s on Barrier CAP between Namyang-ni and Vladivostok reported 12 MiGs coming our way. We were slowly scratching for altitude to be in a position to cover our strike aircraft below and behind our fighter cover. LT Hal Crumbo, my wingman from VF-871, an Osage Indian with eyes like an eagle, spotted them to the northeast about 30 miles out and some 5,000 feet above us. The MiGs approached to within 20 miles with a closing rate of about 500 mph, stayed in formation, made a diving turn to the right and went home. Not a shot was fired, but they scared all of us! They were in position to attack, but didn't make a firing pass. It was the only time we saw MiGs during our entire deployment. We completed our first tour on the line, five weeks of operations with lots of flak, but no personnel losses.

By mid-September we were back on the line for restricted operations, as the port catapult was only partially operational, and it finally died. We flew more sorties, followed by a side trip to Yokosuka for cat repairs, more strikes against troop centers, supply dumps, bridges at Hoeyang, Yongpyong-ni, and Kowan on the west coast. Some interesting missions were close air support over the beaches on the west coast in support of a landing feint by ComPhibGroup Three near Kojo, followed by joint strikes with *Bon Homme Richard* and *Princeton* against mining and railroad facilities at Hyesanjin on the Korean/Manchurian border. We would destroy what we could in daylight, which they would then rebuild overnight, especially rail lines down the east coast to Wonsan—the 'Wonsan Express.'

Another interesting strike developed one night when LCDR Ed Potter's night attack AD-4Ns trapped a railroad engine and a couple of freight cars in a tunnel. They kept the train there until daylight, stalled, with smoke coming out both ends of the tunnel. Firing approaches into the tunnel entrances were made in level flight at 100 feet over water. An early strike was planned using two VA-55 AD-4s with 'Tiny Tims,' a 12-inch rocket packing a 500-pound warhead. The pilot flew straight and level, not bothered with AA, and aimed and fired at the entrance. Bullseye! The rocket entered the tunnel, hit the engine, exploded, and destroyed not only the rail cars, but half the mountain slid into the sea. The second aircraft with the spare rocket was saved for a second tunnel that made a 30-degree turn inside. As the rocket flew in a straight line it would hit one side or the other and detonate. The same result. The whole side of the mountain slid into the sea. The Wonsan Express was out of commission for a while.

VADM J.J. 'Jocko' Clark, CTF 77 and ComSeventhFlt, would probably have given me a reprimand had he known we used $200,000 worth of rockets to destroy a Korean railroad engine, a couple of flat cars, and two small tunnels, worth at best $5,000, but he never knew, I hope. A fun strike. No MiGs were reported while we were airborne.

With five F2H-2P photo Banshees aboard, pre-strike and post-strike coverage flown by LT Tom Nielson of VC-61 and his pilots was excellent.

After six weeks on the line, we again headed for Japan. However, the second line period was not good—bad weather, five planes lost, two pilots MIA and one severely injured. We had a lot of flak and battle damage.

The first two weeks in Yokosuka were spent catching up on paperwork, history, logs, citations, awards, a little R&R, and more work on the port cat. During this period ATG-2 submitted to CTF 77 a modified operational doctrine in which we would hope to eliminate much flak damage. Targets were getting harder to find and were better defended. We recommended run-ins above 10,000 feet, minimum pullouts above 1,500 ft., no low strafing except on CAS, and no repeat runs on targets. With our plan approved by CTF 77, we left for our third tour on the line, arriving 16 November.

Targets were fewer, and we had a non-productive joint operation with *Oriskany* (CVA-34) against Chonjin, followed by attacks on supply dumps and warehouses at Kyongsang with *Kearsarge* (CVA-33) and hammering bridges and railroads at the border city of Hoeryong. A short, bad trip; three pilots lost in eight days, then back to Yokosuka for the ailing cat after it put LT W.J. Keefe and his Banshee, as well as VF-23's LTJG L.T. Frietas, in the water. We recovered Keefe, but lost Frietas.

On 6 December we returned to the line for our fourth, and hopefully last, combat operational period. Jocko Clark wanted lots of awards, so we ruefully submitted many citations in the hope that CTF 77 would be happy; the paperwork was horrendous.

Admiral Clark called for a briefing of CATG 2 and CAG 7, CDR George Brown, in *Bon Homme Richard*, so we helo'd to the flagship. The admiral wanted to change the major thrust from attack missions to close air support. Jocko had a large map of central Korea with combat lines drawn in, about a two or three mile-wide strip across the country. He stressed we were to hit close air support targets with the Marines. This area was code-named the 'Cherokee

Strip.' He kept pounding the table with the flat of his hand, repeating over and over, 'The Cherokee Strip.' After the briefing was over, I asked CDR Gene Shirley, Jocko's Flag Secretary, where Clark got the name for this operation. He answered, 'Because he is a full-blooded Cherokee.'

Transferring our efforts from strike warfare to close air support was not difficult, but required better gunnery and positive target recognition.

Our last strikes were made 30 December against Rashin and the northern railroad yards with CVG-101 off *Kearsarge*. CAS was the only game in town until we would leave the line for the last time. The fourth line period lasted 35 days, some 2,200 sorties were flown with 26 aircraft damaged and three lost, but no pilots even injured. The new doctrine had paid off.

An incident during a CAS mission on one of our last operational days supposedly involved a flight of VF-871 F9Fs, possibly strafing our own forces in the Cherokee Strip. Hutch Cooper's operational team determined that his aircraft were not in the area. When *Essex* left the line 8 December, Cooper and I were flown to Seoul, met with the JOC Korea staff, Marines, and Air Force personnel. ATG-2 was cleared, and we were to return to Yokosuka. RADM Apollo Souchek flew us to Atsugi in his R4D, and we rejoined *Essex*.

A date and time had been set for the final inspection of *Essex* and ATG-2, and for an awards ceremony on the hangar deck. All aircraft had been transferred to Atsugi to be used as spares for future operations. All hands lined up before Admiral Clark to receive more than 500 awards. Mess tables were awash with medals and citations. My name was called, and I stepped up to receive an award. I thought the admiral would pin the medal on my uniform and that would end the ordeal. No such luck; he opened the pin, and instead of attacking the uniform, he attacked me! He stuck the pin straight into my left breast. I could feel the blood running down inside my shirt and tried to ignore the fact I was mortally wounded (with no purple heart in sight) as I stepped back, saluted, and said, 'Thank you, sir.' he looked me in the eye and said, 'Come back here, I have four more for you.' I was shocked when he stuck four more pins into me. Gene Shirley, sitting in the front row, almost fell off his chair laughing while my blood poured.

On 25 January we sailed for the U.S., arriving at Alameda 6 Feb 1953, a seven-month deployment. All wounds healed prior to our arrival."

CDR JAMES D. "JIG DOG" RAMAGE, USN (RADM Ret)
CAG 19 10 Dec 52-28 May 54
USS *Oriskany* (CVA-34)

"Jig Dog" Ramage (from the old phonetic alphabet for J and D) graduated from the Naval Academy in the Class of 1939, then served two years in the gunnery department in USS Enterprise *(CV-6) before reporting to Pensacola in 1941 for flight training. His assignment to the "Big E" was prophetic.*

Winning his wings in February 1942, Jig Dog first went to the cruiser Salt Lake City *(CA-25) as a scouting pilot, but in late 1942, after being thrown out of the ComAirPac personnel office on a number of occasions, he was able to talk his way into assignment to* Enterprise's *air department. Upon returning to the Big E in Espiritu Santo, Ramage went directly to the ship's exec, CDR John Crommelin, to ask for transfer into Air Group 10. Crommelin's response was very supportive, and resulted in LT Ramage's checkout in the SBD with LCDR "Bucky" Lee's VS-10, then temporarily ashore at Bomber One in Espiritu Santo. With only one field carrier landing practice he carrier qualified and was assigned to Bombing Squadron 10.*

On 20 Jun 1944, as skipper of Bombing 10, Jig Dog led an attack against Japanese Carrier Division Two (Ryuho, Junyo, *and* Hiyo) *at the Battle of the Philippine Sea. One of Jig Ramage's proudest accomplishments in life was bringing all of his 12 SBD crews safely back from the infamous "Mission in Darkness." All but one crew landed successfully aboard task force carriers in probably naval history's most confused night recovery. The one crew was rescued after ditching inside the force perimeter after running out of fuel on final approach to landing. Most aircraft had about five gallons of fuel remaining after the arduous long-range mission. Ramage's performance at Philippine Sea earned him the Navy Cross.*

"Jig Dog" Ramage led VB-10 through some of the Pacific War's heaviest carrier fighting. He was CAG 19 at the end of the Korean War and believes his CAG job was great, but "...squadron CO is even better." USN photo, courtesy J.D. Ramage, USN(RADM Ret)

At war's end, Jig Dog was CO of VB-98, the SB2C replacement training squadron at NAS Los Alamitos, CA. His other tours before and after CAG 19 include a stint with the nuclear weapons program at Sandia, NM; CO of VC-3 (night fighters); Commander Heavy Attack Wing One; as well as command of Salisbury Sound *(AV-13),* Independence *(CVA-62), and* Franklin D. Roosevelt *(CVA-42). Selected for flag rank in 1967, he served as ComFair WhidbeyFleet Air Wing Four, then on CinCPacFlt staff before becoming Commander Carrier Division Seven in the Gulf of Tonkin. Next came Chief of Naval Air Reserve, then Commander Naval Forces Caribbean/Antilles Defense Command before his 1 Jan 1976 retirement from active duty.*

At retirement, Jig Ramage "made up my mind I was not going to get involved with the aircraft industry passing out calendars and tie clasps to people who used to work for me." Instead, he has remained active in naval aviation affairs, dedicating himself to supporting the aircraft carrier and its people through his work with the Association of Naval Aviation and Tailhook Association. Retired in Coronado, CA, he has also served as chairman of the Coronado branch of the Red Cross.

"When I left VB-98 in June 1946, I couldn't get a flying job for six years. I was too qualified! Each time I contacted a detail officer I was told that my career needed 'rounding' and that I was in danger of having a reputation as a throttle jock. I responded that this was fine with me. Then came the rebuttal, 'You've already had command of two squadrons. We want to give some of your contemporaries command opportunity.'

So, when I finally got orders to command Air Group 19, I was more than ready. I had a lot to do because I had only 2.1 hours of jet time. My first task was to get myself shaped up, then concentrate on my job. I reported to CVG-19 by letter on 13 Nov 1952, and proceeded directly to the Fleet Air Gunnery Unit at El Centro. My logbook shows two fam flights in F9F-5s, then I went directly into the air-to-air gunnery syllabus. I think you will admit, this was on the job training. FAGU CO, 'Boogie' Hoffman, and his instructors did a great job on our class, and me in particular. Although I had only 30 hours in the F9F-5 at graduation, I felt confident that my personal problem had been solved.

The early '50s were very trying years for carrier aviation. The aircraft were far ahead of the pilots. In 1952 the air groups were transitioning from three prop and one jet squadron to three jets and one prop. There was only the tiny Jet Transitional Training Unit at NAS Olathe, KS, to train a cadre of jet pilots. The Navy had disestablished Air Group 98, the replacement training group, in 1947. It meant that untrained pilots were to train other untrained pilots within the fleet squadrons. The RAG (Replacement Air Group) concept was not brought back until 1958. I saw the problem coming before reporting to CVG-19—experienced jet pilots, lack thereof. I was one of them!

Fortunately, Bob Elder, one of the most knowledgeable pilots in jet aviation, was ordered in to command VF-191. He came from the Naval Air Test Center, Patuxent River, and he knew the talent. He was responsible for getting Ed Holley, Jeff Davis, 'Tiny' Granning, and Al Shepard aboard. At that time, all lieutenant commanders and below were ordered into NavAirPac for further assignment. Having been on that staff two years previously, I was able to monitor the incoming pilot records and request certain people for Air Group 19. This group included a number of excellent Reserve officers who were recalled for the Korean War.

Jet aviation on straight-deck carriers was never easy, particularly at night. Add to this problem the lack of low-level ejection seats with quick-opening parachutes and you have quite a dicey operation. It was claimed that a bailout under 5,000 ft. above terrain would be fatal.

Air Group 19 consisted of VF-191 (F9F-6) under Bob Elder; VF-192 (F9F-5) with John Dinneen; VF-193 (F2H-3) commanded by Deke Carr; and VA-195 (AD-4) led by Bill Elliott. I was particularly pleased that Bill volunteered to come as skipper of VA-195. He had been CO of the Ryan FR-1 Fireball squadron and had jet time, but elected to come along with us in props.

I felt that the CAG job was to be the front man to keep the crap from flowing into the squadrons. I did not involve myself in internal squadron affairs. I relied heavily on Bob Elder for jet tactics. Excellence with primary weapons was our goal, which was validated during our ORI in Hawaii while en route to WestPac. The report noted that we were the best group around the target that they had inspected during the Korean War. We were particularly well-qualified in nuclear weapons delivery using the F2Hs and ADs.

We deployed in *Oriskany* in September 1953, returning in May 1954. Don Griffin was skipper, and a fine one. I reported to him as my commanding officer. He put up with the air group's attempts at humor with good graces. I learned a lot from him.

In recalling a rather tragic, but also humorous, incident, I must tell the Frank Repp story. About mid-cruise, Frank had a ramp strike in an F2H-3. It was in the afternoon. He hit the ramp square, with the tail going into the spud locker on the hangar deck, while the fuselage and cockpit, in a ball of flame, rolled up the flight deck. Flames and smoke poured from the deck as I ran up to the bridge. CAPT Griffin asked the name of the pilot, whom we both knew was dead. I responded sadly, 'A Reservist named Repp.' As we talked for a few minutes, I noticed a movement at my side. There, totally unscathed and looking quite normal, stood Frank Repp. I asked, 'My God, Frank, weren't you in that plane? What happened?' His response, 'I don't know, CAG, but I sure was colorful, wasn't I?' As usual at that time, we had him on the next scheduled hop the following day.

I was relieved of command 28 May 1954 and moved directly across Moffett Field to take over VC-3 (night fighters and mini-transitional training unit)."

CDR GEORGE C. DUNCAN, USN (CAPT Ret)
CAG 5 31 Jul 53-30 Nov 54
USS *Philippine Sea* (CVA-47)

Hailing from the Pacific Northwest, CAPT George C. Duncan, USN(Ret), was born 11 Feb 1917 in Tacoma, Washington. Graduating high school in 1934, George spent one year at the College of Puget Sound before entering the Naval Academy, where he graduated with the Class of '39.

George did his two years of surface ship duty in USS West Virginia *(BB-48) before reporting to Pensacola in September 1941 for flight training. Earning his wings in March 1942, LT Duncan was ordered to* Louisville *(CA-28) flying SOC scoutplanes. The cruiser was in Alaskan waters during the Japanese decoy-feint at the Aleutians during the Battle of Midway, where he "managed to collect some 400 bullet holes from a float Zeke in my seaplane and decided that fighters were more to my liking." But first, he was sent on temporary additional duty to VP-12 in the South Pacific to do aircraft spotting for a cruiser bombardment of Munda Island. During this tropical land-based stint, George contracted malaria and was returned to the States in early 1943.*

After his recovery, George pursued his desire to become a fighter pilot and was ordered to VF training at Melbourne, FL. He then reported to VF-15 as operations officer when the squadron was established 1 Sep 1943. The squadron deployed in the new Hornet *(CV-12) for her February 1944 shakedown and transit to Hawaii. As part of Air Group 15, Fighter Squadron 15 embarked in* Essex *(CV-9) in April for the air group's initial combat tour. VF-15 CO, Charlie Brewer, was lost at Guam during the Marianas Turkey Shoot, and George moved up to the XO slot. Showing the true measure of a carrier fighter pilot, he takes pride in the fact that "Charlie Brewer had developed such a well-drilled escort posture that we came through the deployment without losing a single dive bomber or torpedo plane to enemy air."*

As CAG 5 during a "peacetime" cruise, George Duncan's CVG-5 mixed it up with ChiCom fighters off Hainan, resulting in the downing of two La-7s. USN photo, courtesy CAPT C.G. Duncan, USN(Ret)

George led escorts for LCDR V.G. Lambert's torpedo jocks, where he assisted in the sinking of Japanese carriers, a battleship, and a cruiser. On one sortie he sank a destroyer by strafing alone—all this while bagging 13.5 enemy aircraft, which temporarily placed him right after CAG 15 Dave McCampbell, who ended the war as the Navy's leading ace. George's combat performance earned him the Navy Cross, along with numerous other awards.

Air Group 15 returned to the U.S. in November 1944 to reform and begin training for its next deployment. However, neither George, now CO of VF-15, nor CVG-15 would make it back out to the Pacific before the end of the war. Detached in March 1945, George attended the postgraduate school at Annapolis, then went on to MIT in 1947 to earn an MA in aero engineering. He next served two years on the staff of Commander Naval Air Pacific, before attending test pilot training in Class Three at NATC Patuxent River, MD.

After graduation, George was assigned to the Carrier Branch of Flight Test, where he participated in test programs of the Navy's early jets, such as the Vought F6U Pirate, Grumman F9F Panther, and North American FJ Fury, as well as the last carrier propeller aircraft, such as Grumman's F8F Bearcat and AF Guardian.

Departing Pax River in April 1952, George assumed command of VF-51 the following month to fly F9F-5s from Philippine Sea. *In August 1953 he became CAG 5 in time for a May 1954 WestPac deployment. The deployment would prove to be eventful when CVG-5 aircraft shot down Communist Chinese fighters after being attacked during a search mission over international waters off Hainan.*

The near-inevitable Washington tour took George out of the cockpit in 1955 to become head of the Fighter Branch for the Bureau of Aeronautics. While there he ramroded the Vought F8U Crusader through design and its early development. The highly successful McDonnell F4H Phantom also came under George's aegis, as well as work on the Grumman F11F Tiger, the FJ series, Douglas' F4D Skyray, F5D Skylancer, and F3D Skyknight, as well as McDonnell's F3H Demon.

During February 1957 CDR Duncan became XO of Forrestal *(CVA-59) for her first Med deployment. During May 1958 he became CO of VF-101 at NAS Key West, FL. After a year at Key West it was back to Washington in June 1959 as assistant to the head of the Aircraft Research and Development Group until June 1961.*

George next did a deep-draft stint as CO of Jupiter *(AVS-8) before commanding* Ranger *(CVA-61) in 1962. His last tours were with AirPac staff and in Washington before retiring in late 1967. He began a new career in law in Arlington, VA, where he resided for 30 years in his second retirement until his death. CAPT Duncan logged more than 5,500 flight hours during his career.*

Although George Duncan's tour as CAG 5 came during "peacetime," it also was at a time of maximum international tension. The Cold War was at one of its hottest periods in the Far East, which contributed to the encounter between CVG-5 aircraft and ChiCom La-7 propeller-driven fighters over the South China Sea; a foreboding precursor of things to come in a few short years. Excerpts from CAG Duncan's report follow.

"On Saturday, 24 July 1954, the air group engaged in normal operations from *Philippine Sea* in an area approximately 150 miles southeast of Tourane, Indochina. During that day I learned that a British commercial airliner had been shot down off Hainan.

During the next two days we were ordered to conduct searches for possible survivors from the airliner. We were instructed to remain outside the 12-mile limit and not provoke any incident that could be construed as hostile. In the event of hostile action against us by aircraft, we would return fire and, in this case, we could disregard the 12-mile restriction if in hot pursuit.

On the 26th, I led a flight of 11 VF-54 ADs and two VC-3 F4U-5Ns, along with an AD-4W from VC-11 and two AD-4Ns from VC-35 on the 0900-1130 search. At the initial point, the search flight was positioned with one division of four aircraft at 500 feet, led by VF-54 XO LCDR Paul Wahlstrom, another four aircraft division led by LCDR Bill Alexander at 3,000 feet, and my division of three ADs at 5,000. The communications relay team was about five miles south, as VC-3's LCDR Bud Salsig, with ENS E.L. Southerland, flew air cover overhead at 10,000 feet.

As we neared the northern limit of the search area Wahlstrom called that his low division had reached the northern point and he was turning around. At that time I sighted two aircraft at four o'clock, about two miles and slightly above my division. The aircraft, Chinese Communist La-7s, turned toward my division. I called their position and directed a hard right turn toward them as they initiated a low side run on LT Roy Tatham on my wing. They opened fire from about 3,000 feet, and I could see the flashes from their guns firing tracers from the upper cowl. The bullets passed to the rear of Tatham's aircraft.

The two fighters were approximately 1,000 feet apart as they passed Tatham, then turned and made another firing run, this time on me and my other wingman, ENS Dick Nuebower. I couldn't return fire as my guns jammed, so I then called for Salsig and Southerland to come down and help, as well as for the two divisions below to come up and join the action. I scissored with the bogies twice and got on the tail of the wingman and stayed with him. Tatham made a pass at the leader and scored some hits as ENS Dick Crooks joined in, also scoring hits. The La-7 leader rolled on his back and made about three turns as he spun into the water.

The wingman was fired on by other aircraft from the lower two divisions, then straightened out on about a 20-degree glide toward the shore before crashing into the sea. During the glide one of the F4Us and three ADs all fired on him.

When the shooting was over I ordered the flight to resume formation. I heard over the radio that we were being fired upon by an escort vessel that was in the area. We were ordered by radio from Phil Sea not to return fire and completed our search as scheduled.

Roy Tatham and Dick Crooks were credited with shooting down the lead aircraft. Paul Wahlstron shared the credit for the wingman with LTJGs Dick Ribble, John Damian, and John Rockford, along with Bud Salsig."

CDR C.L. "MIKE" BURBAGE, USN (CAPT Ret)
CAG 19 JULY 55-20 Sep 57
USS *Yorktown* (CVA-10)

"Mike" Burbage's introduction to combat came early and hard in his career. A 1941 graduate of the Naval Academy, Mike was officer-of-the-deck in USS Detroit *(CL-8) at Pearl Harbor during the Japanese attack. He served in the South Pacific and Aleutians until entering flight training in early 1943.*

Receiving his wings in June 1944, Mike returned to combat with VF-5 in Franklin *(CV-13) as operations officer. He was on board when* Franklin *was hit off Japan, resulting in 724 men killed and 265 wounded. He next was XO and then CO of VBF-80 and CO of VF-22 after the war. Several other assignments were followed by his 1955–1957 CAG 19 tour.*

Mike attended the Naval War College, succeeded by tours as XO of Franklin D. Roosevelt *(CVA-42) and with CNO's staff before assuming command of* Canisteo *(AO-99). He returned to* Roosevelt *as her skipper in June 1965 after a three-year stint in the Bureau of Personnel. Tours as chief of staff to ComNavAirPac, then CO Naval Air Technical Training Center, Millington, TN, brought Mike to his retirement date of 1 July 1971.*

Mike worked at the corporate headquarters for Holiday Inn, Inc. at Memphis, TN, until his full retirement in September 1979. He now enjoys traveling from his Memphis home when not hunting, fishing, and playing golf.

"I was CAG 19 from July 55–September 57, one of two CAGs on the West Coast who had a group for two years. We were at the beginning of the transition from the CAG training to the RAG (Replacement Air Group) system. As a result, I flew with every pilot in the air group except the AJ detachment and had the training responsibility for the group. Our training was in four phases, the last being CarQuals.

In two previous air groups I had been in, we never got involved in operating with the carrier until the final phase. I changed that and had all of our people carrier qualified during phase one of our training. This bit me later at the time of the Suez Crisis, as we were deployed with little overall training accomplished.

When the Suez Crisis came up during summer 1956, AirPac was directed to provide a carrier and air group to go to the Persian Gulf. At that time there was an air group next door to us at Moffett Field which had finished all of its training except CarQuals. So, AirPac directed us, on 24-hour notice, to fly our aircraft to Alameda and go aboard *Yorktown* since we were all carrier qualified. I told AirPac we had done no gunnery or bombing but, to them, it made no difference.

"Mike" Burbage's peacetime CVG-19 found how quickly "peacetime" can change to immediate contingency operations for carrier air groups when they were emergency-deployed in *Yorktown* (CVA-10) for the 1956 Suez Crisis, but were halted at Pearl Harbor. USN photo, courtesy CAPT C.L. Burbage, USN (Ret)

We reported aboard *Yorktown* and shortly afterward sailed for the Persian Gulf. By the time we arrived at Honolulu the Joint Chiefs had second thoughts about a carrier group in the Persian Gulf, and we were halted at Pearl. Finally, they decided to bring us back to ConUS, except Moffett Field then had no place for us. We spent the remainder of our time aboard *Yorktown* until we were deployed normally. This arrangement caused a lot of problems as far as training was concerned and as far as being continuously at sea. We did deploy in February 1957 on a regular six-month Far East deployment in *Yorktown*.

The chain of command structure was hazy, but in those days there was far less paperwork than today—we had to be careful to be sure ComFair Alameda and ComAirPac were cut in on our operations, either by info copy or through them. When deployed, we also had a CarDiv commander who had an interest in our doings, so the whole command structure was sort of a mess. We had a good skipper in Ed Colestock in *Yorktown*, and many of our problems were solved by personal contact. Our cruise went exceptionally well. Our other deployed air group ended up in deep trouble by flying over Communist China out of Hong Kong, and some of them were relieved of command.

Our deployment was rather routine other than two prolonged periods at sea. We had one of the first mirrors aboard any carrier. It worked beautifully until BuAer came out and found that due to refraction (aluminum mirror), the meatball stayed dead center from four degrees above to four degrees below the glide path. We had had no problems getting aboard, but after it was fixed, everyone had problems for about a week trying to keep the meatball in the center. I had no deck accidents the entire time I had the group, which was a record.

Our AD LSO had a tendency to give too many signals. I went back there one day and took his paddles and landed all the ADs with no waveoffs. All I did was give them a 'roger,' and when they looked OK to me, gave them a 'cut.' After that, he changed and forwent all the signals.

One other thing I did during my tour was to get all our pilots qualified to fly instruments. We had miserable weather in the Bay area during our early training, but were able to get clearances out to sea from Moffett and, by forcing them to go, we never had a weather related casualty. An air group next door to us lost five pilots to weather during the same time.

I wish I were young enough to do my entire aviation career over, since it was so much fun, but now it is relived in sea stories."

CDR ROBERT M. ELDER, USN (CAPT Ret)
CAG 12 6 April 56-8 Nov 57
USS *Lexington* (CVA-16)

CAPT Robert M. Elder, USN(Ret.), is a tailhooker's tailhooker. Carrier aviation has been a major part of his life for more than 50 years, and his influence on carrier aviation has been remarkable. Combat pilot, test pilot, squadron CO, CAG, carrier skipper, aerospace industry—Bob Elder has done it all

Commissioned and designated naval aviator in May 1941, Bob entered flight training after graduation from the University of Washington, where he majored in aeronautical engineering. His first duty station was with VB-3 flying SBC Helldivers from USS Saratoga *(CV-3). The number of aircraft he flew during his 25-year naval and 46-year total flying career ultimately grew to 140 types. In that time he logged better than 8,200 hours and 948 accident-free arrested landings in 35 different aircraft types on board 23 carriers.*

While serving in Bombing Three, flying SBD Dauntless dive bombers, ENS Elder flew in the battles of Coral Sea, Midway, and Guadalcanal. At Midway, Bob assisted in sinking the Japanese carriers IJNS Soryu *and* Hiryu, *as well as severely damaging the heavy cruiser* Mikuma. *Later, at Guadalcanal, he was credited with damaging the seaplane carrier* Chitose. *He was awarded two Navy Crosses for his WWII actions in the Pacific.*

Bob returned to the States in late '42 to instruct at NAS Deland, FL, before joining the Ship's Experimental Unit (SEU) at NAMC Mustin Field, Philadelphia, PA. While at SEU Bob conducted the first carrier trials of the F8F-1 Bearcat, F7F-1 Tigercat, and FR-1 Fireball. Probably his most unusual project was flying carrier trials for a tailhook-equipped P-51D Mustang on board Shangri-la *(CV-38) in November 1944.*

After attending General Line School, Bob returned to the fleet as XO of VA-5A to fly the second Curtiss Helldiver, the SB2C. However, he soon was switched to VF-5A (VF-51, after August 1948) to fly FJ-1 Fury jet fighters as XO under CDR Pete Aurand. Bob and the "Screaming Eagles" CO made carrier aviation history 10 May 1948 when they became the first U.S. naval aviators to take operational jets aboard an aircraft carrier as they conducted qualifications on board Boxer *(CV-21) off Southern California.*

In January 1950 Elder entered Class Four of the U.S. Naval Test Pilot School at NAS Patuxent River, MD. An honor student graduate, Bob became head of NATC's Tactical Test Division fighter section. After this test pilot tour he returned to the fleet, and in January 1953 assumed command of VF-191. Under CDR Elder the squadron became one of the first Navy units to fly swept-wing jets from axial-deck Essex-*class carriers. Bob fleeted-up to CAG 12 in April 1956, deploying in* Lexington *the following spring.*

Returning to Pax River, Bob served as Director of Flight Test Division where he led the team conducting the initial Navy flight test evaluations of the F4H-1 Phantom II, F8U-3 Crusader III, and A3J-1 Vigilante. Selected for captain, he attended the Naval War College during 1959–60, then commanded Waccamaw *(AO-109), followed by a CNO staff tour in Washington, D.C., before assuming command of* Coral Sea *(CVA-43) in June 1962.*

Retiring from active duty in November 1963, Bob embarked on a 25-year career with Northrop Corporation where he performed variously as chief test pilot, director of flight operations, and assistant to the president for aeronautical systems. He subsequently served as director of military requirements until his retirement from Northrop in 1987. While with Northrop, Bob was the driving force that applied the technologies embodied in the YF-17 into the P630 design, which culminated in today's F/A-18 Hornet strike fighter.

Bob's awards and honors over the years have been legion. Those of which he is most proud are his selection as the Tailhook Association's "Tailhooker of the Year" in 1964 ("And damned proud of it!") and his 1987 induction into the Carrier Aviation Test Pilots Hall of Honor. He was at that time one of nine former naval aviators (three living) to be enshrined. Bob is also a past-president of the Society of Experimental Test Pilots and was instrumental in founding the Tailhook Association in 1956. Bob currently lives at Palos Verdes Estates in Southern California.

The immediate post-Korean War era was a time of challenge and innovation for carrier aviation. Bob Elder's CVG-12 was in the forefront of introducing new aircraft, weapons, and tactics into the fleet.

"Challenge is the adrenaline of carrier aviation or at least for me, defined the history of my career. A career pattern which spanned

Bob Elder's 46-year military and civilian flying careers netted him more than 8,200 hours and 948 arrested landings in 35 aircraft types on board 23 different carriers while flying 140 types of aircraft. USN photo, courtesy CAPT R.M. Elder, USN(Ret)

that golden era of naval aviation during the '40s through the '60s when new combat aircraft and weapon developments for the Navy were coming on at a rate that will never again be duplicated. And, I was indeed fortunate to be deeply involved in both flight test and fleet operations with many of the most provocative and challenging of the developments—which proved to be the format of my readiness for air group command—especially this one!

However, translating lessons learned from personal experience to the command responsibilities and decisions requisite for the training and operational readiness of a squadron or an air group takes on a whole new significance—which only makes the challenge more exhilarating.

I inherited Air Group 12 from my long-time friend CDR Charles N. (Tex) Conatser. The ceremony was traditional, the reconstitution of the air group was not—a more disparate collection of controversial aircraft would be difficult to imagine. Nothing was dull about trying to form-up an air group with the menagerie of aircraft assigned to CVG-12, such as the F7U-3 Cutlass—a mantis-like creature, radical in design, marginal in performance, and difficult to maintain; the F3H-2N Demon—a reasonably well-designed supersonic aircraft with good stability and flying qualities, but cursed from birth with the plague of that era, i.e., another airframe looking for a compatible engine; and the AJ-2 Savage—an aircraft that only a mother, or a low-fuel-state fighter, could love. Large and cumbersome to handle on deck, the Savage was never a favorite of the carrier commander.

The balance of the lineup was considerably more conventional in size, complexity, and maturity of their aircraft, but no less challenged by the many tasks ahead.

CVG-12's greatest assets were its squadrons and magnificent personnel:

VF-121 'The Pacemakers.' Fourteen FJ-3Ms, led by a consummate fighter pilot and tactician, CDR T.S. 'Tom' Sedaker, the Pacemakers were the first Sidewinder-equipped Fury squadron to deploy to the Pacific (through a bit of 'asset management' that I was able to arrange while still head of air group training on ComAirPac staff). As recounted by Tom: 'All or any doubt as to the effectiveness of the weapon was put to rest during our ORI (Operational Readiness Inspection) in Hawaii. Led by CAG, an F6F drone out of Barbers Point was sent to a watery grave thusly: The first missile hit the left wingtip flare; the second hit the right wingtip flare; the third, the engine; and the fourth impacted the falling debris.' A real eye opener.

The Pacemakers were indeed a spirited and inquisitive team with great pride in professional excellence. Challenge a Ready Room One briefing with the query, 'Is there a Pacemaker in the house?' and the resounding response was 'Bet your ass!'

VF-123 'Haley's Comets.' With 14 F9F-8 Cougars, under the able leadership of CDR Richard E. Haley, blended in during the training cycle when the wisdom of deploying VF-126's Cutlasses became questionable. Prior to the cruise the 19 pilots underwent intensive training at Miramar, El Centro, and China Lake, honing their skills for the upcoming WestPac deployment. The squadron made aviation history when four of their Cougars were hooked up simultaneously in the world's first multiple refueling operation from a seaplane tanker—a Convair R3Y-2 Tradewind.

Making the trip to WestPac aboard *Lexington,* the Comets were shorebased at NAS Atsugi, Japan, for the greater part of the cruise.

VF-124 'Moonshiners.' With the powerplant compatibility problem ameliorated by the installation of a higher thrust, Allison engine in the -2 Demon, the Moonshiners became the first squadron in the Pacific Fleet to receive and subsequently deploy with CVG-12 in the Navy's newest Sidewinder-equipped, all-weather interceptor, the F3H-2N.

Two tragic accidents occurred near the outset of VF-124's predeployment training cycle which lowered the squadron's ceiling temporarily, but did not obscure its visibility. The first happened 10 Aug 1956 when their XO, LCDR James W. 'Shorty' Ewing, suffered career terminating burns in a near fatal crash while attempting a deadstick landing following a flameout in the landing pattern at Miramar. That setback was compounded some two months later when their esteemed commanding officer, LCDR Shannon McCrary, mysteriously disappeared from the radarscope during a night controlled instrument approach to Miramar and crashed on the eastern slope of the Laguna Mountains.

The McCrary's were both personal friends, thus the always onerous command responsibility of breaking the news that night to his wife Magee became a gut-wrenching experience for my wife Irene and me.

But, with steady hitters like LCDRs Don Shelton, Ed Porter, and Wally Schirra in the lineup, LCDR Paul Payne took command and quietly set a pace that contributed immeasurably to the accolades that ComCarDiv Five and ComSeventhFleet accorded CVG-12 following our WestPac deployment.

These were the 'Demon Drivers' and the 'Demon Doctors'; Hangar Bay 3 in *Lexington* was Satan's around-the-clock clinic—through superhuman effort those dedicated medicos met the challenge. But I shudder to think what the situation might have been if the decision *not* to deploy the Cutlass had gone the other way.

A unique and amusing event that occurred during our ORI involved a fighter weapons exercise using a Regulus I guided missile as a target. The Regulus blasted off from Barking Sands, Kauai, was picked up and aligned on course at near-sonic speed by one of the specially-configured FJ-3D Furys from our Guided Missile Group One detachment, and promptly pounced on by three Demons from VF-124. The CO, Paul Payne, had the honor of firing first, but his Sidewinder guidance system went ape. Next in line was the XO, Don Shelton, who blew the target to smithereens. But, Wally Schirra, the original 'Gotcha' guy, still maintains that he was in perfect position when Don's Sidewinder whistled by him. RHIP.

VA-125 'The Rough Raiders.' With a thoroughly-seasoned attack pilot, CDR John H. Bahlman, at the helm, the 14 AD-6/-7s of the squadron provided the stability that literally took the monkey off my back during the reconstitution of CVG-12. For that relief, plus some humorous times we shared, I am deeply indebted to him.

While at El Centro prior to deployment, the Rough Raiders rode that old warhorse, 'Alfa Delta,' for a total of 41 Navy E's awarded for proficiency in special weapons profiles, strafing, rocketry, dive-bombing, night-bombing, and loft-bombing—then aptly demonstrated during the ORI that those were no idle honors.

VA-126 'Tough Guys.' Under the unique leadership of CDR George L. Cassell (RADM, Ret), VA-126 established remarkable records in loft- and dive-bombing competitive exercises with the Cutlass. Despite the complexity of maintaining the machine, the squadron met all training commitments; they had the spirit, they met the challenge. However, because of logistic and support problems for the F7U-3, a wise but most disappointing decision by higher authority ultimately prevented their deployment to the Far East with the air group.

I would be remiss not to note that it was George Cassell's wry humor and exceptional gift of expression that served him well in getting the message across to the troops. This did much to create the *esprit de corps* in VA-126 that has stood the test of time.

Leaving them stateside will always be a sad memory that I shall never forget.

VAH-6 Det Golf. Despite previous comments about the AJ aircraft, under the able, no-nonsense leadership of LCDR M.A. 'Mal' Lea, with two AJ-2 Savages, Det Golf was indeed a welcome addition to the air group and during the ORI. The Det complimented our professional standards with near perfect performance in all phases of their specialties. An admirable accomplishment considering the critical surveillance of the PacFleet Special Weapons evaluation team.

GMGRU-1 Det Golf. Guided Missile Group 1 was a unique addition to the air group. With LCDR V.F. Forsberg in charge and two specially-configured FJ-3D Furys, Det Golf provided an extended range delivery capability of a nuclear weapon carried by a Regulus I guided missile.

Notable among the successful operations conducted was the 270-mile strike of a Regulus missile fired from the cruiser *Helena* (CA-75). Taking over guidance 180 miles out, the Fury, piloted by LT R.P. Blount, shepherded the bird to a point directly over the target island where the charge was fired by remote control.

This performance brought a 'well done' from the Flag.

The tactical training and weapons delivery deployments of the squadrons (including detachments) to El Centro, China Lake, and Fallon during the formative stages of integrating the air group sharpened both the individual flying skills and unit combat doctrine, while developing the equally essential elements of competitiveness, cohesive pride in achievement, and spirit that characterizes a truly professional combat team.

The awards for excellence that each of the squadrons racked up in competitive weapons employment exercises during those deployments was proof positive that the training was infectious. As exemplified by four sharpshooters of VF-121 (led by their skipper, Tom Sedaker) that scored a total of *234 hits for 489 20mm rounds expended* on a standard 30-foot towed banner. A *48 percent average*, with 64 percent high (guess who) and a 31 percent for low man on the totem pole. Not too shabby.

As Tom modestly remarked: 'The 20mm cannon could still be a heck of an offensive weapon.'

Day and night CarQuals, using either 'paddles' or the mirror landing system, had progressed quite well under the critical tutelage of our LSOs, LCDR Ray Manhart and LT Jack Nelson; the normal hiccups along the way, but without accident or serious incident. Bless the angled deck, along with some approbation for the 'meatball' landing system.

The deployment of *Lexington*—affectionately dubbed 'Runway 16'—19 April 1957 was to be her second WestPac cruise since conversion to the latest innovations: an angled deck, mirror landing system, and steam catapults.

By departure time I was confident that the squadrons and dets were well trained in their basic roles and missions; indeed, ready to take their act to sea as a combat team.

The upcoming ORI for ship and air group, by CinCPacFleet in Hawaii, would be the first impartial test to quantify that assertion. The months at sea to follow would challenge our ability to maintain the requisite level of combat readiness, which was never an easy task. But we met both challenges with commendable success as noted by post-deployment accolades from ComSeventhFleet and ComCarDiv Five.

And, besides that, we had a hell of a lot of fun!

When Commander Carrier Division One, RADM Herbert D. Riley, sent the following message dated 17 Oct 1957, little did we realize that this deployment was to be the final one for Air Group Twelve, and the planting of the roots that would ultimately branch into the Combat Readiness Air Wing (CRAW) concept that has revolutionized the effectiveness of naval aviation today:

> THAT TAPE AT THE END OF THE STRETCH IS NOW IN SIGHT AND IT HAS BEEN A PLEASURE TO WATCH CHAMPIONS COME THROUGH ACCORDING TO FORM X WHEN RUNWAY SIXTEEN AND CARRIER AIR GROUP TWELVE REPORT QUOTE ALL DOWN CMM TURNING FINAL UNQUOTE OFF PT LOMA THURSDAY CMM THEY HAVE EARNED THE RIGHT TO DO SO WITH THE CLEAR KNOWLEDGE AND SATISFACTION OF HAVING COMPLETED AN ARDUOUS DEPLOYMENT IN A UNIFORMLY OUTSTANDING MANNER X YOUR RECORDS IN THE SEVENTHFLT WILL NOT SOON BE FORGOTTEN WELL DONE X RADM RILEY

My closing comments to the air group might well have been: 'Runway 16 in the wind, Foxtrot at the dip, double-nuts turning final—CAG arriving.' Thank you all for everything you gave me.

As a footnote, I might add that the CNO 1957 Aviation Safety Award was presented to *Lexington*, then passed on by the Commanding Officer, CAPT John A. Gannon, to me with the comment: 'Accept this plaque with pride—Air Group 12 earned it, they deserve to have it.'

It has a prominent niche in my 'I love me' memorabilia room at home."

CDR RICHARD HAROLD MILLS, USN (CAPT Ret)
CAG 3 4 Sep 57 - 23 Oct 58
USS *Saratoga* (CVA-60)

Born 28 Jun 1920 and raised in Joplin, MO, Dick Mills entered naval service as a seaman second class 6 Feb 1941 at NRAS Robertson, MO (Lambert Field). Prior to enlisting, he attended civilian pilot training and two and one-half years of college. He was appointed an Aviation Cadet 1 May 1941 at NAS Pensacola. He completed flight training as a carrier attack pilot at NAS Opa Locka, FL, 23 December following his commissioning and designation as a naval aviator during the preceding two months.

After advanced carrier training at NAS San Diego and Norfolk, Dick became a member of Bombing Squadron Six in which he served during the battle for Guadalcanal in 1942 while flying from USS Enterprise *(CV-6). He later returned to the Pacific theater during WWII with VB-15 in* Essex *(CV-9). As commanding officer of VA-45 he deployed to the Med in* Coral Sea *(CVB-43) and later, during the Korean War, embarked in* Lake Champlain *(CVA-39). In 1958 CDR Mills returned to the Med in* Saratoga *as Commander Carrier Air Group Three during the Lebanon Crisis.*

Dick's shore duty tours include Advanced Air Training Command instructor; on the staffs of the Bureau of Aeronautics and Chief of Naval Operations; Supreme Allied Commander Europe; and Bureau of Personnel. His command tours at NAAS Whiting Field and NAS Key West were followed by assignment to the Joint Strategic Planning Staff at Offutt AFB, Omaha, NB.

CDR Richard H. Mills (CAPT Ret) led CVG-3 from *Saratoga* (CVA-60) during the 1958 Lebanon Crisis. USN photo, courtesy CAPT R.H. Mills, USN(Ret)

During his naval career Dick attended General Line School, Command and Staff Course at the Naval War College, Industrial College of the Armed Forces, and he holds a BBA degree from George Washington University. He retired from the Navy 1 Jul 1972 as a captain and holds the Navy Cross for his actions at the Battle of Leyte Gulf, where he successfully bombed a Japanese carrier.

Following his naval retirement, Dick settled in Naples, FL and worked in the construction industry building custom homes until his complete retirement in 1982.

CAG Mills led Air Group Three during the initial 1958 Lebanon crisis.

"During my naval career I deployed several times with squadron and air groups, and I must say my tour as CAG 3 in the Med was a bit unusual—*stressful* might be a better word for the occasion. Had there been another ship and air group available and an admiral not as understanding as RADM Clifford S. Cooper, I'm sure we would have been *beached.*

The air group training had been seriously curtailed as a result of matériel problems and modifications to the aircraft. Additionally, *Saratoga* had a low experience level of the crew. The air group and ship had a very short time to develop into a team and, frankly, we were not ready.

The last four months of our training period, which included Christmas '57, before embarking in Sara in January 1958, were spent correcting serious problems and modifying aircraft. The air group was composed of the F3H, F8U, A4D, A3D, AD-6, AD-5N, AD-5W, and F9F-8P.

I assumed command of CVG-3, and the next day, VF-32 with 12 F8U-1s, flew to NAS Guantanamo Bay, Cuba, for a 30-day concentrated training period. A message was awaiting them upon arrival in Gitmo: 'One-time flight authorized to NAS Cecil Field. Upon arrival, F8Us grounded until further notice. ComNavAirLant.'

I learned of this message when the CO of VF-32 called me Saturday evening to find out what was going on. I advised him to stand by while I tried to get in the picture.

After several exchanges with Chance Vought in Dallas, AirLant in Norfolk, and VF-32 in Gitmo, it was agreed that AirLant would leave the squadron in Cuba to make a change to the fuel system.

On Sunday morning, 8 September, VF-31 launched an F3H, our all-weather fighter, for Dallas to pick up the parts and deliver them to Gitmo. After a 20-minute flight, the Demon flamed out while flying in clouds and the pilot safely ejected. The F3Hs were then grounded until further notice, and replacement parts were safely delivered to Gitmo by CAG's AD-5N.

In less than two days, after I became CAG 3, the F8U day fighter had been grounded to correct a fuel system deficiency (also, its armament system had not yet been declared fully operational), and the F3H all-weather fighter had been grounded and then restricted to flight where moisture would not be a factor. Additionally, the AD-6s were undergoing modifications to provide them with a tanker capability, and new A4Ds were being accepted. A busy period followed, but not much training.

Soon after Sixth Fleet's devastating message of August 1957 regarding A3D carrier operations, Heavy Attack Squadron Nine reported to me for operational control as a unit of CVG-3. During a visit to NavAirLant I had been advised: 'You know, Dick, VAH-9 will succeed, or else!' As I recall, this was the first time a full squadron of A3Ds would make a normal deployment aboard ship as a unit of the air group.

Believe it or not, we finally got organized enough to go aboard *Saratoga* for training in January 1958—our first operations on our ship as a group. As you might guess, we 'flunked the ORI.' *Ship and air group.*

With few or no alternatives, CVG-3 and *Saratoga*, with RADM Cliff Cooper as CarDiv, set sail from Mayport, FL, 1 Feb 1958 for an eight-month tour of the Med. It was reported that the *Saratoga* carried what the Navy called the most formidable naval air group ever to visit Europe!

By the end of May 1958 we had grown up—if not the best, we were damn good! It had not been easy.

We arrived at Gibraltar 13 February and relieved *Roosevelt* (CVA-42). From that date through May we were at sea 74 days and in port 42. Most of the days at sea we trained; we still had our problems. At the end of February, when leaving Naples, Italy, *Saratoga* damaged a propeller. Early in the year all fuel tanks of the A3Ds had to be changed because of salt water in the ship's fuel system—quite a logistical undertaking. All F8Us had a change to the landing gear to prevent further damage from carrier landings.

The end of May we left Crete a grown up, fully-ready, and top-notch ship and air group; we could well have been called the 'Top Guns.' We gave flawless aerial demonstrations for their Royal Hellenic Majesties and SecNav Gates while we established new records of all types for flight operations. CVG-3 set new sortie and flight time records on 'Admiral Cooper Day.' All of this led up to our show of strength over Jordan and the Lebanon crisis.

About 0400 15 July, *Saratoga* and other ships of the Sixth Fleet made an unscheduled and emergency sortie from Cannes, France. At maximum sustainable speed, the fleet headed for Lebanon.

Although I personally have never seen the message, nor a copy thereof, I am positive that it was sent—very much a 'Cat Brown'-type (Sixth Fleet) message, and very appropriate at the time.

'TO CNO FM SIXTH FLEET X SIXTH FLEET UNDERWAY AND READY CMA WHICH SIDE ARE WE ON?'

As I recall, we steamed for a day and a half at max sustainable speed when CAG 3 was launched with 28 jets and 13 props for a show of strength over Jordan.

I had three general orders when I was catapulted from the ship: '(1) This overflight had better be good; (2) Your planes have the right to defend themselves—an overflight of Israel has not yet been obtained; and (3) If a message is received in flight the AD-5Ns will land under your air cover and fly the King of Jordan and his family to the safety of the Sixth Fleet.'

ComCarDiv Six was especially interested that the capital of Jordan—Amman— was acutely aware of our presence and that the planes were positively identified as U.S. Navy planes. Apparently several months earlier an overflight of Jordan was scheduled and the planes had missed the target. Although CAG 3 was not involved in the 'miss,' CarDiv Six was advising me to be accurate.

The overflight of Israel was approved and uneventful, while the show of strength over Amman went perfect. Our A3D navigators hit Amman on the button, and we did indeed rattle the rafters. I know the King and his family could not have slept through the demonstration of naval air power, but he must have felt safer, as we did not receive the message to pick him up.

In that our flight over Jordan had been launched at max range, our recovery aboard *Saratoga* was of concern to all. I'm positive that an early deck crash would have put several jets and crew in the drink as a divert field was not too close. Fortunately, we had a perfect recovery and a very complimentary message from Sixth Fleet.

We began the Lebanon crisis in support of the U.S. Marines in July 1958 and continued the action until we were relieved in early September. Around-the-clock combat air patrol, reconnaissance, and patrol missions were daily events. The AD tankers were also used for range extension of the photo jets and were on station for emergencies. The air group flew 8,886 hours while flying 4,200 sorties in support of Lebanon—didn't fire a shot or drop a bomb! Some small arms fire did minor damage to our planes, but was of little consequence.

We sailed from Gibraltar 21 Sep 1958. The air group was launched for the beach on the 30th, and *Saratoga* steamed into Mayport 1 October. CAG 3, VF-31, VF-32, and VA-34 returned to NAS Cecil Field. The photo jets and VA-35 went to NAS Jacksonville. VAH-9 returned to NAS Sanford, FL, and the all-weather units with the early warning detachments returned to NAS Quonset Point.

Commander Leo B. McCuddin relieved me as Commander Carrier Air Group Three 23 Oct 1958 at NAS Cecil Field. I departed with a satisfied feeling that the group had done an outstanding job under some very adverse situations. All in all, we had a good cruise."

CDR E.L. "WHITEY" FEIGHTNER, USN (RADM Ret)
CAG 10 20 March 59-18 March 60
USS *Forrestal* (CVA-59)/USS *Essex* (CVA-9)

Edward Lewis Feightner enlisted in the U.S. Naval Reserve 16 June 1941 following his graduation from college and shortly thereafter went into the Aviation Cadet program. "Whitey" was designated naval aviator 3 April 1942 at NAS Corpus Christi, Texas. His first tour was in VF-10 flying from USS Enterprise *(CV-6) and later* VF-8 in Bunker Hill *(CV-17). Whitey participated in numerous battles of the Pacific war including both carrier-based and land-based actions at Guadalcanal, as well as the Battle of Santa Cruz, then Palau and Woleai Atoll. His skills as a fighter pilot brought him nine victories over enemy pilots.*

Returning to the States in November 1944, he next served with VF-98, then VF-21 until November 1946. He next was flag lieutenant for Commander Carrier Division Three and then ComCarDiv Two. Whitey went through test pilot training at NATC Patuxent River, MD, 1948–49 and remained at Pax River until December 1951 as a project pilot.

During the next nine months he was a member of the Blue Angels flight demonstration team where he flew the F7U-3 Cutlass, two of which had a brief service life with the team. From September 1952 until December 1954 Whitey was in VX-3 before becoming commanding officer of VF-11 January 1955-February 1957. He was next OinC of the Jet Transitional Training Unit, NAS Olathe, KS, before temporary duty as prospective CAG 10 while attached to VF-101 October-December 1958. He assumed command of Air Group 10 in March 1959 and transferred to CVG-4 the following year. Naval War College and staff tours were followed by command of Chikaskia *(AO-54), then* Okinawa *(LPH-3) in February 1966.*

After further schools and staff jobs RADM Feightner became Deputy Director for Plans and Programs at Naval Air Systems Command. He retired from active duty 1 July 1974, then formed FMF Inc., a consulting firm, where he serves as president and lives in Washington, DC.

A young LT Whitey Feightner beams from his VF-10 F6F-5 Hellcat emblazoned with his nine victory flags over Japanese pilots during the war in the Pacific. USN photo courtesy RADM E.L. Feightner, USN(Ret)

"My CAG 10 tour included deployments to the Med in both *Forrestal* and *Essex*.

Command structure between CAGs/carrier COs and the Type Commanders was (to put it kindly) fluid and unique to each particular situation. For instance, ComNavAirLant ordered me to the Med to relieve CDR A.O. Morton, CAG 10 aboard *Forrestal*. ComSixthFlt would have no part of it. He refused to allow a change of CAGs during a deployment. As a result, I spent three months as a Deputy CAG and finally took over as CAG after we returned to the States. On board *Forrestal,* CAG was a department head and was treated as such. In *Essex,* CAG was treated as 'Mr. Air' and was consulted on every phase of ops. The main purpose of the ship was to assist the air group in accomplishing the air mission. How is that for a novel idea? Needless to say, we had a great cruise. This deployment was one of the last during which the CAG flew with every squadron, each one of which sported a spotless aircraft with the double '00' on its nose.

Ours was not the regular CVG-10 air wing. This group was the brainchild of CAPT Mike Handley of the AirLant staff, who had been charged with forming a group specifically tailored for the *Essex* deployment to the Med. In spite of ample and knowledgeable advice to the contrary, he insisted on assigning an F4D-1 fighter squadron to the group. The F4D was too much airplane for the catapults and arresting gear of an *Essex*-class carrier. Nevertheless, AirLant insisted that the F4D was going and that the problems would be resolved. The arresting gear was modified only after the squadron executive officer was killed when a crossdeck pendant broke during a landing. The catapult was modified, and a new holdback fitting was designed after another pilot was killed when a holdback failed as the pilot applied full power for takeoff. Special takeoff procedure was required on every catapult shot, because the pilot was violently thrown forward and backward during the launch. As a result, the aircraft over-rotated at the end of the power stroke, such that the fuel drain valve was damaged by the catapult shuttle and the aircraft invariably caught fire as it left the deck. To avoid this situation, all launches were made with the pilot grasping his right leg with his right hand and allowing the stick to float free until the aircraft was airborne, at which time he could again grasp the stick and start flying the plane. Can you imagine asking pilots to live with such a procedure for an eight-month cruise, day and night, and in all kinds of weather? We not only did it, but we did it successfully for the entire cruise. Most of the time the ship was unable to get enough wind over the deck to permit launching the F4Ds with external fuel, so many of the flights were very short (30–40 minutes) because of limited internal fuel in the aircraft. This routine, of course, played havoc with scheduled cycle times and deck handling procedures. A F2H-3 or -4 squadron would have been more effective and saved all the hassle.

The air group for this cruise consisted of: VF-13, 12 F4D-1s; VF-62, 12 FJ-3Ms; VA-106, 12 A4D-2s; VMA-225, 12 A4D-2s; VA-176, 12 AD-6s; and VFP-62 with three F9F-8Ps."

CDR FRANK W. AULT, USN (CAPT Ret)
CAG 10 18 MARCH 60-22 MAY 61
USS *Shangri-la* (CVA-38)

Frank Ault is a 1942 graduate of the Naval Academy. After two-and-one-half years in cruisers, he went through flight training and was designated naval aviator in March 1946. A lifelong attack pilot, his first flying tours were with VT-75 in SB2Cs (later VA-4B in ADs) and VC-5 in nuclear-delivery capable AJ Savages.

In addition to being CAG 10, he commanded VA-216, USS Rennville *(APA-227) and* Coral Sea *(CVA-43). Ault is probably best known for his 1969 work as author of the "Ault Report," a comprehensive study of the U.S. Navy's operations and tactics during the first half of the Vietnam War. From WWII to mid-Vietnam, the Navy's fighter air-air exchange ratio declined from 18:1 to an alarming 3.5:1. Ault's exemplary work resulted in sweeping changes to the Navy's fighter training, weapons handling and employment, along with introduction of better tactics and maintenance. Additionally, he is credited for helping to identify problems which led to the establishment of the Navy Fighter Weapons School ("Topgun"). By the end of the Vietnam War the Navy's kill ratio had risen to better than 12.5:1.*

After his 1971 retirement, Ault served as an aircraft industry executive before founding International Prowess, Inc. Since mid-1980, he has been working closely with the Department of Defense, Coast Guard, and Customs Service on drug interdiction programs. During his CAG 10 tour, life may have been quieter, but training for future wars is never simple, and the next one was just over the horizon for him.

"I relieved Whitey Feightner as CAG 10 and was relieved, in turn, by Bob Stegg. CVG-10 was based ashore at Cecil Field and made two deployments in *Shangri-la*; one to the North Sea for NATO exercise in spring 1960, and a Mediterranean deployment in fall-winter 1960-61.

CVG-10 had '00' on three aircraft, a VF-62 F8U, VA-106 A4D, and VA-176 AD-6. I did not put '00' on a VF-13 Skyray because I did not fly the F4D after completing the RAG syllabus in Key West. The flight characteristics and cockpit configuration of the F4D were just too different from the other aircraft types in the air group, all of which I flew regularly.

I don't recall any particular events which distinguished the CVG-10/CVA-38 operations from any others of the period. Our operations were of the standard Cold War variety; NATO exercises in the Norwegian Sea and Bay of Biscay, as well as in the Med. In November 1960 we were suddenly sent to the Caribbean to act as a deterrent force to prevent movement of arms and equipment between Cuba and the Dominican Republic. I was in the hospital at Jax at the time with a bad back, but the medical people rigged me a metal corset so I could fly. I would note that during the Med cruise we left VF-62's Crusaders at Rota, Spain, because we had a full deck without them. They joined the ship for refresher work whenever we were in the Western Med. Concomitantly, they impacted significantly on the manners and morals of the Spanish population near Rota.

On the command front, I made myself quite unpopular with ComNavAirLant and the Replacement Air Group (RAG) CAG by complaining about the lack of night carrier capabilities of the RAG product. The problem was not solved during my tour.

Two humorous recollections: ComSixthFlt was VADM George Anderson (later CNO), a stickler for military protocol and uniform regulations. Knowing this, I had all my COs in uniforms with new gold, ribbons, etc. for our official call on the flagship. Just out was the new 'Command at Sea' insignia, which was about an inch-and-one-half in diameter. Frankly, it looked like it came out of a Wheaties box. We were all wearing them when we made our official call on Anderson. He looked at my chest, pointed, and said, 'What is *that*? Feeling I was ahead on brownie points, I said proudly, 'That's the Navy's new Command at Sea insignia, sir!' George roared, 'Take it off! I won't have anyone in the Sixth Fleet wearing the symbol of our enemy (the star)!' End of bid for brownie points.

The other incident occurred during workup for our Med deployment, for which we added VMA-225 in A4Ds, which replaced VA-12. Only the skipper and exec of the Marine squadron had ever operated from a carrier. On the first night of landing quals the ops officer of the squadron made an arrested landing, raised the hook,

A Cold War CAG, Frank Ault earned his fame as author of the "Ault Report," which brought about sweeping changes in tactics and operations that greatly improved naval aviation's performance in the Vietnam War. His aircraft, as well as his helmet, were adorned with "00" to identify him as CAG 10. USN Photo, courtesy CAPT F.W. Ault, USN(Ret)

added throttle, and taxied smartly off the angle. (We got him back.) About 15 minutes later I got an outraged call from the Marine CO asking me to come to VMA-225's ready room. There I found that my ops officer, LCDR Dave Scherer, had written on the blackboard: NOTICE TO MARINES: WHEN CLEARING THE ARRESTING GEAR, DON'T FORGET TO TURN RIGHT. Although he was an A4 driver, Scherer didn't get to fly with the Marines that cruise.

While I am, admittedly, a member of the 'old school' and Air Group 10 was not engaged in combat during my tenure, I felt that the command structure was ideal: simple, direct, and effective. A Super CAG, in my era, would have been tits on the boar, and I have no idea what he could have done—or would have done—to contribute to the improvement of planning and operations. We have learned the hard way, in both the military and the business worlds, that the more people you have in the chain of command/management, the harder it is to pass the *authentic* word and get things done—and the harder to affix responsibility when things go wrong. There is an old saying: 'Eagles don't come in flocks; you have to find them, one at a time.' The Ault appreciation of that axiom is that a flock is likely to contain a lot of crows and buzzards. The former will squawk about everything while accomplishing little, while the buzzards will be glad to pick at your bones after the crows have driven you to distraction. End of philosophical diatribe."

CDR HOWARD E. GREER, USN (VADM Ret)
CAG 10 11 Sep 61-5 Sep 62
USS *Shangri-la* (CVA-38)

Coming out of the Naval Academy class of '44 (graduated June 1943 because of wartime acceleration), "Howie" Greer's first fleet assignment was in USS Princeton *(CVL-23) as a ship's company officer participating in Pacific campaigns from Tarawa to the Philippines. He reported for flight training in October 1943 and was awarded his wings of gold in November 1945.*

For the majority of his career, Greer served in fighter squadrons (flying 11 different types of fighters) and fleet staffs, deploying to the Sixth Fleet in the Mediterranean five times, and seven times to the Seventh Fleet in the Pacific.

Greer commanded VF-64 before becoming CAG 10 in September 1961; afterward commanding Ponchatula *(AO-148) and* Hancock *(CVA-19) before becoming Commander Naval Air Reserve. He also was Commander Carrier Group Three, Second Fleet and Naval Air Forces Atlantic.*

Howie saw extensive service during the Vietnam War, holding four Seventh Fleet Staff positions, including assistant chief of staff (1964-66), as well as his ship commands and CarGru Three. While serving as Yankee Station commander, he directed all naval air operations during Linebacker I and the mining of the port of Haiphong harbor.

Retiring from active duty in 1978, Greer formed and became president of Greer Industries, Inc., where he worked several years as a near full-time consultant for General Dynamics Corp. in the weapons field. Now living in Honolulu, Hawaii, the industrious admiral remains active in aerospace consulting, as well as in public speaking before military and civilian audiences.

"I relieved CDR Bob Stegg, who was acting CAG 10, to fill the gap between CDR Frank Ault and myself. I am from the Naval Academy Class of '44, and all of my squadron commanders were from '45—they kept me on my toes, and I checked my six every few seconds.

During fall 1961, we participated in CarQuals and Second Fleet exercises along the East Coast. In these exercises ComCarDiv Six RADM Toby Hogle was embarked, and he had as chief of staff, CAPT Roger Mehle. I point this out simply leading to the statement that during our Sixth Fleet cruise we had no flag officer embarked. In other words, except for a very limited period, I reported only to the *Shangri-la* commanding officer. That relationship was extremely cordial, and neither encumbent ever told us 'how to do it.' CAPT Herbert Houck was CO until mid-March when CAPT Robert Rynd relieved.

In December 1961 I was informed that, in accordance with a policy decision to place more Marine squadrons aboard ship, VMF-251 would replace VF-62 for the cruise to the Mediterranean. VF-62 was the highest performing squadron in the group and was commanded by a superlative leader, CDR Joe Moorer. Needless to say,

Howie Greer mans an A-4 Skyhawk on board *Shangri-la* (CVA-38) during 1962 Med cruise. Typical of most CAGs, Greer, a career fighter pilot, flew nearly all aircraft types in his air group. USN photo, courtesy VADM H.E. Greer, USN(Ret)

I was initially disappointed. Soon I was to learn that VMF-251's average number of carrier landings per pilot was less than 15. They reported in December and logged a few landings aboard *Shangri-la*. We were to operate them day and night!

On 7 Feb 1962 we sailed from Mayport for the Med with VMF-251, LTCOL Ernie Poor; VF-13, CDR Maury Esmiol; VA-106, CDR Stu Nelson; VA-46, CDR Cal Calloway; VA-176, CDR 'Beetle' Forbes; and a VFP-62 det, LCDR Jim Kauflin.

We were scheduled to participate in fleet exercises immediately upon reporting to Sixth Fleet, and we were weak in recent carrier landings (VMF-251 grossly so). It was therefore necessary to conduct CarQuals with a full deck and often with no bingo field while crossing the Atlantic in February weather. Night landings in rough weather on a 27 Charlie carrier in F8Us and F4Ds were challenging! We bent a few, but didn't hurt anyone.

I flew with VF-13, VA-106, and VA-46, but did not continue with the F8U since their availability was low and need for carrier work quite high. At this point, I would like to interject the fact that VMF-251 came up to speed and was a top-notch squadron before the halfway point in the cruise. They should never have been asked to deploy on short notice with their low experience aboard ship; however, they met the test.

The 'Shang' sounded water hours as soon as we cast off lines at Mayport and stayed in that state until we completed the cruise. RADM 'Chick' Hayward was CTF 60 embarked in *Independence* (CVA-62). Prior to each exercise, I was summoned to be overhead at a designated time to attend a briefing aboard Indy. Chick always added a footnote that my shower was ready—a ritual which preceded my entry to the briefing room.

We were relieved by *Enterprise* (CVA(N)-65). About two weeks prior to rendezvous for turnover discussions, we informed the 'Big E' that we had the German Defense Minister embarked and that he was to be transferred to the 'E' for transportation to Naples, their first port of call. At anchorage he was sent by barge to *Enterprise* and was rendered full honors. After being shown to the flag cabin, he shaved his mustache, changed to his Marine uniform, placed his tweeds in his ditty bag, and joined a small liberty party. A few hours later we sent a message to *Enterprise* informing them if they were searching for the 'Defense Minister,' alias 1st Lieutenant USMC, that he was safely back aboard *Shangri-la*.

We returned to Mayport on 28 August, having visited many interesting ports and logged a lot of flying (for those days). I flew 419 hours, 75 at night, and logged 173 traps, 44 at night.

CDR KENT L. LEE, USN (VADM Ret)
CAG 6 22 June 62-7 March 63
USS *Enterprise* (CVA(N)-65)

Kent Lee enlisted in the Navy 15 Aug 1940 and became a V-5 Aviation Cadet 15 Aug 1942. He was designated naval aviator 7 Aug 1943 at NAS Corpus Christi, Texas. He went through SB2C training at VB-100, and in June 1944 reported to VB-15 in USS Essex *(CV-9) in the Pacific. With the need for increased fighter protection as a result of the* kamikaze *threat, Kent was temporarily assigned to VF-15, where he participated in numerous strikes against enemy positions and shipping at Marcus and Wake islands, the Marianas, Bonins, Palau, and Philippines. He was credited with shooting down a Japanese Betty bomber off Formosa 12 October.*

Terminating his temporary role as a fighter pilot, Kent returned to VB-15 before joining VB-151, then VB-17 in September 1945. A short post-war tour with VB-5A was followed by assignment to Columbia University as an ROTC student, then General Line School. He served in Korea on the staff of Carrier Division 15 and CarDiv 17 before getting back in the cockpit with VA-115 flying missions in Korea from Philippine Sea *(CV-47).*

Several more tours under instruction were followed by VX-3, Fleet Training Center, Norfolk, VA, then command of VA-46 in May 1958. In September 1959 Lee began a lengthy association with the Navy's nuclear program when he became Technical Aide Nuclear Physics, Power Branch in the Office of Naval Research. Two more staff tours occurred before his command of CVG-6 in July 1962. Nuclear propulsion training was followed by command of Alamo *(LSD-33), then aide to the Assistant Secretary of the Navy (Research and Development). Next came command of* Enterprise *in July 1967,*

Kent Lee was CAG 6 in *Enterprise* during the Cuban Missile Crisis, which brought the U.S. and the Soviet Union to the brink of nuclear war. USN photo courtesy Naval Institute Proceedings.

where he was also Commander Task Unit 77.1, with the responsibility for detailed planning and execution of extensive and sustained combat air strikes against enemy positions in North Vietnam.

Several Washington tours were next before Kent's duty as Deputy Director of the Joint Strategic Target Planning Staff, Offutt AFB, NB. He returned to Washington in August 1973 to become Commander Naval Air Systems Command until his retirement 31 Oct 1976. Admiral Lee is now retired on a farm near Gordonsville, VA. The following information provided by him is excerpted from his oral history, given to the Naval Institute, which graciously gave permission for its publication here.

"I assumed command of Carrier Air Group Six in June 1962. We were the first permanent air group of *Enterprise*, a brand new ship, just back from Guantanamo, Cuba. I relieved CDR James Holbrook; Vince de Poix was the captain, and Max Harnish the executive officer. Later on John T. Hayward became the carrier division commander.

I had the group only about a month when we moved on board *Enterprise* for some exercises. We spent August and September 1962 in the Mediterranean. The idea was to show off the Navy's first nuclear powered aircraft carrier in that region.

We came home for about ten days in October, then were unceremoniously loaded back on board to head south for the Cuban Quarantine. We loaded our complete air group aboard with the exception of the A-5s, which we left ashore. We took some extra A-4s aboard—VMA-225—for a total of one F-4, one F-8, one A-1, and three A-4 squadrons for the Cuban operation. We had an extra A-4 squadron instead of the A-5s, which had as its mission the delivery of nuclear weapons against the Soviet Union.

We went down to Cuba and were keeping track of what was going on. We got all the photography we needed and made plans to attack the missile sites. We had a lot of practice, usually at sea, loading weapons and dropping them on practice runs and attacks. We put together teams to attack the Cuban missile sites, which we studied assiduously.

Survivability was our main concern. We wanted to destroy the missile sites, but we did not have any guided standoff weapons in that period. Our plan was to go in low, beneath the umbrella of the missiles, beneath their altitude capability, and attack the site and pull out. That was our basic plan; go in low using strictly conventional, non-nuclear weapons.

We had very little intelligence on the Cuban antiaircraft capability. What the intelligence folks did paint was an ominously grim picture. We always tend to make the enemy ten feet tall. We had pictures of surface-to-air missile sites that the Russians were putting in for the Cubans. We also had pictures of the ballistic missile sites they were putting in. We made plans to attack all of those. We were just standing by on a day-to-day basis to hit them.

Every day we would stream a sled behind the carrier as our target and practice our bombing. On 26 October I went flying in a VA-66 A-4. We were to make runs on the sled. After I'd been airborne maybe half an hour, one of my division pilots said that I was smoking and burning. Sure enough, I was. I came down to perhaps 10,000 feet about 20 miles south of *Enterprise*. My fire warning light came on, but I found it very difficult to leave that airplane. I'd been flying around for 20 years or so, never had a problem, and there it looked like I was going to have to leave this one. That was a difficult decision—to eject or not.

I slowed the plane down to about 200 or 250 knots, somewhere in that range, and at 10,000 feet decided that I'd really better go. I could move the stick all around and there was no response. I knew the control lines had burned through. I said to myself, 'Here goes.' I reached up and pulled the face curtain, and there was a tremendous thud. The face curtain comes down over the pilot's face and ignites the ejection seat. You really don't know what's going on. You feel this tremendous thud, the world kind of blacks out for a minute, and you feel the parts flying every which way.

When I was fully oriented, I was at about 9,500 feet or so, drifting down in a parachute to land in the water, south of Cuba.

I then began to take stock of all my survival lessons. I remembered that, above all, you never release yourself from the parachute until your toes touch the water because you have no depth perception over an ocean. Pilots have done that thinking they were right at the water and they released themselves and plunged 200 or 300 feet to their death. Then, when your toes do hit the water, you release your parachute. If you don't, it will act like a sail and pull you through the water and drown you.

I hit the water feet first and went down pretty far; it takes an eternity to bob to the surface. But I came up by and by and inflated my Mae West; I also had a life raft with me. Before I ejected, I had alerted my division mates, who had alerted *Enterprise*, so it wasn't long before they sent a helicopter after me. I was in the water maybe half an hour.

After I ejected, the airplane was still burning fiercely. It just rolled over and plunged into the sea. I was about 20 miles south of *Enterprise*, and there was nothing down there. I hadn't wanted to get the plane up around the formation; it might hit a ship. After I landed in the water, I realized I was not injured, but I think every joint and muscle in my body was sore. It's a wild ride, believe me.

I was helicoptered back to the ship and piped aboard by CAPT de Poix—'CAG RETURNING,' BONG-BONG, BONG-BONG—and they made a big joke of it.

One of the problems with carrier duty is how to spend your time at this constant state of readiness without wearing too thin. We tried to fly as much as possible—got everybody in the air and kept our planes ready to go. We put out a sled every day, but it's difficult to be in a high state of readiness for two months, just operating south of Cuba, with no ports. It's not the easiest thing in the world to do, especially when you face a potential danger in which you know you're likely to lose some people.

We didn't realize that we were not going to have to attack Cuba until John F. Kennedy had settled with Nikita Khrushchev. We stayed down there a week or ten days after the settlement to make sure that the Russians removed their missiles, which was part of the agreement. We had airplanes observing this (removal). In late November 1962 *Enterprise* once again headed for Norfolk. In view of what we later learned in Vietnam about the effectiveness of surface-to-air missiles, I'm just as glad things turned out the way they did.

In 1976, after five tours of duty in Washington during my naval career, I wanted to get out of the place for retirement. Besides, Washington is not a healthy place for a retired admiral. If the nineteenth hole doesn't get you, the politicians will.

In early 1977 we found a farm in Albemarle County, near beautiful Charlottesville, VA. We bought it and moved in 6 Oct 1977. Our three daughters were married here on Logon Farm, and we now have nine grandchildren. We attribute that good fortune in part to Logon Farm. It was a good start.

The years have gone quickly since retirement, and life has been more than happy and pleasant at Logon Farm. Even my wife, a city girl, has fallen in love with the place. What more could a sailor ask for?"

CDR DONALD B. EDGE, USN (CAPT Ret)
CAG 14 24 March 64-5 March 65
USS *Constellation* (CVA-64)

Don Edge is a 1945 graduate of the Naval Academy who received his wings of gold 7 June 1947 after his mandatory "blackshoe" tours in the cruisers USS Canberra *(CA-70) and* Montpelier *(CL-57).*

His first flying tour was with VA-94, and he flew combat missions in Korea with VC-3 in F4U-5Ns. CDR Edge later commanded VF-213 and assumed command of Attack Carrier Air Wing 14 on 24 March 1964, just prior to its deployment to Southeast Asia in Constellation *(CVA-64). During this deployment the Gulf of Tonkin incident occurred, pulling the U.S. further into the abyss of the Vietnam War. CAG Edge's air wing conducted the first retaliatory strikes against North Vietnam as they launched attacks 5 Aug 1964 on coastal PT boat bases and petroleum storage areas at Vinh. CVW-14 was awarded the Presidential Unit Citation for its performance under the leadership of CAG Edge during this deployment.*

Don Edge later commanded Kawishiwi *(AO-146) and NAS North Island.*

He is now retired from the Navy as a captain living in Coronado, CA, where he enjoys a relaxed life playing tennis and serving on city committees.

"I assumed command of CVW-14 24 March 1964, upon its return from Hawaii, where Bob Gallatin had taken the group through the ORI (Operational Readiness Inspection). I flew out to get some flying, and to become acquainted with members of the squadrons. Upon the day that we returned to North Island, we held a change of command ceremony on the flight deck. There wasn't any other time to do this between getting ready to deploy and our 5 May departure. At least we thought it convenient, and guests would get to see a real carrier.

A lot of time was spent in going to *Constellation* to coordinate the move on board ship, assignment of the CAG and squadron spaces, and those other matters that would allow for a smooth coming aboard for deployment. As the CAG, you were considered a department head on the ship and were accorded all the responsibilities thereto.

We settled into the WestPac routine with an initial line period of approximately 59 days. Much of this time was in recce, the escorting of planes into Laos and the *Plaines de Jars*, where *Kitty Hawk* (CVA-63) had lost a VFP-63 photo plane and a VF-111 F-8 escort to ground fire, with the *first* of the many rules of engagement (ROE) that were to follow. Very simply, those first missions of escort were that you could not fire weapons unless fired upon. (A great way to give the advantage to the enemy.)

During our July-August Hong Kong visit, we had been doing our shopping and R&R, when, on about the fifth day in port, something altered the shore leave. Standing port and starboard watches, only one section was allowed to go ashore, and they were cautioned that there might be a general recall. They were to be prepared to return and be aboard no later than 1600. For whatever the insider reasons, another two sections were allowed to go ashore after lunch. Sure enough, the recall came and everyone was on the ferries or wala walas headed for Green Island where the big boys or hazardous ships always anchored.

Connie steamed away handsomely on 3 August and, for the first time, we began to hear rumbles that a U.S. ship might be, or could be, endangered. The skipper, Fred Bardshar, had all boilers

CDR Donald B. Edge was CAG 14 in *Constellation* during the August 1964 Gulf of Tonkin incident, which led to deeper U.S. involvement in the Vietnam War. CVW-14 aircraft, along with CVW-5 in *Ticonderoga* conducted the first retaliatory raids against North Vietnam. USN photo, courtesy CAPT D.B. Edge, USN(Ret)

on the line, and we began to think of a night launch in order to send support if ordered. We hurried to get past the southern tip of Hainan, since this would be the maximum distance for a strike, plus the wind had to be coming out of the west for both the launch and recovery. And so it happened.

The launch was after midnight, with all planes recovered successfully after an uneventful flight into the northern Gulf. I began to think that nothing had been seen or heard and that the events of the early morning were completed. Not so! About four or five in the morning, word came that a strike would be executed against the Vinh oil fields by *Ticonderoga* (CVA-14), and an additional three northern targets would be given to the Connie. Since the majority of the pilots had flown that night, I had a great concern in turning around many of those same pilots within a matter of hours. I called the COs together and asked that they be selective about those pilots who would be on the next mass launch. Morale was such that everybody really wanted to go, and most of the pilots from all of the fighter and attack squadrons were on the reprisal strike. (One footnote: President Johnson notified the American public of the strikes *before* they were ever launched, giving the Vietnamese an opportunity to be unnecessarily alerted.)

A 64-plane strike from Connie and Tico was launched against four major PT boat bases along the North Vietnam coastline and their fuel depots at Vinh. The A-1s went to Vinh, and the A-4s went to the more distant northern targets. Although the strike was opposed by heavy flak, generally events went well, except for the loss of LTJG Richard C. Sather, of VA-145, in a dive against small-boy fire. We also lost the Navy's longest-term POW, LTJG Everett Alvarez, from VA-144. Fortunately, he was returned in the second airlift from Hanoi in 1973.

I was never convinced that there had been an incident requiring reprisal strikes and, when several days later, the same radio chatter from the northern Gulf began to reappear, I volunteered to launch in an F-4 to personally investigate. For an hour of recce, neither my RIO nor myself saw anything that looked like hostilities on the high seas. We returned to Connie, and that ended the so-called attacks by enemy boats.

Though line periods in excess of 50 days were the norm in those days, there were never any more incidents. Mostly, we did a lot of recce flights; photo birds to Saigon, escorts for the photos and air shows for SEATO, PacFlt, etc. All sorties for the cruise returned safely, and other than Alvarez and Sather, no pilots were lost. This fact was in itself considered remarkable."

CDR H.P. "HANK" GLINDEMAN, USN (RADM Ret)
CAG 15 8 Jul 64-29 Apr 65
USS *Coral Sea* (CVA-43)

"Hank" Glindeman is a product of the U.S. Naval Academy, Class of '46, which graduated 6 Jun 1945 because of WWII expediency. He was first assigned to USS Pennsylvania *(BB-38) and subsequently had tours in* Quincy *(CA-71), staff duty in Tsingtao, China, and in* Furse *(DDR-882) before reporting to Pensacola for flight training in December 1948.*

Winning his wings of gold in June 1950, Hank went to VS-23 until November 1952, than back to the Naval Academy for a two-year instructor tour. He returned to the cockpit in 1955 and transitioned to jets flying with VF-143, then VA-55. A tour with VA-126 as operations officer preceded the Naval War College, followed by a two-year CNO staff job. Hank was XO, then CO of VF-154, December 1962–July 1964 before becoming CAG 15 on 8 July 1964.

Fying from Coral Sea, *CAG Glindeman led the initial strikes against North Vietnam in February 1965 following the Gulf of Tonkin incident and retaliatory strikes of August 1964.*

Leaving CVW-15 in April 1965, staff jobs at ComNavAirPac and First Fleet were followed by his command of Passumpsic *(AO-107), then another CNO staff job. Hank assumed command of* Ranger *(CVA-61) in September 1971 and was in the Gulf of Tonkin when the cease-fire was signed 28 Jan 1973.*

Following Ranger, *Hank was promoted to flag rank and became Chief, Fleet Coordinating Group at Nakhon Phanom, Thailand, until October 1974, after which he went to the staff of the Secretary of the Navy. In August 1975 he assumed command of Carrier Group Seven. In October of the next year Hank wore the dual hats of Commander Carrier Striking Force Seventh Fleet/Com-*

Hank Glindeman served as CAG 15 at the onset of the air war against North Vietnam. His frustration of micro-management from Washington of the conduct of carrier strikes was a precursor of the mis-handling of the Vietnam War by politicians. USN photo, courtesy of RADM H.P. Glindeman, USN(Ret)

mander Carrier Group Five before returning to the States in November 1977 as head of the Naval Safety Center.

After having nine commands in his naval career, Hank retired from the Navy 1 Sep 1978, and now lives in Clearwater, FL, involved in the production of environmental clean-up materials.

"As major events occur in one's life, most of us do not say to ourselves, 'Aha, I'm in the right place at the right time.' More likely, as we reflect back on various major events we do find an answer for our good luck, by recalling the saying, 'I was in the right place at the right time.' This chapter is written about my tour as Commander Air Wing 15, 1964-1965. However, I'm also a believer that you must set goals for yourself. From the time I was commissioned an ensign in the Navy, my goal was to have a command. In the aviation community especially, being a commanding officer was a real achievement, because you had competed with the best and won.

The first screening for command is an anxious moment for all aviators. To find out you have screened for command is a truly exhilarating moment. I made it and was assigned as XO to Forrest Peterson in VF-154, the Black Knights, flying F-8D Crusaders out of Miramar with CVW-15. I had the job only about six weeks when my predecessor was selected to enter the nuclear power training program.

Next career goal after squadron command? Really didn't have any in mind. I didn't have to wait long, however. At the end of CVW-15's *Coral Sea* WestPac deployment, Skipper CAPT Charlie Roemer informed me that my orders had just come in to relieve Bob Bascom as CAG 15. Wow! I was pleasantly surprised, as I hadn't expected orders because I still had two months left on my CO tour.

My initial feeling was euphoric—two commands!—how lucky could a guy get? Also, because I already knew the people in the wing I really felt doubly blessed to get Air Wing 15. My XO in VF-154 at the time was Bill Ellis, a great naval officer who, after his CO tour, got Air Wing 21.

I went back through the RAG, getting checked out in the Phantom and Skyhawk as a P-CAG. Survival school again—special weapons, refresher, day/night CarQuals, all the good stuff.

The CAG billet was a plum and required greater management as well as stick and throttle skills. Being CAG under the base loading concept required you to coordinate with each Commander Fleet Air (ComFair) at the individual bases and/or locations. CAG needed to know and understand the geography of each area. ComFair Miramar was different than ComFair Alameda, Whidbey, or Lemoore. Each area had its own set of problems. For example, air traffic control departures and arrivals at Miramar, with its interface with civilian traffic at San Diego's Lindbergh Field and North Island, were in a constant state of change. The advent of afterburner jets generated many meetings with base commanders, FAA, unit commanders, type commanders, etc., on the impact of high performance takeoff training. Other problems arose with the Boardman Range in Washington State that was utilized by the Whidbey A-3 squadrons. As CAG, you had to be aware of these problems and spend time seeking solutions in support of your squadrons.

The base loading concept with the squadrons widely scattered was excellent for material and maintenance support, but was not conducive to providing an opportunity to train as an integrated unit in the same manner in which you would fight. That opportunity came when the air wing went to Fallon for a weapons deployment. There the wing came together as a fighting unit. That was where jets had to learn timing with the A-1 propeller aircraft and coordinated strike tactics. These weapons deployment training periods came late in the training cycle and lasted from three to four weeks. Some dissimilar aircraft type training was done prior to the whole air wing going to Fallon and operating as a unit, but it was very limited (tanking, some AEW coordinated training with the VF community) until you got the entire wing together.

Some units hadn't structured their flight schedules with the same people flying together. I felt then and feel now that good combat tactics can be made better if a four-plane division always has the same four pilots. People flying together get used to each other and will perform better under stressful combat conditions if they train together.

My tour as CAG, flying for the first time in two-seater F-4Bs, was an eye opener for me. I had flown single-engine, single-pilot aircraft for the majority of my career, handling radios, armament switches, etc. by myself. The RIO turned out to be a welcome adjunct to flying and increased the effectiveness many times over of the weapons platform.

As deployment time came near, coordinating with the carrier became a priority. Different airplanes, more people, more yellow gear than the previous deployment created new problems. The CAG's billet was as a head of department, under the XO. The marriage of the air wing to the ship is one that seems to be forced down each others' throats. By that I mean that the ship has been without an air wing aboard for many months, and the air wing has been ashore where its spaces are more generous, can go home most nights, and life just seems easier. CAG and his staff have to establish their rapport early with the ship and be the point of contact for those inevitable problems which arise. I worked closely and frequently with the carrier XO early on to establish the ground rules and space assignments. Strong leadership at the unit level in educating the new young first-cruise officers and men is a must.

I was fortunate in that my XO, CO, and CAG tours were all on the same carrier, *Coral Sea.* I went through Bob Elder, Charlie Roemer, Pierre Charbonnet, and half way through George Cassell's tours as *Coral Sea's* CO. Each was different, each was excellent, and each had a good understanding of our common goals.

August 1964, with its sudden eruption of the Gulf of Tonkin incident, put a whole new light on things. As *Coral Sea* and CVW-15 put to sea out of Alameda in November 1964, severe salting problems in a main condenser arose, which necessitated off-loading the air wing to NAS Barbers Point when we arrived at Pearl Harbor. We catapulted off the ship while tied up to the pier at Ford Island (having downloaded fuel first, of course). The launch course put the aircraft headed straight for Aeia, a suburb of Honolulu, in the late afternoon. There was much furor at the Pearl City Tavern's Monkey Bar over the noise level encountered, I'm told.

We operated out of Barbers Point for about four weeks while the ship was repaired. During that stay I received a summons to go to CinCPacFlt headquarters to talk with ADM Tom Moorer about our upcoming deployment to WestPac. I took with me the A-4 skipper, Pete Mongalardi (who relieved me the following year). Upcoming operations? Why does a four-star admiral want to talk directly with the CAG? He asked about our weapons training and whether we were ready for combat. No war had been declared, he added, but as kindly and firmly as he could, he predicted combat operations in Southeast Asia. Admiral Moorer went on to add that, if it happened, it would be a non-standard operation. Things were so political that he was sure we would find ourselves at the point of the spear, so to speak, with unusual rudder orders and constraints. He stated that if combat operations commenced, I would think he and his staff had gone off their rockers. He quickly added that this was not the case, and he assured me he was behind us 100 percent, but politics were going to be playing a major role and driving things. It was quite an experience for me. I thought that calling me in was a bit unusual but, to this day, I really appreciate his council and the setting of the stage.

On 15 Jan 1965, CAPT George Cassell took over as *Coral Sea's* CO, and we sailed from Honolulu westward to Subic Bay in the Philippines. Upon arrival at Subic we off-loaded a few airplanes to ease the deck load, took aboard ComCarDiv One and his staff, and shoved off for the Gulf of Tonkin with only 24 hours in Subic and no liberty.

Now I had a new set of people to work with. ComCarDiv One was RADM Eddie Outlaw. I had known him during a tour in Op 50 a few years before. CAG almost becomes divorced from the CO of the carrier with a staff aboard because the day-to-day tasking flows directly from the staff to the air wing. The CO generally attends the admiral's morning briefing, but frequently his CO's duties require him to stay on the bridge. CAG is a must attendee. CAPT Lucien Powell was chief of staff, and CAPT 'Jeep' Streeper was the division operations officer.

Our first line period in the Gulf of Tonkin, coupled with transit time from Pearl harbor and one day in Subic, was a long time between stops. CarDiv One was the operational commander, and finally *Coral Sea* was headed back to Subic for some in-port time. But this was not to be—At The Right Place At The Right Time?

On 7 February I was awakened by the admiral's orderly about 0200 and told he wanted to see me up in Flag Plot. When I arrived, Admiral Outlaw was in his bathrobe and slippers. I'd never seen an admiral before in bathrobe and slippers, one of the unusual things that forever remain in a person's memory. He said we had received orders to proceed to a launch point and to get ready to bomb Dong Hoi Barracks in retaliation for the North Vietnamese attack on the Pleiku Barracks, a U.S. installation in the central highlands of South Vietnam. The TOT (time over target) was not yet fixed. He asked me when was the earliest we could launch without having to tank any airplanes. CVW-15 had A-4C and A-4E Skyhawks. The A-4C was the limiting factor on range with a payload.

Admiral Outlaw had turned us around (with accompanying escorts), and we were proceeding at best speed back to the Gulf of Tonkin. I took a compass and struck an arc on the surface chart in Flag Plot equal to the maximum un-refueled range of the A-4C from Dong Hoi Barracks. The admiral asked me if I was sure and did I want to consult with anyone. I said no. When he was finished with me I left Flag Plot to wake up my squadron COs and staff ops officer to brief them. We all spent the rest of the night passing the word, getting ordnance broken out, checking the weather and doing our strike planning.

That is how Flaming Dart One, the first of seven years of nearly daily offensive operations, got started. Flaming Dart Two occurred 11 February against Chan Hoa.

Admiral Moorer was correct. The early months of offensive operations against the North Vietnamese were particularly frustrating. First and foremost, our TOT was dictated in our tasking directive with no consideration as to the weather. The Gulf of Tonkin in the early spring frequently has severe fog, called a 'crackin fog.' Very low, very dense, but generally burns off by 1000 or 1100 from about DaNang north to Haiphong. Everyone was angry and dismayed that we had to be over our target, floundering ineffectively and at great risk to ourselves, for no reason. We (from Admiral Outlaw on up the chain of command) tried on many occasions to let the on-scene commander have some flexibility as to TOT, aircraft composition, target selection, and ordnance, to improve our military effectiveness. I came to realize the truth in Admiral Moorer's words back in his office at Pearl Harbor—it was a strange conflict indeed.

In addition to times over target, the selection of targets, composition of the strike aircraft themselves, and selection of ordnance were being made, not by anyone on the ship, but came from Washington. The Rules of Engagement (ROE) were constantly being changed, and if you broke one, the penalty was severe.

As CAG, I became a buffer between angry squadron COs and their pilots and the system. Admiral Outlaw was a jewel; a CAG himself in WWII and winner of a Navy Cross, he was as frustrated as anyone.

Our mission was not clearly stated to us. By the end of February 1965, when I saw Soviet freighters loaded with trucks going up the Gulf of Tonkin toward Haiphong, I went to Admiral Outlaw and recommended we mine Haiphong Harbor. We had some mines on board and plenty more on an ammunition ship in company with us. The admiral jumped at the idea. He forwarded a mining concept plan up his chain of command. Approval was recommended all the way to Washington.

Some time later the reply came back, in essence saying, 'you guys don't understand the problem. We're only retaliating, not declaring war.' The irony of the mining recommendation is that I finally got to do it in December 1972, albeit, vicariously as CO of *Ranger*, when, during Linebacker Two, U.S. forces finally were authorized to mine Haiphong Harbor. No further ships entered Haiphong after we mined it. Just think, seven years earlier we could have done the same thing. How history is molded and changed because of the political winds. From the outset, the manner in which we conducted those early combat operations was the genesis of the disillusionment of our nation as a whole to Vietnam.

The paperwork associated with operations in Vietnam was staggering. The Operational Reporting (OpRep) system was almost diabolical. Designed to give the whiz kid operations analysts all the information they could possibly need to design combat operations, but did nothing to win the war.

As CAG, I resented not being able to use all the weapons we had on board, i.e., Walleye, IR guided bombs, napalm, etc., for the appropriate targets. I resented being directed to reduce a strike from 32 aircraft to 18 because somebody in Washington thought that 32 was too large a number of aircraft for the target assigned. I had to reduce the number—what do you cut out? SAM suppression, MiG CAP, SAR, aircraft, target bombers? I didn't like that call from Washington at all.

Two other areas when I was CAG were bothersome: coordination with other forces in the theater and publicity (news media). Initially, in the Gulf of Tonkin, Navy Air didn't know what the Air Force and the Marines were doing and vice versa. More than once I locked up on what turned out to be USAF planes flying through or near my target. Only the visual identification criteria of the ROE prevented me from shooting down a friendly. By the end of my CAG tour a communication and organizational setup was established to assist in solving the mutual interference problem.

It appeared to someone that the Navy was not getting its share of publicity. I don't know who that somebody was, but I was ordered to fly to Saigon after a strike and to be there to attend and be part of the daily 1700 debrief ('5-o'clock follies') provided the press corps. So, 30 Mar 1965 I flew the strike on Back Long Vi Island, refueled in the air and flew to Tan Son Nhut AB in Saigon for the press briefing. I flew the strike with a pair of khaki trousers and shirt rolled up and tucked in the corner of the cockpit. It was not a good day. By the end of the strike I had two COs and one operations officer shot down. One CO was picked up that afternoon by a submarine I had asked Admiral Outlaw to preposition on the south side of the island. The other CO was found 48 hours later and rescued. We lost the operations officer, who was last seen in a high negative-g maneuver with no ejection observed.

The press conference was anticlimactic. I related the success of the mission—two rockets went into ammunition storage areas with huge secondary explosions, and a radar site was also taken out. I avoided direct answers about aircraft losses. My one and only press conference. Instead, reporters such as Peter Arnett were flown to the carriers to get the Navy story.

Vietnam was a strange, self-defeating conflict. By comparison, Desert Storm was executed by the on-scene commander in absolutely jig time—100 hours, with outstanding inter-service cooperation. Hopefully, that is a lesson learned from Vietnam by the politicians; a clear, reachable goal, adequate forces, and then let the professionals go at it.

CDR M.S. SNOWDEN, USN (CAPT Ret)
CAG 5 12 Jan 65-13 Dec 65
USS *Ticonderoga* (CVA-14)

Graduating from the Naval Academy with the Class of '45, ENS Macon S. "Mac" Snowden was assigned to USS Caperton *(DD-650) with TF 38 off Japan just prior to the end of WWII. After a subsequent tour in* Knight *(DMS-40) in the Pacific, he went through flight training and won his wings of gold at NAS Jacksonville, FL, in July 1948.*

He first flew the F4U-5 Corsair from Coral Sea *(CVB-43) with VF-21, then F9F-2 Panthers from* Franklin D. Roosevelt *(CVB-42) in the Atlantic. His first shore duty tour was assistant professor of naval science at the University of Illinois 1951-53. Mac then returned to the fleet with VF-41 in* Randolph *(CVA-15) until 1956 when he reported to the Commander Naval Air Atlantic staff as atomic weapons officer for ships.*

In 1957 Mac underwent training as a test pilot, and upon graduation became a project test pilot in the Armament Test Division of NATC Patuxent River, MD, until the following year. He then attended the Armed Forces Staff College before returning to Pax River to complete his test pilot tour there in 1959.

Tours with VF-124 (replacement pilot training) and VF-91 as executive, then commanding officer in Ranger *(CVA-61) were followed by assignment to* Constellation *(CVA-64) as assistant air officer, then air officer during 1962-64. Mac was next assigned as CAG 5 in January 1965 and deployed to Vietnam in* Ticonderoga.

Next came a Washington assignment with the Chief of Naval Operations staff (Op-05) as the F-4 Phantom, F-8 Crusader, and

Mac Snowden led CVW-5 from *Ticonderoga* during the early stages of the Vietnam War. USN photo courtesy CAPT M.S. Snowden, USN(Ret).

F-111B program coordinator 1966-68. During this time he wrote specific operating requirements for the F-14 Tomcat and was promoted to captain. Returning to school at the Industrial College of the Armed Forces, he wrote his Masters thesis on "Comparison of Soviet and American Management Techniques in the Development of Fighter Aircraft."

During 1969-70, CAPT Snowden was commanding officer of Chilton *(LPA-38) deployed to the Mediterranean, then returned to the Office of CNO (Op-05) as Deputy Director of the Air Capable Ship Study Group. He then commanded* Guadalcanal *(LPH-7) during 1971-72 before again returning to the Op-05 staff as Head, Aviation Plans Branch, then Head, Aircraft and Weapon Requirements Branch 1972-74. After logging more than 4,000 hours in the air and 503 carrier landings, CAPT Snowden retired in 1974.*

Since his retirement from active duty, Mac has served as Assistant Director of Administration for Montgomery County, MD, and during 1975–80 as military advisor to a member of the House Committee on Armed Services. He continued his active support of naval aviation through his 1980-86 positions of executive director and treasurer of the Association of Naval Aviation.

Mac's second retirement came in 1986, and since then he has served as a member of the Board of Directors of Carteret General Hospital and on the administrative board of a local church. He resides in Pine Knoll Shores, NC.

First as air officer in Constellation *and then CAG 5 in Tico, Mac was present during the early phases of the U.S. Navy's involvement in the Vietnam War.*

"Following my relief as air officer in *Constellation* on Yankee Station, I returned to the West Coast and went through refresher training before assuming command of CVW-5 on 12 January 1965. We deployed to WestPac in *Ticonderoga* 10 October, stopping in Hawaii for a short training period.

One incident in Hawaii caused a real flap. We had a Spad driver who dropped a live 500-lb. bomb on the beachline of the wrong island. It was a moonlit night, and through the clouds it looked like the spot he was supposed to hit, but it wasn't, it was an inhabited island. Senator Patsy Mink raised a holy stink in Washington. The headlines were bigger than those for Pearl Harbor—'NAVY BOMBS HAWAII'—in two-inch letters. Man, was that a flap! Probably *one* of the reasons I didn't make flag rank!

I preceded CVW-5 to Cubi Point and took the COD out to *Bon Homme Richard* (CVA-31), then over to *Midway* (CVA-41) where I flew sorties with VF-191, -194, and -111. I rejoined CVW-5 on its arrival in Tico at Cubi, then we proceeded to Yankee Station to commence strike ops 8 November. On 13 December I was relieved in Yokosuka, Japan, during an off-line period and reported directly to OpNav for duty.

As for the chain of command structure, we were still under the 'old' system (Super CAGs had not been heard of as yet). If the CV was operating independently, details were worked out between CAG and CV operations, then presented to the carrier skipper. My observation was that if a flag was on board, liaison was more of a CAG-Flag Ops link, and the CV CO receded somewhat to ship's management, but he was still briefed as things progressed. This (method) was not the usual way of doing things, and was an indicator of the strength and knowledge, or lack of it, of the CV CO.

It was a rather universal view that management of operations came in large measure from Washington or Pearl Harbor. Target selection, in my opinion, was poor. Rules of engagement were, in some cases, illogical. Weaponeering was occasionally very poor. For example, weapon selection and loading instructions came from without—in one case we were instructed to load 19-shot rocket pods on A-1 Skyraider centerline stations. This action is pretty hard on propellers.

Another case of illogical thought was not permitting attack on targets which had been previously struck—even if the primary target at the time was destroyed and you wished to use your weapons on something that needed to be hit as an alternate. Example: we struck a large intersectional bridge and obtained a number of direct hits with Mk 84 (2,000 lb.) bombs. Because of defective fuzing, the bombs largely exploded prematurely or passed through the structure, exploding in the valley below without much damage to the bridge. Approximately two days later as part of a two-air wing strike on a large bridge in North Vietnam, we had to pass over the bridge we had previously attacked. If that day's target was found to be eliminated on our arrival, we were not allowed to use the first bridge as an alternate target. Instead, we were directed to proceed back to the ship and jettison our bombs at sea while en route. An exception to this was that we could attack it *if* we observed vehicles crossing it. Needless to say, there were *always* vehicles crossing it. This was also during a time when we had a bomb shortage. Hard to believe, isn't it?

During predeployment to Vietnam I had set up 'Operation Singing Sword.' (The name was taken from the comic strip 'Prince Valiant.') CVW-5 was probably the first wing to attempt extensive wing training exercises in preparation for a tour at Yankee Station. We flew a number of mass A-4 flights from Lemoore, joined by F-8s from Miramar, to various target areas on the West Coast and east of the mountains. Coordinated attacks were the routine, and opposition to the flights was provided by VF-126 TF-9Js out of Miramar.

Later in the training cycle we deployed the entire wing plus detachments to NAS Fallon, NV, for strike training, emphasizing all aspects of live ordnance. We also provided cross-training in that VF pilots taught fighter tactics to VA pilots in the classroom and in the air. Conversely, the attack folks taught the VF guys all the recent tricks in bombing. Remember, at this time F-8s were largely pure VF and some had acquired rocket tubes, but there had not been much in the way of bombing. We had the bomb racks and we used them. All the ground crews, as well as the pilots, had a real workout and were fully prepared for deployment."

CDR JAMES B. STOCKDALE, USN (VADM Ret)
CAG 16 18 Feb 65-9 Sep 65
USS *Oriskany (CV-34)*

James Bond Stockdale, U.S. Naval Academy Class of '47, is the epitome of leadership and inspiration, as well as the ultimate naval aviator. Admiral Stockdale's performance of duty during eight years as a prisoner of war in North Vietnam earned him the Medal of Honor, and he has the distinction of being one of the most highly-decorated officers of the naval service.

After graduation from the Naval Academy, Jim first served in USS Carmick *(DMS-33) then* Thompson *(DMS-38),* Charles H. Roan *(DD-853), and as executive officer of* PCS-1392 *before entering flight training in 1949. Earning his wings in September 1950, his initial fleet squadron tour was with VS-27 as LSO and safety officer. He received test pilot training at NAS Patuxent River in 1954 where he remained as an instructor until January 1957. Returning to the fleet with VF-211 and then VF-24 in early 1960, Jim became the first naval aviator to log 1,000 hours in the F8U Crusader.*

In September 1960 he received postgraduate education at Stanford University where he studied history, political science, philosophy, and economics. In a few short years to come, these studies would serve him well in a way he could never imagine.

Again returning to the fleet, Jim joined VF-51 in September 1962 as XO before fleeting up to CO a year later. On 4 Aug 1964, he led a flight of four Crusaders from Ticonderoga *(CVA-14) to repel an attack on* Maddox *(DD-731) by North Vietnamese PT boats; the event now known as the Gulf of Tonkin Incident. The next night, he led another flight against purported PT boat attacks on* Maddox *and* Turner Joy *(DD-951) off the coast of North Vietnam. Although this event led directly to the 10-year quagmire of U.S. involvement in the Vietnam War, Jim adamantly believes that there was no attack that night by the North Vietnamese. During the mission, he made repeated passes as directed by the on scene commander and, although he expended his full load of rockets and ammunition, the "Screaming Eagles" skipper never observed any trace of a target.*

Returning briefly to the States in November 1964, CDR Stockdale took command of Air Wing 16 just in time for its April 1965 WestPac/Vietnam deployment in Oriskany. *CVW-16 sustained heavy losses in the rapidly-escalating air war, and Jim was shot down 9 September while flying a VA-163 A-4E Skyhawk. He became the senior Navy prisoner of war, confined in Hanoi's infamous Hao Lo prison. CAG Stockdale was tortured 15 times as he spent two years in leg irons and four years in solitary confinement—a result of his assumption of command of the POWs and his underground organizational leadership in defiance of his captors' orders. Jim was awarded the Medal of Honor for deliberately inflicting a near-fatal wound to himself in order to convince the North Vietnamese that he was willing to die rather than to capitulate to their demands.*

Promoted to captain while in prison, Jim was released 12 Feb 1973 and underwent extensive medical treatment for his injuries received upon his ejection and subsequent torture. Although these injuries left him permanently disabled with a locked left knee joint, then-RADM Stockdale was able to continue his naval career as he assumed command of Antisubmarine Warfare Wing Pacific in January 1974. He later was president of the Naval War College before he ended his 37 years of active duty with his 1979 retirement as a vice admiral.

A strong believer in academic excellence, VADM Stockdale first entered the civilian world as president of the Citadel Military College in Charleston, SC. He is currently a Senior Research Fellow at the Hoover Institution on War, Revolution, and Peace at Stanford University, Palo Alto, CA. He is an active writer, and has published several works concentrating on the central theme of how man can rise in dignity to prevail over his adversities. His most recent books are "A Vietnam Experience: Ten Years of Reflection" *and* "In Love and War," *co-authored with his wife, Sybil. In 1992 he was a vice-presidential candidate on the Independent ticket with Ross Perot. He also holds 11 honorary doctoral degrees.*

Jim describes some of his CAG 16 experiences, as well as command relationship with his carrier CO and his final mission.

"My *Oriskany* CO, CAPT Bartholomew J. Connolly III, treated me every bit as much his 'right hand man' as he did the ship XO. For instance, I had authority to clear (sign off for transmission) all *Oriskany* messages that dealt primarily with air wing matters. For

CDR James Bond Stockdale, CAG 16, callsign "Double-O Seven," became one of the Navy's most highly decorated aviators and was awarded the Medal of Honor for his leadership as an eight-year POW in Hanoi. USN Photo courtesy VADM J.B. Stockdale, USN(Ret)

example, it was I who decided whether a shot down pilot was MIA or KIA; I sent the message from the ship to Washington, and CAPT Connolly just read it on the message board the next day.

In operational matters, I was similarly delegated complete authority. Whether or not I ran it by the captain before executing it was a matter of taste and style. Another example being the matter of standardizing the landing pattern procedures of all squadrons; I wrote the pages that got in 'the book.' I did not get this approved by CAPT Connolly because I felt sure he didn't want to bother with it. On the matter of clearing the Crusaders to carry two 2,000-pound bombs (one under each wing), I *did* consult the captain. The reason being to get the idea worked out and on line as soon as possible because it was primarily a pilot-saving safety initiative. (My Marine squadron came up with the idea as a way to take down the Thanh Hoa bridge in a hurry, rather than send flight after flight over to bomb it with little 500-pounders and losing a pilot every week or so.) I knew if we 'went by the book' on this procedure, we would have to involve the BuAer guys, because we would be making an exception to all the handbooks they had published. I knew getting approval would take months and run a 50/50 chance of being turned down flat. (With these bombs and a full fuel load, the planes were too heavy to catapult, so we had to launch them with bombs and about a one-third load of gas, then air-refuel them off the bow.)

In other words, to do what I wanted would stick the captain's neck out a mile. What actually happened was that I kept him abreast of the 'test project' we ran as we tried these procedures out on a series of hops off the ship. Once we agreed it was practical, he said to go with it. Without mentioning it, we both agreed to let Washington go to hell.

In terms of my autonomy, with the relationship between CAPT Connolly and me, I had the best of both worlds. We saw eye to eye on the philosophy of command at sea. I was always welcome to a private session with the captain any time I asked for it, and to benefit from his wisdom and advice in all matters concerning my Air Group Commander duties.

As for my relationship with higher authorities, I can't remember a single instance of my having the need to communicate with any senior command off the ship. With regard to the embarked CarDiv commander and staff, either the captain or I was welcome to sit in on the morning meeting of the admiral and staff; we shared the privilege and compared notes afterward. When I had a difference of opinion with the staff, I let the captain of the ship know it, and he fought my battles with Flag.

In a nutshell, from what I know about the Super CAG concept—to let the air wing commander have unimpeded authority over the way the wing is to fight in the target areas—for all practical purposes on the *Oriskany*, I had that authority in the rank of commander as CAG 16.

One unusual event that took place during my CAG tour was the Air Force kept borrowing our A-1s to augment their search and rescue forces in trying to find and rescue shot down pilots in western Vietnam. At times, half or more of our A-1s were 'TDY to Udorn' (Thailand). Our A-1 skipper was getting upset over half his forces being out of his control, and he had a point. The guy in charge of SAR forces over in Udorn had been a recent National War College classmate of the admiral and, at the admiral's suggestion, CAPT Connolly and I arranged for me to be launched to Udorn in a 'Spad' just after midnight when night carrier operations were over. It was just daylight when I got there and, at the Officers' Mess breakfast, I found the brigadier general in charge, explained Admiral Cousins' and our concern about the situation, and he freed up the planes—provided they would do a little SAR up and around Dien Bien Phu on the way home. I joined them on that trip, and we got back to *Oriskany* about noon.

By the end of the summer of 1965 we had already lost eight pilots. But our spirit was high, particularly when, on 10 September, we prepared to throw eight *Crusaders*, each with two 2,000-pound bombs, against the Thanh Hoa bridge to break its back once and for all. We were just beyond mid-deployment and, after a long and costly line period, were scheduled to repair to Hong Kong for R&R the next morning.

This air group finale of the line period got to be a rather big operation; each Crusader had to have its own A-4 tanker to top-off its necessarily light fuel load before it crossed the beach. Then there were flak supressors, along with high fighter cover and 'Fuds,' as well as an EA-1 to zap the DRV fire control radars in the target zone. I think there were 37 airplanes in my formation when I, as strike commander, leading an A-4 flak suppression division, barked out the 'Kilo' signal to the *Oriskany* CIC as we headed for the bridge.

The first bad news came from our A-1, way up ahead: 'The weather is zero-zero over the bridge.' What a letdown! There was nothing for me to do but to direct all elements to proceed against secondary targets, and I had neglected to mention to Wynn Foster, XO of VA-163, who found himself on my wing, what exactly our alternate target would be. I remember we loitered offshore listening to VA-163 CO Harry Jenkins' division comparing notes on their detections of SAM radars, on their just-acquired gear, flying in the soup, down near Vinh. That SAM was the southern-most one yet detected, I believe.

I knew of a place where we could normally get rid of a bomb load into a worthwhile target without prior arrangements, and with surprisingly little flak: a railroad siding four or five miles west of Highway 1, about 15 miles south of Thanh Hoa. The ceiling was low up there and, since Wynn and my planes had 500-pound bombs with an option switch for snakeye or steep dive delivery, we went in under the clouds. We proceeded up the coast past our target, then swung west to a nice '180' turning point, and swept in to line up on those beautiful boxcars.

I didn't even have my oxygen mask snapped across my face. We were coming in flat, in trail, right on the deck at about 500+ indicated, I, and undoubtedly Wynn, both totally relaxed. I had pulled this foul-weather snakeye maneuver on this same train parking area a couple of days before and got no flak.

Then I heard it!! 'Boom! Boom! Boom! Boom!' I knew right where to look—the noise came from two o'clock. And there it was, firing at me point blank, 200 yards at three o'clock. It was, I think, a 57mm mobile cannon. I watched it belching fireballs directly at me and, by the time I recognized all this, every red light in my cockpit was on and pulsating: fire! Hydraulics gone! Everything unwinding!

I had to get that oxygen mask up to give Wynn a Mayday. That took an instant, and by that time the plane was out of control with me just along for the ride. The stick didn't work, and I was heading for the deck. The confusion of 'g' forces interrupted my grabs for the curtain, so I went for the alternate handle between my legs. I remember craning my head back to see the chute blossom, and a strange feeling coming over me as I watched the chute unravel that 'something about that ejection was just not right.'

I later figured I had 'sensed' something awry with my bones; I had forgotten to grab my right wrist with my left hand, and it was hours later before I realized I couldn't raise my left arm. Instantaneous wind blast flail, of which I was never conscious, had broken bones in my left shoulder and back. But I had a lot more going through my mind than checking bone structure. Even when, seconds later, I touched down right on the main street of a little country town named Tin Gia (just after I floated over highway one), longing for that open ocean I could see less than a mile ahead, I could hear feet pounding down the street. Then I caught them out of the corner of my right eye at three o'clock—the town's roughnecks, sprinting forward, just about to gang tackle this 'American Air Pirate.' It was maybe two minutes before the police whistle started to blow. The pummeling and gouging and twisting suddenly stopped, and I was sat up in the street and motioned to disrobe by the pith-helmeted official.

Eventually I saw my left leg—broken at the knee and sticking out at a right angle toward nine o'clock. Crippled for life! I knew it in an instant. What a letdown.

You see, between the time I pulled the ejection handle and came to rest in that main street, I had become a man with a mission. I can't explain this without unloading a little emotional baggage that was part of my military generation's legacy in 1965.

In the aftermath of the Korean War, just over 10 years before, we all had memories of reading about and seeing early television news accounts of U.S. government investigations into the behavior of some American prisoners of war in North Korea and Mainland China. There was a famous series of articles in the 'New Yorker' magazine that later became a book entitled 'In Every War But One!' The gist of it was that in the prison camps for Americans, it was every man for himself. Since those days, I've come to know officers who were prisoners of war there, and I now see much of that reporting as selective and a bum rap. However, there were cases of young soldiers who were confused by the times, scared to death, in cold weather, treating each other like dogs fighting over scraps, throwing each other out in the snow to die and *nobody doing anything about it.*

This situation could not go on, and President Eisenhower commissioned the writing of the American Fighting Man's Code of Conduct to guide future POWs.

It is written in the form of a personal pledge. Article Four: 'If I become a prisoner of war, I will keep faith with my fellow prisoners. I will give no information or take part in any action which might be harmful to my comrades. If I am senior, I will take command. If not, I will obey the lawful orders of those appointed over me and will back them up in every way.' In other words, as of the moment Eisenhower signed the document, American prisoners of war were never to escape the chain of command; the war goes on behind bars.

As an insider, as CAG aboard *Oriskany*, I knew the whole setup—that the North Vietnamese already held about 25 prisoners, probably in Hanoi and, that as the only wing commander to survive an ejection, I would be their senior—their commanding officer—and would remain so, very likely, throughout this war that I felt sure would last *at least* another five years. And here I was starting off crippled and flat on my back.

All told, the disaster of this business of my being impeded in my duty by being permanently crippled, was only a temporary setback from things that were important to me. Being cast in the role of sovereign head of an American expatriate colony that was destined to remain autonomous, out of communication with Washington, for years on end, was *very* important to me. Months later, I was 42 years old, still on crutches, dragging a leg, at considerably less than my normal weight, with hair down near my shoulders, my body unbathed since before I was catapulted from *Oriskany* and with a beard that had not seen a razor since I arrived, when I took command (clandestinely, of course—the North Vietnamese would never acknowledge our rank) of about 50 Americans. That expatriate colony would grow to more than 450—all officers, all pilots or NFOs. I was determined to play well the given part. And all Americans continued to refer to me as 'CAG' until after I got home.

I was 'CAG' from February 1965, when on the eve of our departure for WestPac, I took the job at a ceremony on the flight deck of *Oriskany* alongside at North Island, until February 1973, when I arrived back at North Island after an eight-year absence. CAG—the boss of American flight crews—for *eight full years*. Can anybody top that?"

CDR WESLEY L. MCDONALD, USN (ADM Ret)
CAG 15 10 Jul 65-25 Jun 66
USS Coral Sea (CVA-43)

Wesley L. McDonald graduated from the Naval Academy in the Class of 1946 and, before entering flight training in 1948, was a member of RADM Richard E. Byrd's South Pole expedition, "High Jump." He received his wings in 1950 and served in several carrier fighter and attack squadrons. Wes led the first retaliatory strike into North Vietnam following the Gulf of Tonkin incident in 1964 while serving as the commanding officer of VA-56. Other command tours included Carrier Air Wing Fifteen, USS Hermitage *(LSD-34) and* Coral Sea. *After promotion to rear admiral in 1972, he was assigned duty as Commandant, Thirteenth Naval District. This tour was followed by assignment as Commander Carrier Group Three, where he deployed to the Western Pacific as a carrier group commander in the final stages of the Vietnam War. He continued duty in the Pacific as Commander, Naval Air Force Pacific representative embarked in* Midway *(CV-41), homeported in Yokosuka, Japan, in 1974. He then served as Chief of Naval Air Training in Corpus Christi, Texas, for one year before being reassigned to Washington, D.C., in 1975 as Deputy Chief of Naval Personnel.*

In July 1977, he was promoted to vice admiral and assumed command of the U.S. Second Fleet. Concurrently, he commanded the NATO Striking Fleet Atlantic. Following this tour, he was named Deputy Chief of Naval Operations (Air Warfare). In that position he established policy for the conduct of naval air warfare and served as an advisor to the CNO for all matters involving naval aviation.

McDonald was promoted to the rank of admiral and assumed the duties of Supreme Allied Commander Atlantic, Commander in Chief U.S. Atlantic Command and Commander in Chief U.S. Atlantic Fleet in September 1982.

On 1 Dec 1985 ADM McDonald retired from that assignment, after having served in the U.S. Navy for more than 42 years.

Since retirement, Wes McDonald has been actively involved as a consultant on defense matters and in other civic and organizational endeavors. He has served as President of the Association of Naval Aviation, as well as a member of the Board of Trustees of the Armed Service YMCA, and Vice Chairman of the U.S. Navy Memorial Foundation. Wes' memories are of his Vietnam service.

Wes McDonald led CVW-15 from *Coral Sea* (CVA-43) during grim days of the Vietnam War, but can find humor in his experiences. USN photo, courtesy CAPT R.R. Rausa, USNR(Ret)

"I have tried to reconstruct the following vignettes from my graying memory bank, and have heavily relied on some memoirs from CDR Jim Morin. (Jim was my VA-155 'Silver Foxes' CO—I flew with his A-4 squadron since my past squadron [VA-56] had A-4Es.) To set the stage, I had been selected for the CAG job following my VA-56 tour and had been programmed to become CAG 5 during fall 1965. As a filler for some of the time available from my change of command at VA-56 in February 1965 to the CAG 5 job, I was assigned as ops officer of CAG 12, the replacement training RAG at NAS Miramar, CA. I left my family in Lemoore to finish the school year, then to move down to the San Diego area while I trained and moved into CVW-5, a Miramar-based air wing.

While in the midst of this training in early July, CDR Pete Mongilardi, CAG 15, was killed in action on the first day that his wing went back into combat. *Coral Sea* and CVW-15 had been extended for an indefinite time at the last possible moment—in fact, they were on their last day in port prior to heading home when they were ordered back to Yankee Station.

I was still current in the A-4E and, after a few hurried bounce periods, flew out to *Hancock* (CVA-19) as she was transiting to Alameda from San Diego. CDR Randy Prothro was the air boss and, when I checked in overhead, he gave me an immediate 'Charlie' and said the deck was all mine. First time that ever happened in my career of bagging traps! I established my pattern, made one touch and go, followed by eight traps, gassed, and was told to bingo to Miramar since I didn't need any more landings to deploy. Two days later I was en route to Cubi Point where a waiting A-3 plucked me back into the air and gently deposited me on the flight deck of *Coral Sea.* CAPT George Cassell, the hard-charging CO, called me to the bridge, gave me some sage advice, and said we were heading to 'Indian Country' the next morning. I relieved CDR Bob DiLorenzo the next day. Immediately I was embroiled in the terribly frustrating business of fighting a politically constrained war, which was being directed most heavily by the political/military leadership half a world away with little, if any, input from the on-scene commanders. Selection of targets, ordered from Washington, was particularly galling to those of us daily flying in harm's way, especially when we knew, in most cases better than Washington, the target damage status and lack of importance or utility to the enemy.

Nevertheless, every once in a while, we would be assigned a significant objective and, in September, we found ourselves heading for a relatively small bridge near Kep airfield, located in Route Pack Six. It was on a main supply route northeast of Hanoi in the Thai Nguyen neighborhood. We planned to use 12 A-4s; eight 'Echos' from VA-155 and four 'Charlies' from -153, with four F-4s from VF-151 for cover. We coasted-in northeast of Haiphong and commenced flying at low level once we crossed the beach.

Since this was SAM country the low level was more appealing, even though it put us in some heavy, but less lethal, ground fire. I was leading, and the two A-4 squadron COs, Jim Morin with a division from -155 and Ken McArthur with four Cs from -153, followed in a loose open combat spread. I hit the first two checkpoints, but missed a difficult break in the road and we went 'DR' (dead reckoning) until we could hopefully spot some landmarks to update our position. Every two or three minutes, I would pop up with my wingman to about 1,500 feet, look around and drop down to about 300 feet again. I did this two times with no luck, but on the third try my wingman, George Blosser, called out a bridge at 10 o'clock. I looked at it and replied that it was not the target. George, maybe hoping for some relief from the high visibility I was subjecting him to, pleadingly said: 'But CAG—it's a bridge. Can't we get it?' Suddenly, we were within sight of the Thai Nguyen Steel Factory, and we could have really had a go at it, a major industrial target—but it wasn't on the approved target list. We ducked it, got back on track, found the bridge and attempted to drop this little 60x15-foot structure. Ken McArthur had to bingo his division before we got there, while Jim and I got some hits with our divisions on and around this less-than-tremendous target. Another stressful, frustrating day in North Vietnam just trying to do something meaningful. But a day that I've never forgotten because of that plaintive voice of George as he pointed out 'his' bridge, sorta saying, 'Jesus, CAG, let's get it and get out of here!'

One other mission sticks in my mind during this truly demanding tour for *Coral Sea* and CVW-15 officers and men. It was to occur on the last day of the cruise—the ship and air wing were finally slated to be relieved and head home. You could feel the exhilaration, tempered by an undercurrent of caution, actually filling the ready rooms. We were assigned a no-missile threat target—Dong Hoi—the very place it all had started for the air wing in February 1965. I briefed the flight in the evening prior to the early launch the next morning and guys started turning in early. Unknown to them, *Oriskany* (CVA-34) pilots had found a truck park hidden near some karst ridges near Hanoi, and the admiral decided to send us against them at first light. I heloed over to *Oriskany* that evening, was briefed, and then returned to *Coral Sea* to awaken and brief the flight all over again. Having been briefed for a 'piece of cake' last-day mission, this change to a high-threat environment wasn't exactly greeted enthusiastically!

I got up early, wolfed a quick breakfast and headed for the Silver Foxes' ready room with as much visible enthusiasm and bravado as I could muster. The last combat mission on a nine month combat cruise—and no longer the easy one anticipated a few hours earlier. Exhilaration and caution were replaced by an atmosphere of tenseness. I looked around as I reiterated some briefing points and was struck by Bill Kish's (one of Jim Morin's JOs) complexion—he was ghostly pale—and I mean that in every sense. Bill just felt he was going to 'get it' on that last flight, but he hung in there. Since the weather was forecast to be 'iffy,' a weather recce was launched with John Chambers, XO of VF-151, flying it. Prior to his launch, as Jim recalls in his memoirs, all the ready room intercoms lit up with such comments to John in Ready 1 as, 'Free beer for a year if the target's obscured,' or 'A case of Chivas if you scrub the mission for weather.' Everyone tried to be cool and ready to go if the decision to launch came back, but all hoping it wasn't to be. After about 30 minutes, John radioed back that the entire Delta was under a low overcast and that the strike group would be unable to get to the target. With suppressed emotions of relief and feigned exclamations of disappointment such as, 'Aw, shucks,' 'Gee Whiz,' and 'I really wanted to end up with a bang,' CVW-15 pilots manned up for the '65 cruise farewell at Dong Hoi—where it all started, as I said earlier.

Now, almost 30 years later, I truly am amazed at the professionalism and dedication of those young men. A marvelous group of America's finest—many of whom were killed in that war, leaving all of us who served with them with a deep sense of great loss. Some of those who came home, continued to serve (many still do), while the remainder went on to other walks of life. But, wherever they are, whatever they are doing, I salute them all, collectively and individually. They make me proud to remember that I was privileged to serve with them when our nation called. 'Good on you!'"

CDR KENNETH A. BURROWS, USN (CAPT Ret)
CAG 10 23 June 66-12 Jan 68
USS *Intrepid* (CVS-11)

Ken Burrows entered the Navy 2 July 1945 via the V-5 program. He became a fighter pilot with VF-113 in time to fly combat missions in Korea from USS Philippine Sea *(CV-47).*

His other early experiences include flight instructor, CIC officer Cape Esperance *(TCVE-88), VF-112 and a Carrier Division Six staff assignment, before switching to the attack community. After A-4 Skyhawk training with VA-44 1962-63, Ken became XO of VA-172, then CO, while making three Med cruises in* Franklin D. Roosevelt *(CVA-42). He assumed command of CVG-10 23 Jun 1966 where he made two cruises in* Intrepid *flying 135 combat missions over Vietnam. Departing CVG-10 on 12 Jan 1968, Ken next became JCS action officer on the staff of CNO, then operations officer for Commander Carrier Division One, where he made two more combat deployments to Southeast Asia in* Bon Homme Richard *(CVA-31) and* Midway *(CVA-41). He retired as a captain from the staff of Commander Naval Air Forces Pacific 31 May 1972.*

CDR Ken Burrows, CAG 10, poses with his A-4 Skyhawk on board *Intrepid* (CVS-11). USN photo, courtesy CAPT K.A. Burrows, USN(Ret)

Following his Navy retirement, Ken became a manager for Computer Sciences Corp. where he was directly involved with the design of Naval Tactical Display System programs for carriers. After his 1984 retirement from the computer field he became involved in an estate property development, which he "sold in 1990 in order to take life easy in Las Vegas, NV."

"Both of my CAG cruises were with an all-attack air wing. I relieved Tom Hayward shortly after the wing arrived on Dixie Station in 1966. The wing was well equipped and trained for the mission of supporting our ground forces in South Vietnam. About two weeks after I took command I realized the wing needed more training on tactics if we were to assist in the air operations over North Vietnam from the Gulf of Tonkin. From that point on, I directed that flights practice anti-SAM tactics and anti-air tactics, when feet wet returning to the ship from air support in South Vietnam. This decision proved to be very wise, because within two months we were flying missions from the Gulf. Our only ECM gear was called 'Little Ears.' Those devices were almost useless. In fact, the connecting cords were so long they were a real hazard in the cockpit.

The biggest lessons learned from the first cruise were:

1. Have proper ECM gear installed in wing aircraft before departure from the States.
2. Collect aircraft overhead ship well above any in-bounds when large numbers were assigned to strike the enemy as a group. This method made for rapid collection of forces without the use of radio transmissions.
3. Maintain silence on the radio unless a transmission was critical.
4. Conduct training in the States as close to the real wartime situation as feasible.
5. Learn how to manage fuel on every mission. (That means save.)

During our first cruise we lost no aircraft to SAMs. We had three combat losses, all to AAA. We recovered two pilots and had one KIA. On our second deployment, with heavier operations over the North against increased defenses, we lost 12 aircraft; three to SAMs and nine to AAA, resulting in two MIAs, three POWs, one KIA, and six recovered pilots. Our tactics were to go in high, no low-level attacks, make a single coordinated pass at the target and get out.

Strict radio discipline was mandatory in combat, and it was my custom to have strike leaders do all the talking. All Alfa strikes were led either by myself or a squadron commander. My radio callsign was 'Earthquake,' and each strike leader used this during airborne comm.

My most memorable mission was during the rescue of an air wing pilot from Haiphong Harbor. My fighter support F-4 Phantom aircraft, flown by LCDR 'Tooter' Teague, and I dodged one SAM that landed next to a Russian ship. Later, I was charged with attacking the ship. The downed pilot was picked up by a rescue helo under the noses of our adversaries.

Although we were considered an all-attack air wing (three A-4 squadrons and one A-1 squadron), our air assets also included three RF-8 photo birds, three F-8C fighters and detachments of EA-1 jamming aircraft and E-1Bs. LCDR Tom Barrett, O-in-C of the photo detachment, was one of the most can-do aviators I have ever served with. With Teague's F-8Cs as escorts, we could count on timely BDA (battle damage assessment) photography at all times. The three A-4 squadrons, commanded by CDRs Les Hofto (VSF-3), Kelly Carr (VA-15), and Dick Wigent (VA-34), were the backbone of our offensive capability. Our A-1 squadron, VA-145, was one of the last active duty 'Spad' squadrons to perform combat missions in the Navy. Under the leadership of CDR Walt Schutz, this squadron performed daily SAR missions in support of all air units flying over North Vietnam.

After my tour of duty as a squadron commander, I was sure that nothing could be more rewarding. I was wrong. The CAG job was even better. Perhaps I was lucky. For the most part, CAPTs John Fair and Jeff McVey were the COs of *Intrepid* during my cruises. They let me run the wing, and I kept them advised about what we were planning to do and the results of our efforts. The team effort on the *Intrepid* was the best CV/Wing effort I saw in the Navy."

CDR JOHN G. WISSLER, USN (RADM Ret)
CAG 3 6 Jul 66-4 Jul 67
USS *Saratoga* (CVA-60)

John Wissler entered the Navy V-12 program in 1944 as a student at Marquette University, Milwaukee, WI. Upon graduation in 1948, John was commissioned an ensign, and his first assignment was in the Navy's public relations field. In 1949 he reported to Pensacola for flight training where he won his wings the following year. John next underwent night fighter training at NAS Key West, FL, and in 1951 reported to VC-4 at NAS Atlantic City, NJ, for his first fleet assignment.

Accepted for test pilot training in 1954, Wissler took over service testing of the McDonnell F3H Demon the following year at NATC Patuxent River, MD. During 1956-57 he attended Line School at Monterey, CA, after which he became assistant navigator in USS Lexington *(CVA-16).*

Returning to the research and development field in 1959, John became an F8U Crusader project test pilot with VX-3, and the following year shifted to VX-4 as the project test pilot in charge of operational testing of the F4H Phantom II. During 1962-63 he was a student at the Naval War College, after which he became XO then CO of VF-31 in Saratoga. *In July 1966 he assumed command of Air Wing Three, a position he held until July 1967 when he was ordered to the Industrial College of Armed Forces as a student.*

In a switch of areas of expertise, John became the A-7 Corsair II test program manager at Pax River during 1968-69, then became Director of the Weapon Systems Test Directorate 1969-71. A 1971-72 Washington, D.C., tour on the Chief of Naval Operations staff (Op 508 program) followed, and in 1972 John became CO of NAS Cubi Pt., Republic of the Philippines.

Returning to Pearl Harbor, John was CO of the Navy's Human Resource Management Center during 1974-76 before he became Deputy Commander of NATC at Pax. Leaving Patuxent briefly, he was the Atlantic Fleet Inspector General during 1978-79 before returning to Pax in 1979 as NATC Commander where he retired in 1982. After his retirement from active duty, John was involved in management of a family ranch near Madera, CA. He served on the faculty of Fresno State University 1985-89 and was an active member of various civic and non-profit boards until his death in 1998.

John Wissler's CAG tour in Saratoga *came during the turbulent times of the Vietnam War. His story relates the turmoil of those days and the effect the war had in the Navy's carrier operations in the Atlantic and Mediterranean areas.*

"During 1966-1967 the Sixth Fleet was facing a serious, new challenge. For the first time since WWII the Soviets were embarked on a massive buildup of their Mediterranean fleet in support of increasing Arab pressure on Israel. The apparent objective of the multi-dimensional buildup (ships, submarines, land-based naval air) was to checkmate the Sixth Fleet prior to the resumption of hostilities between the Arabs and Israelis.

For the carrier air wings in the Med, the new threat changed the emphasis from Vietnam-type air-to-ground and air-to-air training to anti-ship offense, as well as fleet air defense against air-to-surface missile bombers. That threat, combined with the ongoing primary concern with SIOP (single integrated operations plan) commitments against Soviet targets in eastern Europe, placed Sixth Fleet

Retired as a rear admiral, John Wissler led CVW-3 during turbulent days for the Navy's carrier air wings in the Med. USN photo, courtesy RADM J.W. Wissler, USN(Ret)

air wing operations and training in sharp contrast with Seventh Fleet Vietnam activities.

The markedly differing missions of the two fleets were played against a substantial shifting of limited assets (aircraft, personnel, ordnance, ECM systems, logistic support) from the potential threats of the growing Soviet Fleet to the real-time, ever-increasing land war in southeast Asia.

The period is an interesting background study relative to the current difficulties facing the Defense Department in trying to downsize and modernize, while retaining the capability to meet two simultaneous regional conflicts, each with different scenarios.

CVW-3's composition and training through two deployments during this period reflected the shifting environment, drawdown of assets, and periods of relatively tense interactions with the Soviet Fleet. An unexpected event was a quick reaction launch in support of *Liberty* (AGTR-5), then under prolonged attack by Israeli sea and air forces for reasons still not fully explained.

While we were making do with whatever was not being sent to Vietnam, we had a lot of squadron repositioning. Essentially, during my CAG tour, I had two distinctly different air wings.

The air wing composition for the 1966 deployment reflected the typical mid-60's aircraft assignments: two F-4B Phantom squadrons, three A-4C Skyhawk light attack squadrons, and detachments of E-1B, RA-5C, and EA-3s.

Upon return from the 1966 deployment, CVW-3 lost its attack squadrons as AirLant played musical chairs in a frantic attempt to stand up an additional air wing (CVW-17). As an alternative to deploying with an undersized wing, I located untasked S-2 and H-3 anti-submarine squadrons and formally requested their assignment to CVW-3 to develop combined CVA/CVS operations. My proposal was rejected, but did serve the purpose of prompting immediate action to replace the missing attack capability. (*Saratoga* finally did test a mixed air wing some years later.)

The 1967 air wing, reflecting the impact of shifting assets, retrogressed to an early '60s composition. The three A-4C squadrons were replaced by one 20-plane A-4B squadron and one 20-plane A-1H Skyraider (propeller-driven) squadron. The 1967 wing had no realistic night attack capability, no modern early warning aircraft, and minimal tanking capability (A-4C buddy store-2,000 lbs. fuel giveaway). ECM systems literally consisted of six small chaff dispensing pods (the total Atlantic Fleet inventory) shared by the entire air wing.

Air-to-air missiles were limited to early series Sparrows and Sidewinders. Air-to-ground ordnance centered around the smaller sizes of the Mk 80 unguided bombs, unguided rockets, and 20mm ammunition for the attack aircraft. An added constraint of considerable concern was that the ordnance available was limited to the shipfills in the carriers and supporting AE replenishment ship. Needless to say, practice ordnance was extremely scarce.

The 1966 wing training concentrated on its most pressing challenge: the total absence of viable tactics that had any chance of success to disable modern Soviet missile-armed warships operating as a cohesive battle group.

The problem was that the underpowered, manual bombsight-equipped A-4Cs had to expose themselves for a prolonged period in the SAM envelope while they closed to minimum range where they had some chance of hitting a maneuvering ship with unguided bombs. The critical requirement was to obtain direct bomb hits to penetrate the ship's armor. The poor CEP performance of jets with manual bombsights gave the attack a low percentage of success, while the long flight time in the missile envelope, combined with essentially zero ECM support gave the defending battle group a much higher percentage of success.

After much experimentation CVW-3 settled on a Rube Goldberg tactic using a compound strike force of about four F-4Bs leading an attack group of about eight-twelve A-4Cs, all under the control of a nearby E-1B. The formation would approach the target ship at very low altitude. At five miles, the F-4s would split into sections, zoom quickly to a shallow firing position, loose a total of 16 Sparrows in an air-to-ground mode, followed by pressing in for a Zuni rocket attack at high speed. Hopefully, while the target ship was recovering from the Sparrow/Zuni shrapnel among its antennas, the A-4C attack group would have sufficient respite to laboriously climb for a close-in dive bomb attack.

An obvious weakness with the tactic was that it could only be delivered against one ship at a time in the enemy formation. The E-1 had the difficult problem of identifying the formation, picking a target ship that had minimum mutual support from the rest of the battle group, selecting the approach corridor, and then controlling two dis-similar groups of aircraft, all without ECM support. Needless to say, even this tactic, which was the best the air wing could develop with the technology available, was a dicey event. The low potential for success probably was a key ingredient in the subsequent development of the current generation of guided air-to-surface anti-ship missiles.

As a matter of interest, a couple of years later when I was testing the A-7D/E, my team worked up an off-the-shelf answer to the anti-ship problem. We took an ASROC rocket-propelled homing torpedo, mounted it on an A-7, and developed the preliminary mathematics to loft it from a low altitude launch point about a dozen miles from the target ship. The concept was to get the torpedo within homing range before it entered the water. We were preparing for airborne proof of concept testing when NavAir stopped the effort, ostensibly because (if successful) it might jeopardize funding for the Harpoon anti-ship missile then entering development.

During 1967, another disturbing activity of the Soviet Fleet began: the placement of warships in or immediately adjacent to the U.S. carrier formations. In addition to observing our operations on a continuous basis, they also posed a close-in threat that could strike without warning.

The carriers, with longer legs and tanker support, frequently resorted to prolonged day-long high-speed runs until the Soviet ships fell off for lack of fuel.

The assignment of the obsolete A-1 to CVW-3 inadvertently provided an interesting counter to the close-in Soviet ships. 'Spads,' with long endurance and ostentatious ordnance, were ideal to place on station orbiting slowly over the threatening Soviet warships, effectively countering their counter.

As Arab/Israeli war clouds approached, the U.S. Air Force found itself completely cut out of the action when their European bases within tactical range were restrained by their host countries from supporting Israel. The Sixth Fleet carriers, *America* (CVA-66) and *Saratoga*, had the only tactical aircraft available in the eastern Mediterranean; an outstanding example of the continuing need for sea-based aviation in polarized regional conflicts.

The jet portion of the air wing received a new assignment: dealing with some 25 Soviet ASM-equipped Badger bombers based at Cairo International Airport, Egypt. The Badgers gave an additional dimension to the Soviet naval threats, which became more ominous as events spun rapidly toward the 1967 Arab/Israeli War.

A large scale alfa-style strike plan was developed to surgically destroy the *Badgers* on the ground. Full dress rehearsals were conducted in the Aegean Sea. The close proximity of the Soviet ships was an additional complication in keeping the true intent of the rehearsals secret.

A similar strike plan was worked up in the event that CVW-3 was tasked with destroying Arab guns covering the straits to the Gulf of Aqaba south of Israel. The plan had a very low probability of success because of the maximum ranges involved and the proximity of Egyptian fighter bases.

The extraordinary progression of the Six Day Arab/Israeli War drastically changed the tactical situation. What the Soviets had anticipated would be an Israeli defeat with a Dunkirk-type finish, requiring Sixth Fleet support (to be checkmated), became an Arab rout. The Soviets suddenly became the checkmated fleet.

When progress of the war became obvious, the Sixth Fleet began to relax. CVW-3 elected to keep about a quarter of its aircraft in an armed ready status until the situation had totally cleared. Shortly thereafter, *Liberty* (AGTR-5) reported that it was under continuous air and sea attack. The ship was about 500 miles away (a distance that precluded sending unrefueled jets), so the venerable A-1s were again pressed into service. Within half an hour they were underway with orders to protect *Liberty* against all attackers. About an hour later, word was received that the attackers were Israeli and that the attacks had ceased. The A-1s were recalled just before getting beyond radio range. It is interesting to conjecture the exchange if they had continued on to cover *Liberty.*

A sidelight of the A-1 launch was that they were diverted to Suda Bay, Greece, to unload their ordnance, vice jettisoning (and losing) their scarce heavy ordnance loads before recovery.

The end of the 1967 Arab/Israeli War brought a semblance of calm, although the Soviet Fleet continued to expand its presence in the Mediterranean."

CDR GENE TISSOT, USN (RADM Ret)
CAG 14 23 Jan 67-5 Feb 68
USS Constellation (CVA-64)

Ernest E. (Gene) Tissot was born in Upland, CA, in December 1927. After graduating from North Hollywood High School he entered the Navy as an apprentice seaman in June 1945. Following completion of the Aviation Midshipman Program he was designated naval aviator in June 1948. He was commissioned ensign in December 1948 and selected for rear admiral in March 1973.

Gene studied at Stanford University, receiving a Bachelor of Science in Engineering and election to Phi Beta Kappa; at the Naval Postgraduate School, receiving a Master of Science in Aeronautical Engineering and the outstanding Aeronautical Engineering student award; at the Naval Nuclear Power School, Bainbridge, MD, and at the Naval Nuclear Power Training Unit, Idaho Falls, Idaho. He also attended the Armed Forces Staff College.

His first squadron tour was VF-192, in which he flew 50 Korean combat missions. Other assignments were Naval Air Advanced Training Unit 1 and VF-33, as well as several staff positions. During the Vietnam War he flew 250 combat missions as a squadron and air wing commander.

Gene's commands include VA-192, Attack Carrier Air Wing 14; USS Thomaston *(LSD-28);* Enterprise *(CVA(N)-65); Fighter Airborne Early Warning Wing, U.S. Pacific Fleet; and Attack Carrier Striking Force, U.S. Seventh Fleet (CTF 77).*

In October 1967 he became the third naval aviator to achieve 1,000 arrested carrier landings. This accomplishment was without accident over a 20-year period flying 11 different aircraft types from 15 aircraft carriers.

Gene Tissot led CVW-14 through a 1967 Vietnam War cruise and also was the first U.S. naval aviator to reach 1,000 carrier arrested landings in operational aircraft. USN photo, PH3 Stewart, courtesy RADM E.E. Tissot, USN(Ret)

RADM Tissot retired from the Navy in 1981 while serving as Assistant Chief of Staff for Plans and Policy, Commander-in-Chief, Pacific. He was Corporate Executive Director, Far East for the Northrop Corporation from 1981 until 1991. He and his wife, the former Mildred Ellen Patton of Oakland, CA, are now enjoying retirement in Auburn, CA.

Gene describes his impressions of being a CAG during the Vietnam War.

"For a carrier pilot, the ultimate experience is to be *the* CAG—it's as simple as that.

CAG—the greatest appellation in the Navy! To be called 'CAG' means you are the senior active pilot in your carrier and in charge of an entire carrier air wing. What a thrill it is! And an even greater thrill is when your air wing has a combat tour, for combat is what every carrier aviator who is worth his salt desires.

From that bright day in January 1967 at NAS Miramar when I saluted and relieved Fred Palmer, until an equally lovely day a year later—on the same podium—when I was relieved by Ken Enney, *I had the time of my life.*

What transpired in my year as CAG was the great adventure of leading a magnificent air wing in combat in North Vietnam during the hottest time of the air war: the summer and fall of '67.

A carrier air wing is the most effective tactical air fighting force ever devised. By virtue of being based in a carrier, all the players—and this includes the ship's company as well as the air wing—really get to know each other. The pilots and aircrewmen fly together every day, and they also live together under the same roof and go on liberty together. This close personal, as well as professional, contact leads to a degree of teamwork that can be achieved in no other way. This is in direct contrast to Air Force tactical air, where each wing is at a different base, responds to frag orders, and only meet in the air.

I had the luxury of having eight squadrons in CVW-14, with no detachments. The 12 squadron commanders I had were superb (two of the 12 were relieved due to normal rotation, and two were shot down and became POWs). Bob Dunn, Bob Holt, Bob Kirksey, Leo Profilet, Ed Bauer, Robin McGlohn, Doc Townsend, Billy Lawrence, Tom Replogle, Pat O'Gara, Jack Eckstein, and Tom Stewart were a talented lot. Four of these men went on to achieve flag rank, with three making three stars. We had two A-4C squadrons, VA-146 and -55; two F-4B squadrons, VF-142 and -143; and a single A-6A unit (VA-196), along with RA-5Cs (RVAH-12), E-2As (VAW-113) and KA-3Bs (VAH-8). CVW-14 was assigned to *Constellation*, commanded by CAPT John Thomas. No carrier commander ever guided and supported his air wing better.

After a workup including a weapons deployment to MCAS Yuma, AZ, we embarked in *Constellation*. Following the mandatory ORI in Hawaiian waters, I took two of my squadron skippers and flew ahead so we could visit the guys on the line in the Tonkin Gulf and learn from them how the air war was being fought *at the time,* since tactics were continually being refined. Even though I had been a squadron commander in 1965, flying over North and South Vietnam, I was well aware that tactics can change in a hurry, and just having been there before does not make one an expert. Air wings that declined to send representatives ahead to learn from their peers did not initially fare nearly as well as those that did. Invariably these wings suffered higher losses until they learned—the hard way—what worked and what didn't.

So, I chose Billy Lawrence as the F-4 representative, Leo Profilet for the A-6, and I had the A-4 background. We visited *Enterprise* and *Kitty Hawk* (CVA-63) on the line, and each of us flew missions off each carrier. From *Enterprise* I flew with my good friend, Pete Sherman, CO of VA-56. Pete was shot down a short time later and did not survive. From *Kitty Hawk* I had the privilege of flying as 'tailend Charlie' in a routine raid on the Thanh Hoa bridge in what CAG Hank Urban called a 'Thanh Hoa Thumper.' Needless to say, the damage inflicted was not noteworthy.

Arriving in the Tonkin Gulf, CVW-14 flew its first combat missions 28 May 1967. On 8 November we flew our last, having made 121 Alfa strikes in North Vietnam, 106 of which were in the Northeast Sector.

Our combat took place during the most intense flying of the entire air war. Our pilots never flinched, and performed in a uniformly fine manner. Although all of our attack and fighter pilots got their share of excitement, our Intruder pilots flew the most demanding missions. They would routinely fly up the river to Hanoi at 200 feet altitude at night to brave the incredible AAA and SAM barages that would be thrown at them.

We had our losses. Six pilots and crewmen were killed in action, and our ten missing all became POWs. Of our many planes shot down, even the outstanding SAR network that was in place was able to recover but one aircrew. That's how tough it was to be on the ground in North Vietnam.

Training is always of major importance. When you are a CAG it is your job to ensure that the CAGs who follow you, the present squadron commanders and executive officers, are well trained. In my case it was easy—we had *on the job training.* For strike leaders I chose only squadron commanding officers and, with our large number of Alfa strikes, each was able to lead several.

Our missions ran the gamut from routinely boring to very exciting. The weather we encountered was sometimes excellent, sometimes terrible, with everything in between. The opposition we faced ranged from none to intense flak, and sometimes a few and sometimes a lot of SAMs. The missions we flew included armed reconnaissance, strikes against bridges and lines of communication, and strikes on airfields, as well as exciting raids into the highly-defended Hanoi area. We were scheduled to fly the first Navy strike on Phuc Yen airfield in Hanoi, home of all their MiGs. This was *the* plum mission of the war, and even though we suited up and briefed for this strike twice, each time the weather forced a cancellation. We were very disappointed when our line period ended and CVW-15 got to make the strike.

Now came the perquisites of a CAG. For each major strike I selected the strike leaders and usually took the most challenging missions myself. I could fly either the Skyhawk or Phantom as I desired—the only pilot in the wing who flew more than one type of airplane. Since during my CAG training I hadn't had time to carrier qualify in the Intruder, I only flew it later on and in a more benign environment. I did not fly the A-6 at night, as I was not qualified for this most demanding mission. On a deep strike or an armed reconnaissance I liked the Skyhawk, as it was the most effective bomber for the mission. This airplane was also the one I was most at home in, having flown it as a squadron commander during a combat tour in 1965. When there was a chance of a MiG encounter I chose the Phantom because, like all fighter pilots, I dearly wanted to bag a MiG. I really enjoyed flying the F-4, with its speed, power, and weapons-carrying capability. Two previous tours in fighter squadrons had given me a solid VF background.

We had little opposition from the air. The MiGs were afraid to face us and, considering the situation, I don't blame them. Our pilots shot down two MiG-21s and two MiG-17s. I outsmarted myself on the day the MiG-21s were bagged by taking another CAP station that I thought offered a better chance and ignoring the CAP station that actually had the encounter.

During summer and fall '67, carrier air wings spent most of their effort flying Alfa strikes into North Vietnam. When the weather permitted we flew three a day, launching our strikes at 0800, 1200, and 1600. The opposition knew our schedule and was usually well prepared for us. With all the early warning they had, the North Vietnamese would have been just as well prepared regardless of our schedule, so the routine timing of our strikes was not really a detriment.

We made several strikes on Kep airfield, which was northeast of Haiphong. There were always a few MiGs at the field, but there were so many revetments that it was difficult to spot the parked aircraft during the bombing runs, especially since the AAA defenses around the field were very heavy. I got the idea of using a Polaroid camera to discover what revetments were occupied by MiGs. Each bomber had a photo of the airfield with the revetments numbered. I had my F-4 RIO equipped with a hand held Polaroid, and I would take a wingman and fly over the field five minutes ahead of the strike. We would retire out of AAA range, and my RIO would develop the picture, then call out the numbers of the occupied revetments. This approach worked quite well, but the second time I tried it my section of Phantoms came under quite heavy fire and my wingman was hit. I didn't try that type of 'hands on' reconnaissance again.

Another CAG perk allowed me to reach 1,000 carrier landings. As with most carrier aviators, I kept count of my traps. I had a very slow start, flying Bearcats and Corsairs during my first tour. Two full cruises, including CQ, had netted but 143 arrested landings. Then later, when I was flying Furys, because of engine problems my squadron was forced ashore for the best part of a full Med cruise and I garnered only 16 traps. From 1960 on, flying from the angled deck, like everybody else I picked up a lot more landings. When I could see the end of our tour in *Constellation* coming—and the end of my carrier flying—I checked my logbook and found that I would most likely fall one or two traps short of the magic number. I mentioned this fact to CAPT Thomas, and I can still remember his words, 'I can fix that!' So when I was CarQualing in the A-6, John let me go around the pattern a few extra times. I bagged my 1,000th trap in a Phantom during our last line period and ended up with 1,008—thanks to John Thomas. I was proud of my achievement. I was the third USN aviator to reach this mark, and the first two had been in carrier suitability at Pax River. So I was the first U.S. naval aviator to make every one of my thousand traps (save the 12 in flight training) in operational aircraft. And I didn't have any accidents, either.

When I ended my tour and went back to Washington, D.C., for my first tour of duty in the Pentagon, I was satisfied. *I had been a CAG.*

Let me add a footnote: I didn't feel that my tour was really complete until our POWs were returned from captivity. As it happened, I was present in the Tonkin Gulf when this event took place. Then Captain of *Enterprise*, with CVW-14 aboard, we participated in Operation Linebacker II and experienced indescribable joy as our ex-POWs were flown from Hanoi to the Philippines. All our CVW-14 POWs survived their captivity, including the A-6 pilot who was shot down by Chinese MiG-19s and held captive there.

CAPT PHILLIP R. CRAVEN, USN(Ret)
CAG 19 11 Jul-27 Sep 1968
USS *Ticonderoga* (CVA-14)
CAG RCVW-12 11 Oct 1968 – 27 Jun 1969

Phil Craven was born 2 Jan 1924 in Los Angeles, CA, and joined the V-5 Naval Aviation Program in December 1942. He reported to active duty in August 1943 for flight training and, upon completion, received his wings and was commissioned ensign in the U.S. Naval Reserve in May 1945. His initial operational flight training was in the TBF torpedo bomber. First carrier qualification was aboard USS Wolverine *(IX-64) on Lake Michigan. Soon thereafter he reported to a replacement air group at NAS Grosse Ile, MI, and was undergoing combat training when VJ day occurred.*

In November 1945 Phil left active duty and became affiliated with the Naval Air Reserve program at NAS Los Alamitos, CA. In 1951, with the outbreak of the Korean War, he volunteered to return to active duty and, after refresher training in the F4U Corsair, reported to Fighter Squadron 92. He made a deployment to the Western Pacific in Valley Forge *(CVA-45) and particiapted in combat operations against North Korea. After flying some 80 missions in 1953 Phil returned to NAS Alameda, CA, with VF-92 and transitioned with the squadron to the F9F-2 Panther.*

In October 1953 LT Craven attended the Fleet Air Gunnery Unit at NAAS El Centro, CA, for advanced weapons training in the F9F. During 1954 he made another WestPac cruise with VF-92 in Philippine Sea *(CVA-47). He left VF-92 in late 1954 and, thereafter, had tours of duty at NAS Denver, CO; General Line School; Randolph (CVA-15); CVG-6, flying F8U Crusaders; Naval War College; VF-174 (the F8U replacement squadron for the Atlantic Fleet); and then took command of VF-62, an F-8E squadron at NAS Cecil Field, FL.*

During his VF-62 tour Phil made a highly-successful NATO cruise to the Norwegian Sea area in Independence *(CVA-64) and a Mediterranean cruise in* Shangri La *(CVA-38). After completing his tour as CO of VF-62, Phil reported to the Bureau of Naval Per-*

sonnel (BuPers) for duty in the aviation officer assignment section. During his BuPers tour he was selected for a "bonus" command, and subsequently was ordered to report to Readiness Attack Carrier Air Wing 12 (RCVW-12) for training as prospective commander of Attack Carrier Air Wing 19.

On 11 July 1967 CDR Craven took command of CVW-19 at NAS Lemoore, CA, and started training for combat operations in the Gulf of Tonkin. In December 1967 the air wing embarked in Ticonderoga *deployed to the Western Pacific. In January Phil was promoted to captain while the carrier made a port call in Yokosuka, Japan, on the way to the Gulf of Tonkin. Phil is possibly the only CAG to make a full Vietnam combat tour as a captain.*

Returning to the States in August 1968, Phil received orders to Replacement Carrier Air Wing 12 (RCVW-12) for duty as Commander of the replacement training wing. In July 1969 orders were received for assignment to Commander Carrier Group Five/CTF 77 staff duty as operations officer. He remained in that billet for one year and then was assigned to Commander 11th Naval District to participate in the College Degree Program. From August 1970 until August 1971 Phil attended San Diego State University where he earned his baccalaureate degree.

In September 1971 in the Gulf of Tonkin, Phil took command of Kansas City *(AOR-3), a multi-product replenishment ship, where he remained until October 1972 when he received immediate orders to return to the Commander Carrier Group Five/CTF 77 staff for duty as chief of staff. During that period CTF 77 was responsible for the conduct of the last naval air combat operations against North Vietnam. Also during that tour CTF 77 was responsible for providing air cover for the final evacuation of U.S. personnel from Saigon. Just prior to Phil's departure from CTF 77 in July 1975, CTF 77 was tasked to provide air cover from* Coral Sea *(CV-43) for the successful recovery of the SS* Mayaguez *and her crew from internment in Cambodia.*

In August 1975 Phil reported to NAS Pensacola as CO of the Naval Aviation Schools Command. He retired from there and the Navy 1 July 1977.

Following his departure from the Navy, Phil went to work for Cubic Corporation as Director of International Programs. Since his full retirement from Cubic in November 1988, Phil has concentrated on relaxing and his golf game, seemingly a contradiction of terms. His comments on his two CAG tours follow.

Phil Craven is another of the fortunates to have served two CAG tours. His first was CAG 19 during Vietnam and his second as CAG RCVW-12, where he was instrumental in forming Topgun. photo courtesy CAPT P.R. Craven, USN(Ret)

"In January 1968 CVW-19 commenced combat operations from *Ticonderoga*. For the first line period the weather over North Vietnam was bad, and the Marines were under siege by the North Vietnamese at Khe San. The air wing flew hundreds of close air support missions in support of the Marines during our first period 'on the line.' When *Ticonderoga* returned to the Gulf in February, the weather had improved over North Vietnam, and several successful strikes were conducted against military targets in the North. One such strike was against an extensive petroleum products storage area just northeast of the city of Vinh. The wing photo reconnaissance team had photographed that area revealing a large concentration of petroleum products stored in 55 gallon drums under trees. We conducted several strikes against the petroleum storage area over a two-day period and destroyed a considerable amount of the stockpiled products. During the night the North Vietnamese were photographed trying to move drums of fuel by bicycle out of the region. However, most of the fuel was destroyed before they could get it moved. The fires attendant to the attacks burned for two days.

During 1968 CVW-19 aircrews flew more than 15,000 combat sorties over North and South Vietnam with the loss of only three pilots. One pilot was lost to a surface-to-air missle near Vinh while the air wing was striking the airfield. One photo pilot was lost during an aerial reconnaissance mission over the southeastern portion of North Vietnam. Tragically, the other pilot was lost on the last day of combat operations when his aircraft went into the sea directly off the catapult. One other aircraft was lost, but the pilot was able to bail out over the water and was recovered unharmed.

The deployment was successful as a result of the professionalism demonstrated by the pilots and crewmen of the wing. CVW-19 was commended by Commander, Task Force 77 for the manner in which the air wing performed, and it was noted we had one of the lowest aircrew loss rates of the entire Vietnam War.

CVW-19 and *Ticonderoga* returned to the U.S. in late August 1968. I was relieved as commander in September and immediately reported to Readiness Attack Carrier Air Wing 12 at NAS Miramar, CA, to take command in October. RCVW-12 had the responsibility for training all the replacement aircrews for Pacific Fleet carrier-based squadrons.

While Commander RCVW-12 I caused a fighter weapons school to be started, which eventually became known as 'Topgun.' This combat training was based on experience gained from the Fleet Air Gunnery Unit. The school was established using RCVW-12 fighter squadron instructors to develop and present the course. The initial course of instruction was devoted to teaching selected fighter aircrews from all of carrier aviation to more effectively employ their aircraft and weapons in the prosecution of their mission. Tactics were also a part of the curriculum in order to demonstrate to pilots how best to maximize their aircraft's capabilities in a combat situation. The first curriculum was developed by instructors from VF-121 and VF-124, as well as some members of the RCVW-12 staff. The first officer-in-charge was assigned from VF-121. Since aircraft to support the school was not then available, squadrons that wanted to send crews to the school had to send one of its aircraft for each crew they assigned.

Topgun was initially a detachment operating at Miramar directly under Commander, RCVW-12, but it was envisioned that the school would evolve to a point where it would eventually become a squadron with its own commander. It was also anticipated that the school would, at some time, receive its own aircraft and would be manned with the necessary technicians to maintain those aircraft. The Naval Fighter Weapons School was formally established in 1972."

CDR EDWIN D. McKELLAR, USN (CAPT Ret)
CAG 8 20 June-13 Jan 71
USS *Shangri-la* (CVA-38)

Ed McKellar entered naval service in June 1949 as an Aviation Cadet, receiving his wings and commission as ensign in December 1950. His first tour of duty was with Composite Squadron 61 based at NAS Miramar, CA. While attached to this command, Ed made two combat deployments to Korea flying F9F-2Ps and F2H-2Ps from USS Antietam *(CV-36) and* Oriskany *(CVA-34). He flew a total of 125 combat missions during the Korean War.*

Ed was later assigned to the Blue Angels Flight Demonstration Team from 1954 to 1957. He then served two years as catapult officer in Franklin D. Roosevelt *(CVA-42) from 1957 until 1959, when he then reported for duty in VA-76, flying A4D Skyhawks. He made two Mediterranean deployments with the squadron in* Intrepid *(CVA-11).*

After attending the Naval War College in 1961-1962 he reported to the Bureau of Naval Weapons as aide and technical assistant to VADM P.D. Stroop. His next assignment was with VA-125 at NAS Lemoore, CA, where he served as training officer and later as operations officer.

Ed reported as executive officer of VA-192 in 1965 and took command of the squadron in 1966. During this period the squadron deployed to Southeast Asia in Ticonderoga *(CVA-14) for combat operations in Vietnam. He flew 146 combat missions while attached to and commanding the "World Famous Golden Dragons."*

On 20 June 1970 McKellar assumed command of Attack Carrier Air Wing Eight, then deployed to Southeast Asia in Shangri-la. *While in command of CVW-8 he flew an additional 97 combat missions in Vietnam. Following this tour Ed reported as operations and plans officer for Commander Carrier Division Five (CTF 77) where he served until May 1972.*

After five years of sea duty, Ed served in the Office of the Deputy Chief of Naval Operations (Air Warfare) as VADM W.D. Houser's executive assistant from 1972 to 1974. On 16 July 1974 CAPT McKellar assumed command of NAS Alameda, CA. He was relieved of these duties in July 1972, and assumed his duties as Chief of Staff, Commander Cruiser-Destroyer Group Three, homeported in San Diego, CA.

Following retirement from the Navy in April 1980, after 30 years of active duty, McKellar became executive director of San Diego's Aero-Space Museum, a position he currently holds.

Not all routes to the job of CAG are the same. Ed relates the hectic final days of his selection:

"So, you are going to be a CAG. Fine, but where do you start the process? Everyone's case is different, I suppose—West Coast, East Coast, the work-up training syllabus, overall timing sequence

CDR Ed McKellar, CAG 8 in *Shangri-la*, with his beloved A-4 Skyhawk. "In those days, it was easy for me to believe my mother had given birth to me in an A-4 cockpit." USN photo, courtesy E.D. McKeller, CAPT(Ret).

and current level of experience in the aircraft involved—all these and many more factors are figured in. But hell, it's a flip of the coin as to who gets the nod and why. Equally important is what you do with the opportunity once you get it.

My case was somewhat unique in that I was in a stash (temporary) assignment at AirPac in late spring 1970. I had just finished a year's 'Boot Strap' program at San Diego State University—finally earning a degree of sorts at the not-so-tender age of 42. After leaving State in spring, I reported to AirPac to await the start of my CAG cycle. I had been selected to take CVW-19, then embarked in *Oriskany*. Both ship and air group were due back from a WestPac combat deployment in September. I was anxious to start training and delighted to have been selected as prospective CAG 19, because this group was my old air wing from the 1966-67 period when I had been XO then CO of VA-192, flying the A-4F Skyhawk.

In those days it was easy for me to believe that my mother had given birth to me in an A-4 cockpit. I was by no means high man in the fleet, flight-time wise in the Skyhawk, but I did know my way around that great aircraft. And let me say that all of us who ever flew the A-4 owe its designer, Ed Heinemann, a big debt of gratitude for that wonderful airplane.

In any case, I had just relieved then-CDR 'Boot' Hill in an airborne change of command ceremony in December 1966 over the Tonkin Gulf. Those had been heady days—with a lot of laughing and scratching—and, yes, we had ample reason for an equal amount of tears during that Yankee Station tour. Before it was over, VA-192 would leave three of its best in Vietnam. Mel Moore, my XO, and Dick Stratton, my maintenance officer, would both be guests at the Hanoi Hilton; POWs for the next six years. Our third loss was Mike Estocin, my ops officer. He was as brave as they come, flying 'Iron Hand' anti-SAM missions on a daily basis until a SAM finally got close enough to Mike over Haiphong to send him west. Small consolation, but Mike left the Vietnam stage with a posthumous Medal of Honor for his bravery and daring.

Over three years had passed since my XO/CO job at VA-192, and there I sat in the AirPac 30 shop trying to look busy for my boss, then-CAPT Paul Peck. Today I can't even remember what he had assigned me to do, but it obviously wasn't holding my attention all that much, because every chance I got, I would toddle out to Miramar and take one of the A-4 pickup hops with VC-7, towing targets, playing the role of radar blip for some surface training exercise, or flying a tanker so some fighter jock in the RAG pipeline could get his required plug-ins.

Then one day in late May, CDR Ken Wallace, who was at that time air wing training officer at AirPac, walked into the office, sat down, and said, 'How'd you like to have an air wing?' I responded by saying, 'Ken, I thought I had an air wing—remember CAG 19?' In a very blasé manner, Wallace then replied, 'What I mean is, how'd you like to have one today?' 'What air wing are we talking about, Ken?' 'Air Wing Eight,' was his response. 'Are you nuts, Air Wing Eight isn't an AirPac outfit, it belongs to AirLant.' 'I know, but we gave 'em some squadrons to flesh out the wing. Specifically, both of the F-8 fighter squadrons, VF-162 and VF-111, and VA-152, one of the three A-4 squadrons, are all AirPac outfits.' 'Oh, great! It's an air wing with squadrons from both coasts. I'll bet that little or no combined wing training ever took place before deployment, right?' 'Right. They were probably still introducing themselves to one another on the way to WestPac.' 'Oh super! What boat are they on?' 'The *Shangri-la,*' Ken said. 'The Shitty, Shitty Shang Shang? Damn, Ken, that's an East Coast boat and, from what I've heard, she probably didn't even make it to Yankee Station. That ship has major problems brought about from deferred and canceled yard periods. Anything to save money. Hell, the Shang has been on the mothball list for years.'

Long story short; it seems that CAG 8, CDR W.M. 'Stolie' Stollenwerck, had hit *Shangri-la's* ramp on Yankee Station. This attention-getting event had occurred about eight-to-ten hours before AirPac was notified that Shang needed a new CAG. I hasten to report that CAG Stollenwerek probably bruised his ego when he made contact with the ramp, and he also banged up the F-8, but both he and the plane stayed aboard. However, I think this particular incident caused him to rethink his future role in naval aviation and, I can only guess that, on the spot, he decided to give up carrier flying. I never had a chance to talk with him, as he had left the ship before I reported aboard.

But, I'm getting ahead of myself. Ken Wallace and I concluded our conversation that late May morning by his saying that two candidates were being considered for the CAG 8 job—me and then-CDR Pete Easterling. At the time, Pete was in line to take over the F-8 RAG at Miramar, and was on the eve of assuming command. In short, he was current in the F-8 and I was current in the A-4. Take your pick, with CVW-8 we had a small-deck F-8/A-4 air wing and two gents who were ready to go in one type or the other. Obviously, there was no time for cross-training, just pack up and go.

Before the day was out, I was selected as the new CAG 8. I distinctly remember having an interview with VADM 'Bush' Bringle, who was ComNavAirPac at this time. We talked about a number of things, but the principal thread of the conversation centered around the question: how quickly can you get out of there?

I told Admiral Bringle and Ken Wallace that I thought I could be ready in about seven flying days, assuming I could get some dedicated bombing range time at Fallon, some day/night FCLP time at Lemoore, and some CarQual time on *Lexington* (CVT-16) at Pensacola. They both nodded in the affirmative to my requirements, made a few quick phone calls and, before close of business that day, I was scheduled for the fastest RAG training program on record.

The last day of May I drove to NAS Lemoore and commenced flying the next day. Between the first and the fourth of June I logged a total of 13 hours in the A-4F, five and one-half at night. I got in four FCLP periods; 56 passes, half at night; and made 24 Mk 76 conventional air/ground runs at Fallon; seven GCA approaches and one wet and three dry tanker plug-ins. Not a lot of refresher work, but, as I said earlier, my mother had given birth to me in an A-4—it was enough.

I drove back to San Diego 5 June and the following day climbed into the back seat of a TA-4F to fly under the bag to Pensacola via Dallas. That was another 4.7 hours, and 4.6 of it was simulated instruments. All this instrument time was required because I also needed to renew my instrument ticket. (Incidentally, I had a straight A-4F flying on my wing to Pensacola, and that's the plane I used 8-

9 June to get my day/night qualifications aboard *Lexington.*) Fifteen day and eight night traps later I was back in the rear seat of the TA-4F under the hood for the return trip to San Diego.

From then on it was shack-up, pack-up, and shag-ass to Travis AFB, CA. Next stop was Clark AFB in the Philippines, then on to NAS Cubi Point. On 19 June I was aboard a COD out of Cubi, and about three and one-half hours later, landed aboard *Shangri-la*, steaming at Yankee Station.

I will always remember my arrival. The flight deck crew taxied us forward to the fly-one area and tied us down. When I stepped out of that COD after a rather hectic 19-day period, the skipper of the ship, CAPT Herb Poorman, was there to meet me. I will forever remember his words, 'Welcome aboard. Are you a flying CAG?' After what I had just been through—flyingwise, simply to get to Yankee Station—well, the man could have chosen better words, to my way of thinking. I responded, 'We'll see,' and went below to the CAG office spaces.

Lo and behold, who should I meet there but my old A-4 buddy from another cruise and another time, LCDR Nick Langston. 'What the hell are you doing here?' I asked. 'I'm your ops officer,' Nick said. 'God, things are going from bad to worse; now I gotta put up with you again. Well, listen up, Nick, this is probably the only direct order I'm ever gonna give to you. First, a question—does the CO that I just met have regular department head meetings like all the other boats I've ever been on?' Nick responded, 'Yes, so what?' 'Well, this is what: you find out what hour of the day/night he plans to hold those meetings and, when you know the time for the following day, you contact one of the three A-4 ready rooms and you schedule me for a flight during the time the department head meeting is held. Then *you* go to the meeting, and when someone asks, "Where's CAG?" you tell 'em I'm in the air. Did you get all of that the first time?' That's the way it worked out. I never went to a *Shangri-la* department head meeting, but I did get in a lot of flight time from 20 June to 5 Nov 1970—97 combat sorties to be exact. We must have come off the line for the final time on or about 5 November, because that was my last combat flight for the cruise.

After leaving the line, we did the usual screw-around things before departing WestPac. The two F-8 squadrons Trans-Pac'd home, and that was a story in itself. But, before any of us left WestPac, we held one last air wing bash at the Cubi club. T'was an event never to be forgotten, but I must now admit that I did forget how in the hell I got back to the ship that night. No matter, we sailed out of Subic Bay at some god-awful hour like 0300 and, when I woke up, we were well on our way to Australia. There were several memorable port calls; port calls that were separated by countless days at sea. In any case, the Shang arrived back in Mayport, FL, her home port, about the middle of December.

The date must have been 16 Dec 1970, because that's the day I led the Cecil Field A-4s off the ship and back to Cecil.

The following morning I met with my relief in the offices of ComFair Jacksonville, where we conducted a low key and decidedly brief change of command ceremony. Never wanting to linger on the East Coast any longer than necessary, I caught a commercial flight to San Diego that same afternoon.

I spent Christmas at home in San Diego and returned to WestPac after the first of the year because, guess what? My old friend, Boot Hill, had nominated me as his relief on the staff of CTF 77. Somehow he had convinced VADM Hutch Cooper that I'd make a good staff operations and plans officer. Regrettably, we can't ask Hutch if I was good at it or not—he's gone west— but I unquestionably served Boot's purposes as his relief. "That wraps up what turned out to be one of the shortest CAG tours on record."

CDR ROBERT F. DUNN, USN (VADM Ret)
CAG 7 23 July 70-10 Jan 71
USS *Independence* (CVA-62)

Robert Francis Dunn is a native of the midwest, born 15 Jun 1928 in Chicago, IL. He graduated from the Naval Academy with the Class of '51 and first served in USS Nicholas *(DDE-449) before entering flight training in June 1952.*

After designation as naval aviator in October 1953, Bob Dunn reported to Attack Squadron 95 flying AD Skyraiders. He then served other AD tours with VF-194 and VA-196 until returning to Pensacola as a flight instructor during January 1957–February 1960. A staff tour with Commander Carrier Division Seven was followed by VA-36, January 1962-June 1963, then to the Naval Postgraduate School at Monterey, CA, as a student until June 1964.

LCDR Dunn was next the Bureau of Weapons Representative at Aerojet-General Corporation in Sacramento, CA, until October 1965 when he reported to VA-125 for A-4 replacement pilot training before becoming XO of VA-146 in February, then CO January-November 1967. Bob led his squadron over North Vietnam on numerous combat missions while deployed in Ranger *(CVA-61) and* Constellation *(CVA-64).*

During 1968 and 1969, CDR Dunn attended the United Kingdom Joint Services Staff College and then served at the Bureau of Naval Personnel in Washington, D.C. He became prospective CAG 7 in February 1970, then commanded the wing during 1970-71. After a Sixth Fleet staff tour, he commanded Mount Whitney *(LCC-20) and* Saratoga *(CV-60) during 1972-1976. Upon detachment from* Saratoga *he assumed command of the Naval Safety Center in October 1976. After instruction at the Naval Nuclear Power Training Center, Idaho Falls, ID, he commanded Carrier Group Eight May 1978-June 1980.*

RADM Dunn next returned to Washington to head the Naval Military Personnel Command until October 1983 when he became Chief of Naval Reserve until December of that year. Bob next took over as Commander Naval Air Force Atlantic and was promoted to vice admiral 9 Feb 1984. VADM Dunn's last active duty tour was as Assistant Chief of Naval Operations (Op-05), where he was the head of naval aviation from January 1987 until his retirement 1 Jun 1989. Since then he has been an independent consultant to aerospace industry, defense oriented non-profit institutions, and non-defense government agencies. In addition, he serves as a member of the NASA Aerospace Safety Advisory Panel and as Senior Advisor to the U.S. Naval Institute. During early 1993 he served as

Interim Press Director for the Naval Institute's book publishing division. VADM Dunn resides in Alexandria, VA.

As one of the Navy's CAGs deployed to the Mediterranean, Bob Dunn's CVW-7 inevitably was caught up in Mid-East events.

"My P-CAG (Prospective-CAG) training commenced 4 Mar 1970. During P-CAG I carrier qualified in the A-4E Skyhawk and the F-4J Phantom. I also flew the RA-5C Vigilante, A-6A Intruder, and E-2A Hawkeye while ashore.

Change of command was 23 July 1970, on board *Independence* anchored in the Bay of Naples. I relieved CDR Dave Leue. The wing makeup was:

VF-33 12 F-4J CDR Moose Myers; VF-10 212 F-4J CDR Bill Hargrave; VA-65 12 A-6A CDR Pete Easton; RVAH-11 5 RA-5C CDR Dave Nothwang; VAW-122 4 E-2A CDR Jack Hoch; CDR Dan Rice, HS-2 8; SH-3A CDR Joe Purtell; CDR Buck Buchanan VMA-33120; A-4 ELCOL Pete Sullivan; VAQ-129 (Det) 3 KA-3B LCDR Dick Youngjohns

My entire CAG tour was embarked in *Independence* in the Mediterranean. First CO of Indy was CAPT B.B. (Beetle) Forbes, Jr., now a retired vice admiral. He was relieved by CAPT Gerry O'Rourke 19 Aug 1970, while anchored in Malta.

Independence was the flagship for CTG 60.2 throughout the Middle East deployment.

Bob Dunn's CVW-7 deployed in the Mediterranean was inevitably caught up in the events of the Mid-East. USN photo, courtesy VADM R.F. Dunn, USN(Ret)

RADM Doc Abbott, a ComASWGru, was the first flag embarked. After the Naples port visit in July there was no flag for awhile, and the Captain acted as 60.2. Several ComCruDesGrus came and went until we were sailed for a contingency operation in connection with hijacked airliners and a Syrian-Jordanian confrontation. George Cassell, Deputy Commander Strike Force South, then embarked. He stayed for just a short while because the leadership in London did not want a NATO commander in charge of national forces. In any event, we soon began operating in close proximity to CTF 60, RADM Jimmy Holloway. Both he and VADM Ike Kidd, ComSixthFlt, made frequent visits to Indy. After the contingency operation was over, ComCruDesGru Eight, RADM Stansfield Turner, embarked.

For the first time an attack carrier air wing included a full squadron of antisubmarine warfare helicopters. HS-2 had flown cross-country from NAS Imperial Beach, CA, stopping at every gas station along Route 66, to embark Indy. On the cruise I flew copilot with them, including a day landing and a night launch from *William V. Pratt* (DGL-13), an unauthorized landing platform for the SH-3 Sea King.

The light attack role was fulfilled by an embarked Marine A-4E Skyhawk squadron. The Navy squadrons normally assigned were transitioning from A-4s to A-7s, so were unavailable for deployment. The Marines were great up and away and around the target, but around the ship and in the maintenance spaces they had some problems. Most pilots had only the original six training command carrier landings before joining the wing and a couple had none; they had gone through Air Force flight training. The maintenance troops had a tough time accommodating to the confines of shipboard maintaining. But it worked out in the end, and we were all better off for the experience.

The embarked Marine squadron provided close air support to Marines ashore in Greek Thrace during Exercise Deep Express.

Independence and CVW-7 established 'Bravo Station' (known as 'Camel Crossroads' at the time) at 33N/33E, standing by to fly support during the 1970 crisis in Jordan. At one point we were loaded up to fly strikes against Syrian tanks heading into Jordan. We never launched, however.

We planned Operation Magic Sword, the support of the NATO central front with carrier sorties from the North Sea and the Bay of Biscay, despite Air Force reluctance; they believed NATO Europe was 'their territory!'

My CAG tour ended early, as I had been selected for captain and because Ike Kidd wanted me on his staff. Change of command was aboard *Independence* at sea in the Western Med 10 January 1971. I was relieved by CDR Bill Knutson."

CDR L.F. "GUS" EGGERT, USN (RADM Ret)
CAG 9 29 Jan 71-20 May 72
USS *Constellation* (CVA-64)

Lowell Franklin "Gus" Eggert was born in Milwaukee, WI, 6 March 1929. He received his Bachelor of Science degree from the U.S. Naval Academy in 1952 and his Master of Science degree in International Affairs from George Washington University in 1963. He graduated from the U.S. Naval Test Pilot School in 1960 and the Naval War College in 1963.

Commissioned ensign in June 1952, Gus Eggert served for two years in USS Valley Forge *(CVA-45). He then reported to NAS Pensacola for flight training and was designated naval aviator 23 Dec 1954.*

Gus's first squadron assignment was with Fighter Squadron Sixty-Two and subsequently with Attack Squadron One Hundred Six, with which he completed several cruises to the Caribbean and Mediterranean Seas.

After graduation from test pilot training at NATC Patuxent River, MD, Gus served as an ordnance test project pilot and later as a test pilot instructor. Subsequent sea tours have included air operations officer on the Staff, Commander U.S. Second Fleet in 1964-65 and operations officer, CVW-8 in 1965–66, again deploying to the North Atlantic, Caribbean, and Mediterranean Seas. CDR Eggert then commanded VA-192 in 1967–68. During this tour he completed a combat cruise to Southeast Asia flying A-4F Skyhawks.

Returning to the Naval Air Test Center in 1969, Gus then served as the Deputy Director and Chief Flight Instructor at the U.S. Naval Test Pilot School. He then assumed command of Attack Carrier Air Wing Nine in 1971 and led his wing on strikes into North and South Vietnam during 1971 and the intensive periods of combat in early 1972.

Under the strong leadership of CAG Eggert, CVW-9's achievements in the Vietnam War were legion. VF-92 and VF-96 produced a combined MiG kill score of nine, the highest single cruise total of any air wing during the war. VF-96 produced the only Navy fighter aces of the war, and CVW-9's bombers delivered more than 50,000 tons of ordnance, also a record for a single cruise. During CVW-9's nine-month 1971-72 deployment, Gus Eggert's stringent training procedures resulted in no operational losses and the lowest combat losses of all air wings during the 10-year war. CVW-9 was awarded the Presidential Unit Citation for the cruise, and CAG Eggert earned a Navy Cross.

Eggert next served as force training officer on the Staff, Commander Naval Air Force, U.S. Pacific Fleet until ordered to command Denver *(LPD-9) in May 1973. He then assumed command of* Constellation *in September 1974. During this command tour he carried out operations in the Indian Ocean and Persian Gulf.* Constellation *was the first aircraft carrier to operate in the confined Persian Gulf in more than 27 years.*

Gus was then assigned as Director, Air Weapons Systems Analysis Staff (Op-50W), Office of the CNO. He served in that capacity until August 1977 when he became Commander, Fleet Air Western Pacific, homebased at NAF Atsugi, Japan.

Returning to Washington, DC, in July 1979, RADM Eggert served as Director, Command and Control Support Systems and Information Systems Division (Op-942), Office of the CNO and as Director, Department of the Navy Automatic Data Processing Management until May 1981. In June, Eggert assumed duties as Inspector General, United States Atlantic Fleet and the Atlantic Command.

Since his 1983 retirement from the Navy, Gus has been Director, Chesapeake Region, for BDM Management Services Co. Living in California, MD, he remains active with the Navy serving as the president of the Board of Directors for the Naval Test and Evaluation Museum at Patuxent River. RADM Eggert's combat experiences spanned two wars, including two combat tours in Vietnam. He relates one of his memorable missions over the North as Commander, Air Wing Nine.

"My naval career was one I would not change one iota, especially my experiences with my squadron, VA-192, CVW-9 and *Constellation.* It was particularly pleasant to have the relationship of moving from CAG Nine in *Constellation* to become the Captain *with* CVW-9 on board. I remember squadron command as the last time an aviator can be the Boss, and still be 'one of the boys.' As CAG you were accepted by *all of the squadrons*, and they wanted to treat you as one of the boys, but you had too many patches on

"Gus" Eggert earned a Navy Cross for successfully leading CVW-9 through some of the heaviest fighting of the Vietnam War. U.S. Navy photo

your flight jacket to be truly one of theirs. Of course, as the ship's captain you were virtually too high above their heads; you were more fatherly, and *they* were '*your* boys.' In every case we were never boys, but men, of varying ages and experience, each of whom knew he was the best in his trade. We were in a very serious business; we took it seriously *and* enjoyed it.

But, it is stories of my CAG days you are seeking. There are so many, but I will not suffer this machine, nor bore you with all those now ancient tales. Sea stories or war stories are usually best, so here's my contribution.

It was either late 1971 or early '72. CVW-9 was embarked in *Constellation* in the South China Sea. One of the many Alfa strikes (large, coordinated total asset strikes—long before Red Flag or some of those innovative ideas) was to take the air wing to a heavily defended target well to the west of Hanoi, or much further 'inland' than most previous strikes had gone. Since it appeared to be a complex and high-risk mission I exercised my prerogative and led the strike. The briefing covered the usual points of high-density missile and AAA fire, the threat of a MiG intercept, navigation, fuel consumption, and tanking. We also covered communications and the possibility of a downed aircraft, as well as rescue procedures. Now, one of the things I emphasized during our Stateside and subsequent training and combat missions was *absolute* radio silence during 'routine' operations. During launch, rendezvous, en route, normal shipboard recovery, etc., radio silence was to be broken only when there was a threat to the flight or the mission.

The launch, rendezvous, and en route navigation went as briefed, without so much as a 'keyed' mike. We went 'feet dry,' and our vigilance increased. It was stirring to look over my shoulders and see more than 35 heavily-armed aircraft flying in perfect tactical formation.

About 10 miles inland we were passing through what was briefed as a very high-threat surface-to-air missile (SAM) area. The missile tracking and launch detection equipment in my aircraft began to give all the indications of a high threat and imminent launch against our formation. *This* (danger) was always grounds to break radio silence and announce the indications of the attack and any information as to its origin, or even sightings. But well trained CVW-9 remained silent! Unable to understand this reaction, I glanced over my shoulder—and saw everyone of those magnificent aircraft executing a proper, but violent, evasive maneuver! I turned back to the front just in time to see an SA-2 (SAM) pass one to two hundred feet (who's measuring?) over my canopy! The instant revelation that I had a radio failure and not an over-disciplined air wing prompted my 'stick pumping' to indicate a change of the lead and my return to the ship. (They later told me they were all shouting warnings, and one pilot thought I was 'too cool.') The designated alternate flight leader took charge, assigned a wingman to escort me, and continued with the mission.

But that was not the end of the story. As we proceeded out of North Vietnam I was considering the many bombs both the newly-assigned wingman and I were carrying and where we could use them to the maximum effectiveness without simply dumping them into the ocean. All of a sudden, I spotted a large concentration of what looked like 55-gallon barrels piled in an open area. Suspecting them to be a fuel dump for NVN vehicles using the trail south, I selected two of the 11 500-pound bombs I was carrying and dove on the target. The poor wingman, unable to communicate with me, must have thought I had gone berserk, diving at the ground. When my bombs hit and violent secondary explosions followed he got the message, and we expended all of our weapons on that target before returning for an uneventful no-radio recovery. The primary group returning from their successful mission passed the area we had hit and, from the vast fires on the ground, assumed another air wing from another carrier had conducted a strike there. We scheduled another sortie to the same area that afternoon (led by me, since I was the only one who knew exactly where to go), with resultant complete destruction."

CDR RONALD W. KENNEDY, USN (CAPT Ret)
CVSG-59 24 May 1971-12 May 1972
USS *TICONDEROGA* (CVS-14)

Ron Kennedy entered the Navy in June 1952 via the University of Southern California NROTC program. He served in USS Wedderburn *(DD-684) for a year before reporting to flight training. Receiving his wings in December 1954, he joined VS-20 at NAAS Brown Field , CA, flying the S2F. VS-20 merged into VS-21 to form a 20-plane, 60-plus pilot squadron in 1956. While with his first squadron he deployed in* Princeton *(CVS-37) and* Philippine Sea *(CVS-47).*

Ron's other early tours were as a flight instructor in VT-1 flying the T-34, T-28, and TT-1; post graduate training; and a year at MACV, Saigon in 1962. Back in the cockpit, he joined VS-38 for two WestPac cruises to the Tonkin Gulf in Bennington *(CVS-20). After a tour in OpNav Plans he returned to flying as XO and then CO of VS-37 during 1968-1970, making another cruise to the Tonkin Gulf in* Hornet *(CVS-12). Following an assignment as* Ticonderoga *Weapons Officer, he became Commander Antisubmarine Air Group Fifty-Nine in May 1971. After his CAG tour Ron returned to OpNav for two years, followed by command of* Paul Revere *(LPA-248) then* Inchon *(LPH-12). His* 1976-78 Inchon *tour was his only LantFlt duty. He retired as a captain from OpNav in 1979. Subsequent to his Navy retirement, Ron started his own highly-successful business, Kennedy Publishing Company, in Coronado, CA, and is still active in the publishing field.*

"I assumed command of Carrier Anti-Submarine Air Group Fifty-Nine in *Ticonderoga* during a WestPac cruise in a quiet ceremony in the admiral's at-sea cabin, along with two other change of commands that took place at the same time. The admiral, ComASWGru Three, was not happy with me, and would not allow a hangar or flight deck change of command ceremony.

My air group consisted of HS-4 and HS-8 flying SH-3Ds, along with VS-33, VS-37, and VS-38 flying the Grumman S-2E. The other ASW air group at North Island, CVSG-53, was made up of VS-21, -29, -35, with HS-2 and HS-6. Our group was fortunate because we had the only CVS in PacFlt to operate off of.

I previously had command of VS-37 in 1970 when the four PacFlt CVSs were going away and things in the VS community were in a turmoil. At the time CO tours were nine months and commanders were all over the place. There were three of us in VS-37, my XO Jay Ortega, and his relief, George Bruce. Squadron morale was not high because of the loss of the carriers, not much of a mission, and low flight time in a seven-plane squadron. Things improved when Tico, an *Essex*-class 27C CVA, was converted to a CVS in Long Beach in mid-1970.

I was ordered to Tico during its conversion period because of my ASW background and because one of the biggest jobs converting the carrier to a CVS was establishing a torpedo shop. I reported aboard as weapons officer in Long Beach, got the torpedo shop completed, and then went through RefTra, firing the old 5"38 guns. I left Tico in early 1971 with orders as CAG, reported to VS-41 for a short refresher in the S-2 along with an instrument check, and then to HS-10, the SH-3 RAG for helo qualification.

I joined the carrier in Singapore and was to take over as CAG after the next at-sea period. The ship had been operating in Indonesian waters, and one of HS-4's helos had made a very low hover over one of the many islands, making contact with the locals, trading for seashells and a hat. The incident received negative diplomatic play, and I was directed to be the Article 32 officer investigating this incident. The Captain and Admiral wanted to make an example of the aircrew, so the heat was on. I conducted a thorough investigation and found the staff insisting that all aircrew had been briefed to avoid Indonesian islands. Well, they may have briefed the other aircrews, but they missed the HS-4 people. My investigation cleared the crew, but the Captain and Admiral still put a negative endorsement on the report. AirPac agreed with me and overturned the Admiral. Nothing like starting out the job as a hero to the air group, which was watching the entire incident like a hawk.

After my change of command Tico and our destroyers went north to the Petropavlovsk area to play games with the Soviets. We operated in complete radar and radio silence, trying to remain undetected. I guess we were successful, because we never saw anyone, or maybe they just did not care. The fog was so bad up north that there were many days we could not fly. One day we did get our usual launch of six S-2s and two to three helos off, only to have the fog close in. We steamed at max speed away from the fog bank and then had to turn around and run back into it for recovery. We got all the planes down except our plane guard helo before we ran into the fog. The LSO had to talk the helo down from its holding position off the starboard quarter. The weather was zero-zero, and the LSO could just make out the outline, besides listening to the SH-3, at 40 feet. It was hairy because there was no place for the helo to go if it could not get aboard.

When we returned from cruise we had another problem. One day out of San Diego the Tico mascot, a big plaster tiger, disappeared. The Captain was furious and ordered the entire ship searched. I got all the COs together, and we came to the conclusion, after the ship had been searched with no luck, that the officers had not taken the tiger, because when a joke goes bad the object usually turns up. We learned later that some enlisted men had taken it and panicked when the search was announced and threw the tiger over the side. The captain was so mad at the air group that he almost did not let us off the ship. On our first at-sea period after the cruise, HS-4 arrived with a Dixieland band on a flatbed playing 'Hold That Tiger' at pier side. The squadron then presented the Captain with a new tiger for the ship, and our air group made a nice recovery.

After the cruise the big project for the VS squadrons was to convert from S-2Es to S-2Gs by adding P-3-type sensor capability, such as DIFAR and CASS sonobuoy capability, the AQA-7 acoustic processor, and the ASN-50 improved navigation system. The addition of this equipment made in effect a non-digitally integrated S-3A prototype, allowing the S-2 to keep up with the nuke submarines until the S-3A came along. The S-2 conversion of all my aircraft was done at North Island by Grumman tech reps. My staff maintenance officer did a great job of keeping the conversion on schedule and shuffling the aircraft around equally so that each squadron had enough airplanes to maintain their flight time.

We did get visits from the Lockheed tech reps about the S-3A, which came out in early 1974, but most of the discussion was about ergonomics and crew comfort on long ASW flights and instrumentation. We were not really consulted about ASW tactics or such.

During the same time we operated off Tico on several short exercises and a RimPac. In fact, half the time I was CAG we were at sea. We operated with a new ASWGru Three, RADM C.J.

Bob Kennedy was one of the Navy's last Antisubmarine Warfare Air Group CAGs. USN photo courtesy CAPT R.W. Kennedy, USN(Ret)

Seiberlich, in one of the last of the UPTIDE series of exercises in Hawaii. We even made a full-blown movie about the exercise—script, professional camera crew, and all. Unfortunately, the movie was classified secret, so we did not get to enjoy it much. The UPTIDE series began in 1969 and ended in late 1972. This particular exercise and the others we participated in were defensive in nature and emphasized confusing the submarine's sonar picture so the carrier could stay afloat. We did a lot of screw masking, and the helos dropped big buoys that played tapes that sounded like the carrier. We usually operated against one nuke attack boat, and it really caused them lots of trouble in locating us. Of course, the CVA could not be bothered doing the deception and usually got 'sunk.'

I had a good staff, and we worked closely with the Admiral on these exercises. He was very interested and knowledgeable in ASW, and I enjoyed working for him.

The tour was great, and I was relieved by Doug Payne in June 1972. CVSG-59 was disestablished a year later. VS-37 and VS-38 with HS-8 eventually made the first two CV cruises in *Kitty Hawk* (CV-63) as part of CVW-11. I consider my CAG tour and ship CO tours as the best jobs I had in the Navy."

CDR CHARLES L. TINKER, USN(CAPT Ret)
CAG 20 28 Feb 75-25 Feb 78

Probably for the sole purpose of being a fighter pilot, CAPT Charles L. "Chuck" Tinker was born 22 Sep 1933 in Vandalia IL. In December 1954, he enlisted in the Naval Aviation Cadet Program and was designated naval aviator in August 1956.

Fighter Squadron 111, based at NAS Alameda, CA, was Chuck's first fleet assignment. He deployed to the Western Pacific in USS Bennington *(CVA-20) flying the FJ-3M Fury. Subsequent tours were with VRF-32 and VT-24. Following F-8 Crusader training, he reported to VF-162 and completed two combat deployments to Vietnam in* Oriskany *(CVA-34). In January 1967 he reported to VF-124 at Miramar as a combat flight instructor, and later that year was ordered to VF-62 at NAS Cecil Field, FL, serving as operations and maintenance officer. This tour included two Mediterranean deployments in* Shangri-La *(CVA-38). He was then assigned to the staff of the Chief of Naval Operations, Op-342 Carrier Requirements. Chuck entered George Washington University and was awarded a Bachelor of Science degree in Business Administration in June 1971.*

Returning to Miramar, then-CDR Tinker became executive officer of VF-191 in December 1971; commanding officer in June 1972; and again deployed in Oriskany *for his third Vietnam combat tour. In August 1973 he reported to the Staff of Commander Fighter Airborne Early Warning Wing, Pacific as operations officer, and in January 1974 to* Constellation *(CVA-64) as air operations officer and later as operations officer.*

CDR Tinker served as Commander, Carrier Air Wing Reserve 20 for two years and, following selection to captain, returned to Washington as Head, Tactical Air Plans and Programs (Op-508C). He reported to NAS Whiting Field, FL, in August 1980, where he served as commanding officer prior to assuming command of NAS Oceana in Virginia Beach, VA.

A third Washington tour, this time with the Joint Chiefs of Staff, was followed by Chuck's retirement 30 Jun 1986, ending a 32-year career, during which time he gained flight experience in every type of fighter and attack aircraft (with the exception of the A-6) then in the naval aviation inventory. This experience exceeds 10,000 accident- and incident-free flight hours, as well as 1,272 carrier arrested landings.

Currently residing in Pensacola, FL, with his wife, Mitch, Chuck is now a flight supervisor with Vitro Services Corp., a DOD contractor providing flight training and instruction for the Undergraduate Naval Flight Officer program. He states: "I'm still having fun!"

"My fond memories of my two years as Commander, Carrier Air Wing Reserve 20 are filled with good people and exciting times. The wing was relatively new to the Naval Air Reserve, evolving out of necessity as a result of the Vietnam War. In its early days CVWR-20 went through some tough growing pains, but eventually evolved into the organization that I was privileged to be a part of.

The Naval Air Reserve had been around nearly forever (April 1917) and had gone through many reorganizations. None, however,

CAPT "Chuck" Tinker, a veteran F-8 combat pilot of the Vietnam War, commanded Carrier Air Wing Reserve 20 during 1976-78, highlighting a 32-year naval career. Bob Lawson photo

was of such magnitude as the reorganization of the early '70s. As the conflict in Southeast Asia drug on, the need to selectively mobilize some reserve tactical squadrons brought attention to an unfortunate reality—some of these forces simply could not answer the bell. The sad and glaring truth was that the Ready Reserve was ill-equipped, under trained, and anything but 'ready.'

I remember in 1968 being attached to an F-8 Crusader squadron at Cecil Field when a Reserve F-8A squadron arrived to start transition into the F-8H. They were a pathetic group. They were more focused on ways to get released from the Reserves, or to politically spearhead a drive that would result in demobilization of the squadron. Transition to an all-weather, night fighter squadron and deployment in a 27C carrier was not what they were all about. And, as time would tell, they were successful in their efforts to avoid complete mobilization.

Within two months the squadron was sent back to NAS Willow Grove, PA, a sad return from the millions of tax dollars paid to support the weekend warriors of the '60s.

This example is but one of the absurdities that ultimately brought about the critical look and eventual 1970 reorganization of the Naval Air Reserve. Little did I suspect during this period that, in 1976, I would have an opportunity to be directly involved with these forces as Commander, Carrier Air Wing 20.

It may be informative to describe the process and procedure necessary for mobilization of Reserve forces during this period, as it has since been changed. During my tenure, in time of national crisis the president had the authority to selectively mobilize a Reserve unit for up to 90 days. After that time, if the urgency of the situation warranted, he could further extend the active duty period for up to an additional 90 days. The combinded total could not exceed 180 consecutive days. Should there be a compelling need to extend beyond the 180 day limit, then Congress, both House and Senate, had to vote their approval and specify the limits of mobilization.

I reported to this new job with some reservations, a few opinions, lots of questions, and a degree of objectivity. Very soon it became apparent that the CAG staff was manned by highly-qualified, dedicated professionals. The same was true of the squadrons. The squadrons were mirrors of their active counterparts. They were operating the same type equipment, met deployment schedules for weapons training and carrier qualification, and established standards of excellence that were hard to believe. They expected to accomplish these goals because these squadrons were manned by outstandingly capable individuals, all with at least one fleet tour under their belts. Competition to participate in this program and gain a seat in one of these squadrons was extremely keen. It was understood, if you did not participate and perform at the established squadron's standards, you were history. There were two types of people conspicuously missing from these squadrons—nuggets and no-loads. Soon my confidence started to rise, and day by day I became more impressed with the Reserve wing concept. I also realized that there was hard work ahead if I was to measure up to the expectations the wing had of me.

Thus began a tour of duty that would provide me with some of the fondest memories of a 32-year naval career.

There were numerous weapons deployments and carrier evolutions, and all were exciting and fun, but the two evolutions that stand out as the most unusual (and probably the most memorable) occurred aboard *Nimitz* (CVN-68) and on a TransLant from Europe to the United States.

CVWR-20 flew aboard *Nimitz* for carrier qualifications. Brian Compton was the skipper and, after I landed, I went up to the bridge for a courtesy call, as well as to thank him for the deck time. He asked if I would be doing any flying. I explained rather sheepishly that I wanted to qual in three types, the A-7, F-4, and RF-8. He gave me a side-glance, smiled, and told me to go crawl into an airplane. Three days later the air wing was fully qualified. I had bagged 42 arrested landings, a qual in all three aircraft, and had achieved a personal milestone of 1,000 carrier arrested landings. As they say in old Milwaukee—it don't get much better than that! That weekend when *Nimitz* pulled into Norfolk, there was a wardroom party at the Breezy Point O' Club. I sent the Captain a case of champagne on behalf of the wing, with our thanks.

The event that had all the suspense of a TV thriller occurred on a flight that was supposed to be routine and so well planned that *nothing* could go wrong. The truth is, almost *everything* went wrong, and the mysteries of the Bermuda Triangle took on a whole new meaning to me.

Our TransLant saga started when the wing was tasked to provide pathfinder service for a flight of 16 Marine A-6Es on a TransLant from Norway to MCAS Beaufort, SC. I decided to go along on the trip since I'd never made a TransLant, and also to build up flight time in the A-3. CDR Red Wynn was skipper of the A-3 squadron and was going as the TransLant project head.

After an uneventful flight to England, we met up with the Marine A-6s and prepared for the long trip back, with overnight stops planned in the Azores and Bermuda.

The flight from England went well; however, we were delayed in the Azores for three days while a hurricane moved up the Atlantic across our intended flight path. Finally we covey-launched on our way to Bermuda; one A-3 and 16 A-6s. The A-3's primary mission was as pathfinder. However, if necessary, it could provide emergency tanking. The mission profile required a rendezvous with a Marine C-130 tanker 600 miles west of the Azores for refueling of all aircraft and then continue the TransLant to Bermuda. No further fueling was planned for the six-hour flight, since the A-6s would still have a minimum of 2,000 lbs. (20-30 minutes) of fuel remaining upon arrival at Bermuda. As we winged westward it seemed like the flight was progressing normally. Our navigator, a seasoned veteran with prior TransLant experience, was confident that we were on track. Red Wynn and I took turns flying the A-3. At the five-and a-half hour point we made our first call to Bermuda Center on our VHF and UHF radios. No answer. Not a big deal, we would try again later.

At the six-hour point we still had no response to our call to Center. Now I *was* concerned! Now it *was* a big deal! We tried all the radios on every available frequency, and actually made contact with New York Center on HF. New York tried to assist us with a telephone call to Bermuda, but Bermuda could neither hear nor transmit to us. Strange, but true. At that point the A-6s were approaching bingo fuel—the point where they *had* to refuel.

By then the cockpit tension could have been cut with a knife. We were all aware that we should have had visual contact with Bermuda by that time, but from horizon to horizon there was nothing but water—and still no radio contact.

I will always remember looking down at the churning Atlantic Ocean, knowing that long term survival would be very unlikely in the cold October water if we were to go down. Strangely, even with all the critical thoughts racing through my mind, my foremost thought was that the 17 aircraft I was responsible for were flying smack in the middle of the infamous Bermuda Triangle. It was almost surreal.

As we slowed down to commence emergency tanking, the A-6 Intruders slid in behind us. The crew chief activated the deploy switch to stream the refueling drogue. Incredibly, it would not deploy from the stowed position! The crew chief crawled back into the bomb bay, but found nothing indicating a malfunction. 'Tense' no longer described the atmosphere in our cockpit. Not only were we lost, but now I had to tell the A-6 guys that we couldn't provide any refueling, either. My routine TransLant flight had become a flying can of worms.

Frantically, as Red and I tried to decide (guess, would be a better word) which way to turn, one of the A-6s radioed that they had a fix on a radio transmission 135 degrees to our left. We turned in that direction, and in about ten minutes rendezvoused with a Marine C-130 tanker, sent out 150 miles along our 'intended' flight path *just in case* something went wrong. A precaution that may have saved 16 A-6Es.

Later, now safely on the ground in Bermuda and after profuse apology to the Marines (who were great!), we went back and recreated the flight from lift-off to landing. We replotted headings, wind, variation, and considered anything else that could have contributed to driving us more than 100 miles north of our intended course. The answer was not forthcoming, and remains a mystery to this day.

The frustrations of that flight haunted me for many moons, and I still get a chill when I consider what our fate may have been if one of the Marine A-6 crews had not heard the C-130's call and reacted so quickly to get the DF fix on his position. The saga probably would have ended much differently.

Stories about the Bermuda Triangle took on a whole new meaning. I almost feel as if I have had some of the same thoughts of other pilots who ventured into the Triangle but were less fortunate and never returned.

Finally, readiness continued to be the target of leadership in the Naval Air Reserve in the years since the '70s. Today there is no question mark behind the capability to execute their missions. They are a ready force.

During my tour as CAG in the Naval Reserve, my life was blessed and enriched by many special people. It was an honor to know and to serve with them.

CDR J.D. "BEAR" TAYLOR, USN (RADM Ret)
CAG 3 17 May 76-27 Aug 77
USS *Saratoga* (CV-60)

Entering the Navy during 1955 via the Naval Aviation Cadet (NavCad) program, Jeremy D. "Bear" Taylor completed flight training in May 1957 and began a life-long career as a light attack tailhooker. He served tours in four light attack squadrons, while flying A-4 Skyhawks and A-7 Corsairs from 13 different aircraft carriers while compiling 6,000 flight hours. He flew more than 200 combat missions over Vietnam while assigned to VA-113 in USS Enterprise *(CVA(N)-65). Bear flew more than 50 different types and models of aircraft, including ten helicopters, and is one of the Navy's select aviators who has logged better than 1,000 carrier arrested landings.*

RADM Taylor's five command tours include VA-46 (1973-74), CVW-3 (1976-77), Kalamazoo *(AOR-6) (1980-81),* Coral Sea *(CV-43) (1983-84), and Light Attack Wing Pacific (1987-88). He served several staff tours and five assignments on CNO's staff. Bear also served as strike plans officer on the staff of Commander, Attack Carrier Striking Force, during the Vietnam War. In addition to his staff assignments, Bear also is a graduate of the Naval Postgraduate School, Army War College, and Senior Officers' Ship Material Readiness (Engineering) course. During one of his CNO staff tours Bear was heavily involved in the study and implementation of the "Super CAG" program, which brought about sweeping changes in the Navy's carrier air wing system.*

RADM Taylor ended his 36-year naval career 1 Mar 1992 and currently resides in Ogden, UT, where he is an independent defense, leadership, and management consultant. He recently has become manager of the Ogden airport.

One of the Navy's more colorful CAGs, Bear discusses some of his personal experiences:

A life-long light attack tailhooker, "Bear" Taylor (*center*) was one of the Navy's more colorful CAGs. He was heavily involved in the study and implementation of the "Super CAG" program that brought about sweeping changes to the Navy's carrier air wing program. USN photo, courtesy RADM J.D. Taylor, USN(Ret)

"When I relieved Cliff Johns as CAG 3 in *Saratoga* 17 May 1976, it was already a great air wing by any standard. I had to look hard to find an area for improvement. Boarding rates and off-center engagements looked like a couple of areas where a little command attention would pay dividends. So, whenever I got the chance to look the troops in the eye I'd put a little emphasis on the importance of getting aboard on the first pass and on centerline. *Saratoga* skipper Bob Dunn was all for this. The way to reduce into the wind time and garner the most time for respots is to get the launch and recovery wrapped up as quickly as is safely possible. That means get aboard on the first pass. The impact of off-center engagements is felt in many areas. First, if an errant aviator is 15 feet to the right or left of centerline, the arresting wire must be pulled. The really bad lineups can cause the whole cable arrangement to be taken out of commission, including the purchase cable. This process means fewer wires available for the rest of the day's operations, which parlays into more hook skip bolters and sharply degraded operations. There are other costs. Every cross-deck pendant is a long labor of love by the greenshirts of the arresting gear team. Pulling and tossing cross-deck pendants puts them to work on manufacturing new ones. This valuable time detracts from other preventative maintenance or equally important time on liberty or home with families. Another cost—taxpayer dollars. In 1976 a pendant cost about $1,200 in materials. So, landing on the centerline is mandatory. To encourage the 110 pilots in the air wing to work harder I backed up my jawboning with a requirement that the next intrepid aviator who made a second off-center (by 12 feet or more) engagement would be required to carry the potted end of a pendant everywhere he went (except flying) until somebody else relieved him of the 30-pound badge of mediocrity. I don't know if it was the jawboning or the symbol that did the job, but I do know that off-center engagements became rarities in CVW-3.

The boarding rate also got better. Everybody knew I was counting bolters, and it was my standard reminder in every briefing to tell the guy who bolters to 'turn to 120 degrees and don't come home,' or, 'if you can't get aboard on your first pass—go to your stateroom and suck your thumb.' Of course, this attitude put me at great risk. Everybody was waiting for CAG to bolter. The pressure was on. The months and the landings rolled by. The air wing boarding rate improved to 96 percent during the day and 92 percent at night. CVW-3 completed the 1976 cruise and was ready to start the 1977 cruise in March. Came the ORE and, of course, I put high emphasis on the boarding rate. The air wing turned in a flawless performance, and it was well into the third day of the exercise before *we* got the first bolter—CAG. There wasn't anybody on the ship who wasn't waiting to rattle CAG's cage. 'Hey, CAG, how come you didn't turn to 120 and keep going?' I heard from one JO as I made my way back to the ready room. 'Hey, CAG, when are you going to your stateroom to suck your thumb?' said another. My balloon was popped after 88 landings. But, it was well worth it. All nine squadrons in the wing were graded 'outstanding' in the ORE, and the wing continued an accident-free skein that extended for the last 15 months of my CAG tour and well over a year into my relief Pat Kober's tour. It was a great group of tailhooking warriors.

The CAG job is, was, and always will be the best job in the world. Nine squadrons commanded by the best and brightest, composed of more than 250 officers with better than 2,400 troops who want to do what they have to do—make 'the job' pure pleasure. On top of that comes access to more than 80 aircraft of seven varieties—magnificent machines that cry out for mastery and provide incomparable opportunities to 'touch the face of God.' Not only is the CAG in charge of a vital element of the national defense organization, he is also doing more of what he's done best for the first 40 years of his life. He is 'the old man' among teammates who are young enough to be his sons. Consequently, he uses wisdom and experience to remain competitive with the younger, stronger, faster, and smarter charges who are there to follow his lead. The CAG is the last great job where the more senior you get, the more in front you go. Proportionally, we lost more CAGs in Vietnam in combat than any other rank. Why? Because only in tactical aviation does the leader lead from in front where the hazards, as well as responsibilities, are the greatest. And the CAG—at least the good ones—will always be found at the tip of his mighty sword—a carrier air wing."

CDR JOHN J. MAZACH, USN (VADM Ret)
CAG 3 29 Jul 82-21 Apr 84
USS *John F. Kennedy (CV-67)*

Born 10 Jan 1944 in Coffeyville, KS, RADM John Mazach graduated from Vanderbilt University in 1966 and was designated naval aviator in October 1967. His first operational assignment was as one of the original members of VA-87 after replacement pilot training in the A-7B Corsair II.

In December 1970 John was assigned to the Bureau of Naval Military Personnel in Washington, D.C., followed by another sea duty tour in 1973 with VA-105 in USS Saratoga *(CVA-60).*

After assignment to the Air Command and Staff College, Maxwell AFB, AL, in June 1976, Mazach went to VA-15 as executive officer in December 1977 and assumed command of the "Valions" in April 1979.

Back in Washington in June 1980, he was assigned to the Naval Personnel Command. On 29 July 1982 he assumed command of Carrier Air Wing Three, where he served until April 1984 when he returned to shore duty as air wing training officer on the staff of ComNavAirLant. In June 1986 he assumed command of Seattle *(AOE-3) until April 1988 when he became chief of staff for Cruiser Destroyer Group Eight.*

In October 1989 CAPT Mazach assumed command of America *(CV-66) and deployed in December 1990 to Operation Desert Shield/Desert Storm before being assigned in June 1992 as Deputy Director, Plans and Policy Directorate, Headquarters, U.S. European Command. Departing Germany in June 1993, RADM Mazach assumed command of Carrier Group Two and interim Commander Operational Test and Evaluation Force. Following a CNO staff tour, he was promoted to vice admiral and served as Commander, Naval Air Force, U.S. Atlantic Fleet until his retirement in 1999.*

CAG Mazach's CVW-3 deployment came at the height of the U.S. Navy's carrier involvement in the seemingly never-ending string of Lebanon crises.

"The 1983 JFK/CVW-3 deployment was one of the most unusual of 12 deployments I have made. Although not much longer than average, it was completely unstructured as to what the operating or port schedules would be. Work-ups were normal in every respect and, at deployment date, the wing was at a very high state of readiness. That readiness level is a point to which I will return shortly. The wing was also a little 'different' as, just before work-ups started, we were informed that we would deploy without A-7s and essentially become a Grumman air wing with two A-6E squadrons: VA-85 with all A-6Es, and VA-75 with A-6Es and KA-6Ds, plus a full complement of F-14s, EA-6Bs, S-3s, E-2Cs, and SH-3 helos. A superb long-range, deep interdiction, all-weather mix, but a little light on daytime pinpoint accuracy (minus smart bombs, which were still in small numbers and not readily available).

John Mazach was CAG 3 during the 1983 Lebanon retaliation raid against Syrian gunners who had fired at an F-14 TARPS aircraft. USN photo courtesy RADM J.J. Mazach, USN

The actual Lebanon raid was the final in a series of false starts that began some three months before December 1983 when Kennedy was on station in the eastern Mediterranean with *Eisenhower* (CVN-69)/CVW-7 (CAG Joe Prueher). Many joint bombing raid plans existed, albeit to the same target each time, and we actually had a man-up or two only to be canceled at the last minute. Because of the uncertainty of the situation in Lebanon, schedules had been canceled, and JFK's trip through the Suez Canal in early December to join PaCom (Pacific Command) forces in the Indian Ocean had been canceled. The uncertainty of the schedule, plus the paucity of overland low-level training routes (day and night), impacted the overall readiness level of the air wing. I must say that these kinds of problems have diminished in the Mediterranean and, although improvements in training available can always be made, we are light years ahead in this area ten years later.

As to the relationships between CAG, the carrier CO and higher authority, the JFK/CVW-3 deployment was made under the existing command and control at the time, i.e., CAG worked for the carrier CO and the CO basically answered to the embarked CarGru (at this time RADM Roger Box, ComCarGru Six). ComCarGru worked for CTF 60 (RADM Jerry Tuttle, ComCarGru Two, embarked in *Independence* (CV-62). Indy/CVW-6 (CAG Ed Andrews) arrived in the Med following Grenada. CTF 60 (RADM Tuttle) worked for VADM Ed Martin who was ComSixthFlt.

Back to the Lebanon strike. Planning for a retaliatory strike against Syrian gunners who had shot at an F-14 TARPS aircraft started late evening 3 December on board *Independence* with RADM Tuttle and staff in control. I flew via SH-3 to Indy to look at the plan and make changes as needed. *We were very limited, however, in our abilities to effect change.* Targets were already identified, and the method of attack (one strike) was also not an option. The targets were mostly gun tubes, and we eventually didn't even get to pick the time over target. To make a long story short and to avoid pointing a finger at anyone in particular, ten years of progress has resulted in positive change. All of the tactical decisions are made by the tacticians. These positive changes were being made two years after the Lebanon strike when the Libyan strike took place. The chain of command was specific, but did not pull all the strings from higher headquarters.

The lack of flexibility, particularly with regard to timing, caused many other pertebations to the Lebanon strike, including bomb loads and aircraft availability. Sacrificing the ability to provide the right load on the right airplane has a negative impact on any strike plan.

The tactics employed on the strike and their execution were sound in spite of many cries (that) 'Vietnam tactics' would never work. Aircrews were professional in every sense of the word. Targets were almost impossible to see, partially because of the sun angle, which was encountered by being forced to execute at first light. Lower altitudes and the element of surprise were not an option during the planning and execution of this strike since time on target and target selection were dictated by Washington.

The idea of a political 'tit for tat' statement being made by military action has also proven to be a poor employment of forces and has, thus far, been avoided since. I hope this does not change in the near future.

The fact that two tactical jets were lost (one A-7E with CAG 6 aboard and one A-6E from CVW-3) is clearly proof of the above discussion points. The loss of life (LT Mark Lange, pilot of the A-6E) is unconscionable.

To summarize, I think the relationships between the CAG, the carrier CO, and the battle group commander were sound in 1983, and that relationship, although altered or modified by the Super CAG program, is still sound. The 'local' chain of command was not at fault during the planning and the execution phase of the Lebanon bombing raid. Everyone's hands were tied to some extent, and we now do business smarter and with much better vision. I am personally gratified and delighted that we have come as far as we have, and can only guess that good people working hard will meet additional challenges as we implement 'From The Sea.' It's the right thing to do as we downsize and, with time, will prove very effective."

CAPT ROBERT G. "BUBBA" BRODSKY, USN(Ret)
CAG 17 23 Aug 85-23 Oct 86
USS *Saratoga* (CV-60)
CAG 9 9 June 89-13 July 90
USS *Constellation* (CV-64)

"Bubba" Brodsky is one of the few of the fortunate—a naval aviator who has served two tours as CAG. After graduation from Georgetown, Kentucky College, Bob entered the Naval Aviation Officer Candidate School at NAS Pensacola and was designated naval aviator in June 1967. His first fleet tour was with VA-37 as an attack pilot flying the A-7 Corsair II, where he made deployments both to the Med and Vietnam, flying more than 100 combat missions over Southeast Asia from USS Kitty Hawk *(CVA-63).*

Subsequent operational assignments as a junior officer included duty as an instructor pilot with VA-174, a department head tour with VA-105, and operations officer for CVW-3.

Following a 1983-1984 tour as commanding officer of VA-105, then-CDR Brodsky became CAG 17 the following year and embarked in Saratoga *during one of the most volatile periods of the unceasing crises in the Mid-East. During CVW-17's August 1985-April 1986 Med deployment, CAG Brodsky was responsible for the orchestration of CVW-17's successful intercept and forced landing of an Egyptian airliner carrying the infamous* Achille Lauro *cruiseship hijackers. Later, during that same deployment, "Bubba" led his aviators in extensive operations intercepting Libyan fighters and patrol boats in and around the Gulf of Sidra.*

Next came CAPT Brodsky's selection for the prestigious assignment as CO of the Naval Strike Warfare Center at NAS Fallon, NV. His second CAG tour came 9 June 1989 when he became Senior CAG of CVW-9 at NAS Lemoore. During this tour, CVW-9 was embarked in Constellation *for her trip "around the horn" from San Diego to undergo her Service Life Extension Program (SLEP) overhaul at the Naval Shipyard Philadelphia, PA.*

As CAPT Brodsky puts it: "Life as a CO or CAG is great...but the downside is the inevitable transition to the Pentagon that follows." His next tour was with the staff of the Chairman of the Joint Chiefs of Staff in Washington. On the joint staff, he worked political-military issues as a member of the Chairman's Staff Group, and later as Special Assistant to the Assistant JCS. In April 1993, CAPT Brodsky was ordered to the Navy Staff as director of the Navy Presentation Office. He retired 1 Jul 1994 to teach ROTC in Ocala, FL.

Another full-fledged "flying CAG," Bubba has logged more than 5,000 flight hours and 1,100 carrier landings during his eventful career. The following information was adapted with permission from Bob's article in the U.S. Naval Aviation Museum's "Foundation." Bob tells of the Achille Lauro *hijackers' intercept by CVW-17. This mission is a prime example of one of the myriad tasks placed upon carrier aviators that are not found "in the book."*

"From a purely military standpoint, it was a fairly simple mission—intercept and force down an Egyptian airliner carrying the terrorists who had hijacked the *Achille Lauro* and murdered an American citizen. Although the successful capture of the terrorists was not wrought with danger in the traditional sense—no one got shot at—there were several individual 'heroes' who performed

Bob Brodsky has served two CAG tours and was a key player in the success of the intercept and capture of the Achille Lauro hijackers. He flew the A-7 Corsair II for more than 20 years. Photo courtesy CAPT R.G. Brodsky, USN

magnificently in the face of a complex politico-military tasking that had numerous opportunities for failure.

The *Achille Lauro*, a 633-foot Italian luxury liner, had left Alexandria, Egypt, on the morning of 7 Oct 1985 after dropping off most of its 750 passengers for a bus tour of the Pyramids. The ship was scheduled to meet these same passengers that evening in Port Said before continuing on to a port visit in Ashdod, Israel. Remaining on board were a hundred or so passengers, most of them elderly, who had decided to skip the rigors of the tour.

Also on board were Palestine Liberation Front (PLF) terrorists, in the guise of tourists, armed with automatic weapons, hand grenades, and explosives. They seized control of the ship prior to docking at Port Said and ordered the ship's captain at gunpoint to sail for Tartus, Syria.

In a fit of rage following refusal by the Syrian government to allow the ship to dock, the hijackers senselessly murdered 69-year-old Leon Klinghoffer, a retired Jewish-American who was confined to a wheelchair. Klinghoffer's body was callously thrown overboard.

Syria continued to refuse to allow the ship to dock, and another American victim was selected from among the passengers. Before they would kill the second passenger, however, the terrorists received a radio message from the PLF leadership directing them to head back to Port Said.

The government of Egypt and the PLF leadership ultimately agreed to freedom for passengers and terrorists alike after the *Achille Lauro* docked in Port Said. Despite U.S. objections, the terrorists had been promised safe passage—a solution acceptable to the Egyptians and Italians partly as a result of the *Achille Lauro* captain's ship-to-shore communication (made with a gun at his head) that 'everybody is in good health.' It wasn't until he personally boarded the ship that evening that the U.S. Ambassador to Egypt, Nicholas Veliotes, found out about Klinghoffer's murder. He immediately demanded that the Egyptian government 'prosecute the sons of bitches.' While this (attitude) was to be widely echoed in world opinion, the Egyptians claimed it was too late—the terrorists had already left the country.

Although heavily involved in conducting a major NATO exercise in the Central Mediterranean, RADM David E. Jeremiah, Commander Task Force 60 embarked in *Saratoga*, had watched events unfold from the beginning. Actual reconnaissance of the *Achille Lauro* was considered risky since any visual sighting of American tactical aircraft could place the hostages in jeopardy. Nonetheless, a wide spectrum of possible contingency operations was on everyone's mind during the tense period following the hijacking. When word was eventually received that the hostages had been freed and the terrorists granted safe passage, a sigh of relief was felt, and *Saratoga* directed its energy to other events.

After completing the exercise, *Saratoga*, commanded by CAPT Jerry Unruh, was steaming northward through the Adriatic Sea toward a port call the next morning in Dubrovnik, Yugoslavia. Aircraft and aircrew had been postured in an 'Alert 60' status, when an unusual 'secure voice' telephone query came from Sixth Fleet headquarters in Gaeta, Italy, asking for the exact location of each ship in the battle group. Sitting idly in his war room, Admiral Jeremiah joked with his watch officer that the magnets holding the ships must have fallen off the Sixth Fleet staff plot board.

What had actually transpired was some clever intelligence work which indicated to the National Security Council (NSC) staff that the hijackers were still in Egypt and about to be transported out on an Egypt Air 737 jetliner. They were, at that moment, at the Al Maza Air Base near Cairo, waiting to be flown to Tunisia.

Credit for the idea to attempt to capture the hijackers should go to a Navy captain on the NSC staff who reportedly suggested, 'Why don't we pull a Yamamoto on these guys?'—the name, of course, being in reference to the Japanese admiral intercepted by American fighters over the Pacific during WWII.

Aboard *Saratoga* orders were suddenly received from Sixth Fleet to reverse course and to launch the Alert CAP (Combat Air Patrol). 'Launch the Alert CAP' was immediately sounded over the ship's loudspeaker. Even with the lack of amplifying information, a sense of urgency flashed through the ship. Despite the official 'Alert 60' posture, two VF-74 F-14A Tomcats and a VAW-125 E-2C Hawkeye were airborne in just 22 minutes. The answer to 'what for?' would come only moments later.

Admiral Jeremiah anxiously pondered the situation in the war room, the secure phone in his hand waiting for another call from Sixth Fleet that would tell us what was happening. Before the call came his face suddenly lit up: 'The hijackers—I'll bet we're going after them...'

Within minutes, confirmation of our mission came from Sixth Fleet. The terrorists were in Cairo attempting to escape by airliner to Tunisia. The exact takeoff time (or had they already taken off?), the route, altitude the hijackers would fly, and the tail number of their aircraft were all initially unanswered questions. Our tasking had all the markings of the proverbial 'find a needle in a haystack' drill. How to communicate with the airliner, once intercepted, and how to entice the crew to divert to the NATO base at Sigonella, Sicily, would be the most difficult problems. The first order of business was to assign CAP stations and brief what little information was available to the already airborne F-14s and E-2. It would be an understatement to say *Saratoga's* combat information center became a beehive of activity, as leadership from the flag staff, the carrier, and the air wing all converged to work the problem. The decision was made to launch additional VF-103 and VF-74 *Tomcats*, while the E-2 ready room was alerted to man an additional aircraft. Ordnance loads on the F-14s were modified in some instances to include tracer ammunition in their 20mm cannons in case warning shots became necessary. VA-85 KA-6D tankers were put on alert to cover airborne fuel requirements of the F-14s.

Critical to mission success, of course, would be C^3 (command, control, and communications) provided by the E-2C. Adequately explaining mission details and all the possible exigencies to the already airborne E-2 mission commander proved extremely difficult, however.

CDR Ralph Zia, commanding officer of the E-2 squadron, widely respected for his airborne professionalism, had been kept abreast of unfolding events almost from the beginning. As much as anyone, he had a clear picture of what we were attempting to do.

When directed, CDR Zia personally manned the second E-2 and was launched to assume airborne control of the operation. There wasn't a better man for the job. His ability to quickly assess the situation and ad lib solutions to each hiccup in the evolution was the key to success.

After initially being assigned CAP stations south of the Greek island of Crete, the E-2 vectored the Tomcats to intercept all contacts that fit the profile of an airliner following the airways between Egypt and Tunisia. We were never given a reliable takeoff time, so there was no assurance that we had set up CAP stations in time. Another possibility was that the Egyptian aircraft may have been routed through Libyan airspace. All *Saratoga's* aircraft could do was throw out the net, intercept everything that flew into it, and hope we could sort out the Egypt Air 737 if it came our way.

There was no lack of activity. Despite the pitch-black night, the Tomcats, with their running lights extinguished, made two hair-raising intercepts of large transports, themselves with lights out. While a radar rendezvous on an airliner-size target is a relatively simple evolution, a night join-up of an aircraft with no lights is not. The F-14s crept up slowly from below and aft until they had the transports silhouetted against the limited starlight. Once joined, identification of the transports could be made only by shining hand-held flashlights.

As luck would have it, we didn't have to wait long for our real prey. On the fourth intercept of the evening, two F-14s pulled up behind an airliner and radioed the markings and tail number—2843—back to *Saratoga*. That was our man! The Tomcats were ordered to remain in position, keeping their lights out so the Egyptian crew and their terrorist cargo would have no idea they were under escort.

Behind the scenes in Washington, the State Department had already asked Tunisia and other friendly littoral Mediterranean nations to deny landing rights to the terrorists. CDR Zia in 'Tigertail 603' dialed in the Tunis air traffic controllers' frequencies and listened in on the Egypt Air pilot's radio communications as they requested and were denied landing clearance in Tunis. The pilot then communicated with the airport in Athens, but was again denied. Knowing the airliner had no choice but to return to Cairo, it was time for the Hawkeye crew to take charge.

'2843, this is Tigertail 603, over,' CDR Zia radioed.

After several more attempts at communications, the Egypt Air pilot finally acknowledged.

Zia continued, 'Egypt Air 2843, you are being escorted by two F-14s. You are directed proceed to and land immediately at Sigonella, Sicily. Over.'

The Egyptian pilot was dumbfounded. 'Say again. Who is calling?'

Allowing the pilot to believe he was talking with one of the F-14 pilots, Zia repeated, 'This is Tigertail 603. I advise, you are directed to land immediately. Proceed immediately to Sigonella, Sicily. You are being escorted by two F-14 interceptor aircraft. Vector 280 for Sigonella, Sicily. Over.'

After the order was forcefully repeated one more time, advice was passed from *Saratoga* to have the F-14s turn on their external lights. The Tomcat crews watched as the excited Egyptian crew (and perhaps the terrorists) ran to either side of the airliner to peer out the passenger windows at the U.S. fighters close aboard on either wingtip.

Zia now had the Egypt Air pilot's undivided attention. He acknowledged his instructions—'Turning right, heading 280'—while at the same time asking a passing airliner to relay news of the intercept back to Cairo. Startled by the close proximity of the Navy fighters, the nervous Egyptian pilot again came on the radio: 'I'm saying you are too close. I'm following your orders. Don't be too close. Please.'

'Okay, we'll move away a little bit,' Zia responded, still sounding authoritatively as if he were one of the interceptors alongside. The range to Sigonella and then back to *Saratoga* was significant, so CDR Zia and his crew vectored four fresh fighters to assume the escort role while simultaneously coordinating airborne tanking of all the Tomcats.

Although *Saratoga* had been informed that the Italian government had approved the landing at Sigonella, the Italian air traffic controllers had other ideas—vectoring the 737 to land at the civilian field in nearby Catania. The senior F-14 commander, VF-74's CO, CDR Ken Burgess, then got involved, requesting a vector for the flight to Sigonella. His request was denied. He asked again, more forcefully, but was repeatedly denied. The F-14 lead then chose his own heading, gave it to the Egyptian pilot, and directed a descent for the 737 and its four escorts into Sigonella.

After a radio-frequency shift to the Sigonella control tower, the escorting F-14 commander requested permission for the 737 to land. When the Italian air controller refused, the Tomcat declared a low fuel emergency and indicated the requirement for an immediate landing.

The unplanned deviation to his original flight plan, inappropriate approach charts for Sigonella, a dark night with 10,000-foot Mount Etna looming just miles to the north, and four F-14 Tomcats on his wing, all proved unnerving to the Egyptian pilot as he attempted to land. After initially settling well below glideslope he decided to wave off his first pass and executed a wide go-around in the dark, with the Tomcats in tow. Everyone breathed easier when he landed successfully on the second pass. The aircraft was immediately surrounded by American troops, who were, in turn, surrounded by Italian troops—but that's another story.

The fact that the terrorists were taken into Italian custody did little to diminish the elation aboard *Saratoga*. (Abu Abbas was flown to Yugoslavia where he enjoyed diplomatic immunity as a member of the PLO Executive Council.) The crew received a congratulatory phone call from President Reagan. Western news media proclaimed a long-awaited victory over terrorism, while appreciative messages and telegrams poured in. Thousands of letters from school children, literally tons of cookies and T-shirts for every sailor aboard were received in the days that followed.

The real reward, however (one every man aboard *Saratoga* could take with him forever), was the knowledge that they had helped bring these terrorists and cold-blooded murderers of an American citizen to justice. In doing so *Saratoga* helped lend credence to President Reagan's declaration to would-be terrorists: 'You can run, but you can't hide.'"

CAPT BYRON L. DUFF, USN(Ret)
CAG 13 29 Nov 85-16 Jan 87
USS *Coral Sea* (CV-43)
CAG 9 21 Oct 87-9 Jun 89
USS *Nimitz* (CVN-68)

CAPT Byron L. Duff is a native of Illinois, where he graduated from the University of Illinois in January 1967 with a Bachelor of Science in Aeronautical Engineering. He entered the Aviation Officer Candidate Program (AOC) and was designated naval aviator 2 Aug 1968.

Following A-7 Corsair II replacement pilot training, "By" joined the "Valions" of VA-15 and completed two Mediterranean Sea deployments in USS Franklin D. Roosevelt *(CVA-42). He was then selected to attend the Air Force Aerospace and Research Pilot School at Edwards AFB, CA. After graduation in August 1971, Duff reported to NATC Patuxent River, MD, for duty involving test and evaluation of naval aircraft.*

In 1975 By served an exchange tour with Number Three Squadron, Royal Air Force, flying the VSTOL Harrier while deployed to West Germany. He next returned to the U.S. and rejoined VA-15 as executive officer. In November 1982 he became commanding officer of the Valions embarked in Independence *(CV-62).*

One of the lucky few, Byron Duff had the good fortune to serve two CAG tours. USN photo, courtesy CAPT Byron L. Duff USN (Ret)

As one of the early F/A-18 Hornet drivers, then-CDR Duff reported to ComNavAirLant in June 1984 as F/A-18 Coordinator and Hornet Air Wing Training Officer. After detaching in May 1985 and completing Hornet/Intruder training, he assumed command 29 November of Carrier Air Wing 13 embarked in Coral Sea *deployed in the Med.*

CAPT Duff was selected for a second tour as CAG 9 in October 1987, completing a Western Pacific and Indian Ocean deployment in Nimitz. *Upon his relief 9 Jun 1989, he then assumed duties as Deputy Chief of Staff Operations, U.S. Naval Forces, Europe, the position he held until he returned to the States to retire in Cobbs Creek, VA.*

Byron Duff accumulated more than 4,500 hours of flight time in better than 20 aircraft types and is a member of the prestigious "Grand Club," having logged in excess of 1,000 carrier arrested landings. His combat initiation was in Grenada as skipper of VA-15, and then his tour as CAG 13 brought him into the volatile Mid-East arena. By gives us his view of what being a CAG is all about.

"Naval aviation has always lived in reality what existed for others only in their fantasies. Fulfilling the warrior image, hair-on-fire, fearless in full-blower, is not what I'm talking about. The core was, and is, in my opinion, the incredible energy of human endeavor that comes into being as the ship turns into the wind, the aircraft go into tension, and launch to meet the challenge of the day. At the pointed end of that sea-born lance of purpose is the highest of highs. The magic of the people, the mission, the ship and aircraft, is what being CAG is all about...there is nothing else like it in the world.

And to think I was honored with two tours—first as CAG 13 on the East Coast, immediately followed by CAG 9 out of NAS Lemoore, CA. From November 1985 to June 1989 it was CAG Duff—my wife and sons even started to call me CAG. It was great, but I digress.

Being an air wing commander means many things, the most important of which is leadership; specifically leadership in the air. My foundation was laid during my first tour on the wing of LCDR Ed "EOR" Rhodes (CAPT Ret). He had honed his teeth as a "Spad" driver in Vietnam and became a friend and teacher. I'll never forget that dark night (they all are) in the Med when Ed had an electrical failure in his A-7. As his LTJG wingman, it fell on my shoulders to bring him down through the soup. Marshal and penetration were right on the money—except for one detail. I was following the tail of the needle. The F-4 launching toward me from the pointed end in full burner was my first clue. Ed held on, and I succeeded on the second try to get him to a centered ball start. For Ed's part, he laughed in his silent way when he found out what had happened and used the incident as a training tool. There never was any incrimination or pressure. I was his wingman and that was that; he shielded me from the wrath of the higher ups. The team concept was firmly in place.

With less than 900 hours of flight time, I was selected to attend the USAF Research Pilot School at Edwards AFB. It was there, as an immature LT, that respect for every type of aircraft and my own limitations was driven home. On my first ride in an F-104, my brain was still operating at SLUF speeds. It wasn't pretty. I wasn't Chuck Yeager. My learning curve was steep enough to graduate and stay alive as a test pilot—a conservative test pilot.

The next plank was laid as an exchange pilot with the Royal Air Force flying the Harrier. All former limits, personal or regulation, were history. From field ops to treetop combat, it was the punchiest, most aggressive flying I had ever done. With the 'no rules' environment in the air came debriefs where there were 'no survivors.' A mission without mistakes simply didn't happen. Constantly flying on the edge required regular doses of humility. A sense of measured aggressiveness was in place.

CAG Ed Andrews was the next adventure. As CAG 6 in *Independence*, 'Hunyak' was bigger than life. Everything he was, I wasn't. I admired him; he was a great CAG. He led from the first day he arrived at my squadron's (VA-15) 'admat' in Athens, to Grenada, then to dancing with SA-7s over Beirut. The Hunyak stories are numerous and without embellishment—there is no need for exaggeration.

For example, after being towed to the cat with a nosegear steering failure, CAG Andrews launched into the skies of Lebanon. In the face of a wall of missiles and AAA fire, he dropped his bombs on target. Three or four breaks followed (we were losing altitude fast). An A-6 off *Kennedy* (CV-67) blew up below us as CAG took a missile up the tail. He ejected over the bay (his third—that is another story), and landed next to a fishing boat. The wooden skiff contained an old man and his grandson. CAG Andrews worked his way over to the boat, placed his .357 Magnum on the gunnels and said, 'American.' The old man, seeing this one-eyebrowed, swarthy giant of a man, answered by pulling a cross out of his shirt and saying in his best English, 'Christian.' They hit it off immediately. He was rescued, and I had an example of gutsy courage that exceeded previous precepts.

The Lebanon raid also left me a greater appreciation for the ship and crew, the value of good intelligence, and importance of all the air wing's parts working smoothly together. Screening for CAG gave me the opportunity to put all of the above together into a cohesive package. I joined up with Air Wing Thirteen (not an unlucky number) and embarked in CV-43, 'The Ageless Warrior,' in November 1985.

CAPT Phil Gay had formed the wing from a paper concept. Four F/A-18 squadrons (two Navy, two Marine), a big A-6 unit, four E-2Cs, and a six-plane helicopter squadron. A Whale det rounded out the package. Matched with *Coral Sea,* the non-standard air wing was designed to give the aging carrier a few more years of life. From my perspective as the new air wing commander, it was a perfect match. Gadhafy and his so-called 'Line of Death' gave the ship/air wing team a chance to prove it.

Within weeks of assuming command, the Mediterranean began to heat up. It was the middle of winter, but in the war room planning was hot and heavy. From providing concepts and plans for freedom of navigation ops in the Gulf of Sidra to counter SA-5 missile tactics, the air wing planners provided more top secret/coded messages for transmittal up the line than I had seen in my entire career to that point. It was fly and plan, with bits of sleep and eat thrown in for grins. With specific targeting, it began to look and feel like Lebanon all over again. With my bag of lessons learned in hand, in I plunged. I wasn't alone—at least from a theater aspect.

From CinCEur to the deck plates, there was a single-minded determination to cover every possibility. If I had any doubts, several eyeball-to-eyeball give-and-takes with four stars asking, 'Now, are you sure there's nothing else you need?' wiped them away. Sure, there were detractors. At times (more than I'd like to admit), the situation became little more than a 'fur ball.' We shot ourselves in the foot on more than one occasion, many times with the help of interested parties inside the beltway. For the majority of the snafus, Navy and Marine personnel provided the energy and desire to do what others saw as the impossible. It never ceased to amaze me what was accomplished through shear willpower and dedication to the ship/air wing team and mission.

Love of ship, respect for each other, a clear sense of purpose, along with crew bonding combined to mature into an entity of living greatness. Borrowing from the commercial, it was like being in tune to the 'Heartbeat of America.' My pulse races even today with the memories. From *Coral Sea's* skipper, CAPT Bob Ferguson, to the men who manned her eight boilers, all cylinders fired when called upon. During the months of meeting the Libyan air threat over the Gulf of Sidra, five days of dueling with patrol boats and SA-5s and the strike on the night of 14 April 1986, the troops never missed a beat.

You can easily understand my elation at screening for 'Super CAG.' I didn't really know what the 'Super' part was all about; what I did know was that being CAG was super. Many ideas come to fruition simply because the individuals involved 'make it so.' In the face of many doubters, that's what happened. Teamed with *Nimitz* (CAPT Brent Bennitt), a Deputy CAG (CDR Pat 'Killer' Kilkenny), who became a friend for life and a superb air wing, I couldn't lose. My second tour as CAG was equal to the first, even without the combat.

I guess that is what being CAG is all about—ships and airplanes, leadership and mission accomplishment, and, most importantly of all, the men and women with whom you work who repeatedly succeed in accomplishing the impossible. There is no other high like it.

Facing a future with the loss of the above is one of the reasons that I left the Navy; my days of afterburners and being close to the men were over. It was time to get out of the way. Do I miss it? You bet! Living under the turn point of a low-level route is painful—in a nostalgic way. When the jets roar overhead on a full-moon night, I know there is a CAG up there having the time of his life."

CAPT JAY CAMPBELL, (RADM USN)
CAG 2 21 Feb 90-26 Jul 91
USS *Ranger* (CV-61)

Fighter pilot Jay A. "Rabbit" Campbell is a product of the NROTC program, entering active service in 1967 after graduation from UCLA. He was designated naval aviator 3 Dec 1968.

Upon completion of F-8 Crusader replacement pilot training at VF-124, Rabbit made two seven-month combat cruises to Vietnam with VF-191 in USS Oriskany *(CVA-34). He flew more than 140 combat missions in the F-8J during a period of 18 months.*

Jay's first shore duty tour came in January 1972 when he served as an F-8 instructor pilot and replacement squadron training officer. He transitioned to the F-14 Tomcat in August 1974 and joined VF-14 at NAS Oceana, VA, in March 1975, deploying to the Mediterranean Sea and North Atlantic in John F. Kennedy *(CV-67). His next assignment was with VX-4 at NAS Pt. Mugu, CA, in October 1977 as F-14A operational test director.*

In 1980 he reported to VF-213 at NAS Miramar, CA, and deployed in America *(CV-66) as the "Blacklions" maintenance and operations officer. He next became executive officer of the Navy Fighter Weapons School ("Topgun") in March 1982, where he served until December 1983. In March 1984 Rabbit became XO of VF-2, deploying in* Kitty Hawk *(CV-63) for a seven-month WestPac/ Indian Ocean cruise. He assumed command of the "Bounty Hunters" in August 1985, where he served until January 1987.*

CAPT Jay "Rabbit" Campbell, led CVW-2 during Desert Shield and Desert Storm. USN photo, courtesy Herman Schroeder

Returning to the East Coast in August 1987, Jay next commanded VF-101, the Atlantic Fleet F-14 Fleet Replacement Training Squadron. Rabbit then became CAG 2 in February 1990 and deployed in Ranger *for Operations Desert Shield and Desert Storm. CAPT Campbell led CVW-2 in the first pre-dawn strike of almost 4,300 combat sorties, which delivered nearly 4.3 million pounds of ordnance on Iraqi targets. He was selected as Tailhooker of the Year for his 1991 combat performance as CAG 2.*

Jay is a distinguished graduate of the 1992 class of the Industrial College of the Armed Forces, and in July 1992 he became Sixth Fleet Chief of Staff. In 1998 Jay was selected for promotion to rear admiral and command of Carrier Group Seven.

At this writing, RADM Campbell has accumulated more than 5,500 flight hours, including better than 3,000 hours in the F-14. He has made more than 1,200 carrier arrested landings. The following is his story of commanding CVW-2 during the Persian Gulf War, written while he was deployed in the Black Sea with the Sixth Fleet and prepared for the next "contingency operations call."

"I took command of Carrier Air Wing Two in late February 1990 from CAPT Chris 'Boomer' Wilson in a ceremony on board *Ranger* at NAS North Island. The air wing went to NAS Fallon, NV, one week later, completed a refresher CQ/ISE (Carrier Qualification/Independent Steaming Exercises) period in the first week of April, completed RefTra (Refresher Training) in April/May, and wrapped up ship/air wing training in SoCal (Southern Calif.) ops and the ATA (Advanced Training Assessment) from 31 July to 3 August. As we were demonstrating the power projection capability of the *Ranger*/CVW-2 team, Iraq massed at the border and invaded Kuwait on 2 August. From then on our training took on a new meaning, and our emphasis shifted to a new regional focus that dominated our every action during the ensuing battle group finale to our work-ups.

By the end of August, Desert Shield was in full swing, with *Independence* (CV-62) on station and CAPT Jay 'Spook' Yakeley (now RADM Spook) leading CVW-14 in CAP and joint planning. The foundation they built for Persian Gulf action grew, and back in SoCal we fine-tuned our plans based on their experience. We carried them into our September Battle Force Inport Training and the ReadiEx at-sea periods in October and November. This 'fine tune' was CVW-2's preparation for the three-day Battle Group Evaluation which would, in fact, be carried into combat exactly two months later. We had the fortune of using our critical training/readiness period to hone our skills, not for some postulated or imaginary threat, but for a very real and imminent threat, that we were surely prepared to meet and defeat.

With the air wing fully loaded aboard, CVW-2 and *Ranger* got underway from San Diego 8 Dec 1990. The winter transit of the Pacific was filled with the anticipation of the inevitable, the hope for the unlikely, the continued practice to meet any contingency. Our four-day stay for voyage repairs in Subic Bay was accompanied by the harbor filled with more warships than had gathered there since the height of the Vietnam War. It was especially meaningful for me, because CAPT Ernie 'Ratchet' Christensen (now RADM Ratchet, Ret.), CO of *Ranger*, let me have the con and bring her in to pierside at the Cubi Pt. carrier pier—a true testament to the teamwork aboard, and a great thrill for me personally.

Ranger entered the Persian Gulf through the Strait of Hormuz 15 Jan 1991. Late that afternoon, as I recall, we received the execute order for Desert Storm, and we knew we had only the following 24 hours to familiarize the aircrew with Gulf operations before we put our training into actual practice. The code name we were given for the execute was 'Wolfpack.' For all of us involved, however, none can ever recall hearing it called over any circuit as H-Hour arrived. It was also on the 14th or 15th that we got the newly-installed *Ranger* satellite TV antenna homed on CNN for the first time. So, as we began the Gulf War, we were as up-to-date on the carrier as everyone on cable at home. In honesty, as I was manning for our first strike, the second wave of that first night, RADM 'Zap' Zlatoper, Commander Battle Group Echo, and CAPT Christensen were watching Bernard Shaw on CNN as the first Tomahawks were striking their targets in downtown Baghdad.

I manned an F-14 for that first strike, as TarCAP for a strike on the naval facility at Umm Qasr. The weather in the Gulf had changed to winter—something I never expected to see—and we had to contend with mid-level clouds en route, and both fog and cloud build-ups at the coast as we went feet dry. Kuwait City was all lit up as it passed by my port wingtip, and the sky in front reflected occasional triple-A as the strike approached the IP (Initial Point). Since we were the second wave of the war, there was *no* element of surprise and, as our Iron Hand (anit-SAM) and TarCAP crossed the beach, all hell broke loose. I truly believe we hurt their air defense and early warning badly all through this and the next day. Although their gunners and SAM operators may not have been able to track us, they weren't blind yet, that was evident. There was AAA at nearly all altitudes up to about 16,000 ft., and lots of 23mm at lower altitude, coincidentally where our main body of six A-6E strikers were, apparently anticipating low ingress, since their listening posts could hear but their radars could not see. Numerous SAMs streaked into the air, some skimming the marshes and crashing harmlessly into the mud delta surrounding Bubiyon Island. On this first night, all our EW and DECM gear was lit up and ringing like a New Year's Eve celebration, with Singer Low and Singer High (SAM warnings) demanding attention to the strobes popping up. Needless to say, we all kept our aircraft moving, perhaps over-reacting in many cases, as preemptive HARMs were launched, slamming into SAM radars while EA-6B jamming blanked and confused the air tracking and warning radars.

It was no cakewalk. Everyone who flew that night, no matter on what mission, to what target, earned a place in history for one of the most successful attacks ever, in the face of severe danger. I saw the night sky in Vietnam during the Son Tay Prison Camp raid, and the beginning of Linebacker II; they did not compare with the defenses thrown up against us in those early morning, pre-dawn strikes of 17 January over Iraq.

My experiences were all positive during this 43-day war, solely because of the people in our battle group, led by Zap, who worked with a single purpose. Also, we were fortunate. Our only combat loss occurred in the evening of the 18th, during another strike in the vicinity of Umm Qasr. Again, I was flying an F-14 as TarCAP. This time, because the fog had really settled in, our A-6Es were going in to lay some precision mines to try to bottle up that part of the Iraqi navy still up in the Az Zubyar canal and at Umm Qasr. We had flown four strikes up to this point, just 30 hours into the war, mixing high/medium altitude tactics, and this was to be our last pre-planned strike at low altitude. The reaction to our ingress past the IP was much like the first strike, with noticeably less SAM warning or launch, but with a fairly heavy concentration of AAA. My guys reported that the AAA in the area of their mine-laying objective was much heavier than they anticipated, and the number four pilot was not sure how he got through unscathed. We'll never know exactly what happened, as it was night, the aircraft were separated by one minute, in fog and light rain with the AAA blossoming in windscreen and sidepanels. But number two, LTs Tom Coston and Charlie Turner, did not check in feet wet and never responded on SAR common or Guard while I remained as on-scene commander until forced to RTB (Return to Base) with low fuel. At the end of the war they were confirmed KIA, having remained with the plane when it crashed.

The air wing pressed on with a renewed dedication. It showed. Great tactics. Great hits. Long Hours. Never a complaint. By the end of the second full 24 hours, CVW-2 and *Ranger* had been dubbed the 'night team.' We actually flew for more than 76 consecutive hours before the flight deck crew got about a six-to-eight-hour break. Then *Theodore Roosevelt* (CVN-71) entered the Gulf, and we were ready to start into the night routine as part of a three-carrier battle group task force. As I recall, our first gathering of all the battle group commanders and CAGs occurred 20 January aboard *Midway* (CV-41). It was a great chance to both share our own experiences and learn what the other CAGs were doing, while getting the guidance of our Naval Component Commander, VADM Stan Arthur. Our discussions ranged from specific target objectives to 'no drop' areas and campaign objectives.

We were Task Group Zulu and specifically charged with eliminating the Iraqi sea surface threat—the destruction of the Iraqi navy. This mission was in addition to participation in the ATO (Area of Tactical Operations) sorties, though early on we committed all sorties to the process to show we were involved in a joint/combined campaign. With no commitment from Riyahd and competing priorities for Eager Anvil (the formal name of the first phase of the campaign) sorties, we devised our strategy to eliminate the Iraqi naval order of battle. This flexibility was part of the decision process for conducting the mining early on, and after that we began in earnest to account for and destroy every platform (missile patrol boats, primarily). I found I could call CAPT Lyle 'Ho Chi' Bien's (now VADM Ho Chi) guys in Riyahd in the evening, after studying afternoon TARPS (Tactical Aerial Reconnaissance Pod System) footage of harbors and bases to locate their boats. I would then lay on the tasking for an early night strike, usually (but not always) from our own deck to hit 'em before they got under way. It all came to a head during the night hours of 28 January and all of the 29th in what we later referred to as the Battle of Bubiyon. The last of the Iraqi surface combatants made a run for the Shat al Arab waterway in an attempt to dash to the relative safety of Iran's inland harbors, much like the Iraqi air force had been doing for a few days prior.

We caught them under way and destroyed all but one, which, we learned later, made it safely to Iran. It was a great combined effort, using all three air wings' aircraft, U.S./British/French helos, and everything we could muster against them. As Zap was fond of saying, 'The three most dreaded words you could hear in the Iraqi navy were, "get under way.'

I personally settled into a sort of routine, having made an agreement with Zap to continue as his strike warfare commander first and foremost, and flying basically every other day—make that night. I began flying the A-6E by the end of the first week of the war, and alternated between the 'Turkey' and Intruder for the entire war. I never did time it right! Certain in the beginning that I would get a MiG kill, I flew the first three missions in the Tomcat. After the fourth or fifth day, though, it was very apparent they were not going to continue to confront us and, in fact, as I mentioned above, they began to flee to Iran or otherwise take deep shelter. And as for a boat kill, I guess I must have let the JOs get 'em all by the time I was roamin' the northern Gulf.

So, as January came to a close, we had convened a second time with all CCG commanders and CAGs in *Midway*, moved from Eager Anvil—while establishing air superiority—to establishing sea and air supremacy, and began in earnest to prepare the battlefield for the land offensive. That preparation got us flying everywhere—beyond the KTO (Kuwaiti Theater of Ops) into central and eastern Iraq. CAG Rabbit was flying past Shaibah and Al Basrah, up to CAP station Elaine, in hopes of catching a stray Iraqi still trying to fly into Iran. As an anecdote, during the first USN strike, there was a VF-1 section from CVW-2, diverted from a pre-briefed KTO mission after already airborne. On that first flight, squadron CO, CDR Ron McElrath, and his pilot, LT Stu 'Meat' Broce, vectored into a bogie that turned out to be an Mi-8 HIP helo, got tally, and bagged it with a single 'Winder.' That victory turned out to be the only USN air-to-air kill in the Gulf. The date was 6 Feb 1991.

About then we began to see more and more tasking on the ATO for missions into 'Kill boxes.' We had heard of this tasking at our third meeting in *Midway* 2 February, and after the defense of the raid made against Kafji. We were definitely in a third phase of the campaign, even if still cleaning up the other two, with serious attention to Battlefield Air Interdiction (BAI). As the Air Force was talking about tank-plinking, I refined our A-6E night tactics to use both Mk 80 series/Rockeye and LGBs to be most effective against tanks and artillery. We saw great success with LGBs against Silkworm and small missile boat targets, and knew we could optimize hits once we confirmed we had a tank. So, in a modified stream, over about a 60-degree sector, we would bring our Intruders in, spaced by 30–60 seconds, and drop medium altitude release 'Rocks' or Mk 83s, continue outbound on that same basic heading for seven miles, then 'reset' in a teardrop, turn back into the target to get a positive FLIR ID on a tank, and put the LGB on it. I got a couple myself—at least a small contribution to the effort.

I will digress with a couple other observations along the way. My first Gulf Armed Recce flight in an A-6E took me up the coastline of Kuwait, and I was witness to Saddam's attempt to keep his promise to 'set the sea on fire.' About two or three miles off the coast he created a giant oil spill at a tanker receiving point, then ignited it. It was fairly spectacular immediately nearby, and created a sickly overcast of smoke that pitched the night into total blackness for several miles, but its biggest effect was the oil itself, still remaining after the fire was extinguished. As an environmental saboteur he would prove to be a master. He began setting fires in the Kuwaiti oil fields after the second week in February, and each night the orange points of light increased and spread, causing an eerie glow under the black-gray canopy of soot and smoke they created. If your target was anywhere near one of the fires, your FLIR was useless. And if the smoke was too heavy on roll-in or reset, there was no way for FLIR to cut through it (though the smoke was not as bad as clouds).

I managed a few four-and-a-half to five-hour CAP hops, manned the platform with the incomparable LSO, CDR John 'Bug' Roach, for a few night recoveries, spent a lot of time in the War Room and TFCC, and along the way got 23 straight night traps. We watched CNN continuously when not flying, and hosted the press for interviews to try to get our story to the public. Although we had considerable success with publicity in Desert Shield, and some limited success in the first ten days or so of Desert Storm, the press pools seemed hesitant to get trapped at sea if anything broke ashore, and it got harder to schedule them out to the CV. I heard from my family and friends that I was seen and heard on CNN and radio a few times, but a major disappointment for me was how little public recognition our Navy men got for all that they contributed to the victory. By mid-February *America* came around from the Red Sea, and we had four carriers operating in the Gulf. The carriers moved farther and farther north, to cut down cycle and transit time and, though some still thought we were in it for the long haul, we saw that things were happening fast. The Marine briefers came aboard on the 21st, I believe, and we got all pumped up about close air support as the ground offensive became more imminent. Our briefer showed the effects of months of preparation and waiting, using all the key phrases to inspire the aircrew. I left the brief charged by the term 'speed and violence.' There was nothing we wouldn't do for our guys on the ground by the time the briefing was over. In fact, however, they moved so fast once the final phase commenced at 0400 on 24 February, that we never really did perform classic close air support. The Fire Support Control Line (FSCL) swept north so quickly, we kept doing controlled BAI just to stay ahead of it.

This last few days included the 'mother of all retreats,' and the resultant 'Highway of Hell' north of Kuwait City made famous in all the weekly periodicals. My last A-6E sortie of the war was on the night of 25 February, a Kill Box BAI mission just to the west of the highway, with ingress from east to west along the north side of Kuwait Bay. My BN, LCDR Larry Clark, called the hits on the money, but we saw so much moving that night it would have been hard to miss. To show that the hop is never over until you have your LSO debrief and are sipping on a Diet Coke, just to make sure there was no post-mission letdown, the right generator crapped out inbound from our timing control point. No big deal at the time, we got our gear all reset, got the bombs on target, and that was that. But the CSD (Constant Speed Drive) was in the process of lunching itself and, by the time we were back in marshal, had made a few turns and started down on our CCA, I lost oil pressure and had to shut it down for a single engine landing. It was uneventful, but I began to wonder—CAG isn't supposed to be dragging single engine emergencies back aboard at night—he's supposed to be bagging day traps. Right?! Ah, well.

My last mission during the hostilities was the night of 27 February, on 'Bong' CAP station, in central Iraq. Flying with LT Paul Monger, who was also my RIO on the first mission of the war, we roamed all over the central section, from the Saudi border, past Tallil Airfield, up to the outskirts of Baghdad and all around for four hours. It was almost serene. A sliver of a moon when it peeked through the high broken cirrus layer, lighting the marshes of the Euphrates, and not a sign of activity on the ground beneath us until we approached Baghdad. Upon recovery I learned at 0100, 28 February, that the end was near. And at 0500 the President announced we would cease all hostilities in three hours. Just like that! At 0800 (0500Z) 28 February, the war came to an end.

We continued to fly recce and surveillance all through March and most of April. All the shooting I saw on those long CAP missions was horizontal—tracers skimming over the Iraqi surface—shooting at each other. *Ranger* remained on station to become the only carrier in the Gulf, with *America* first to leave, then *Midway* and TR. *Midway* went back to finish out her homeport duties in Yokosuka, while TR ran up to the Med and performed CAP for Provide Comfort over Northern Iraq. But, before we all scattered, in a moment of brilliance, all four carriers joined in formation, flanked by *Aegis* cruisers, for a historic picture of what may well be the last time a task force of that size is used in combat. It was a sight to behold, with the striking power of four carrier air wings crowding the flight decks, forward deployed, self sustained, ready and versatile. There is really nothing like it. You've heard the phrase, 'It's great to be king.' Baloney! It's great to be a Combat CAG!"

CAPT WILLIAM J. "Fox" FALLON, (VADM USN)
CAG 8 8 March 90-12 Aug 91
USS *Theodore Roosevelt* (CVN-71)

VADM William J. Fallon was born 30 Dec 1944 and raised in Merchantville, NJ. A graduate of Villanova University, he received his commission through the NROTC Regular Program in May 1967 and was designated naval flight officer upon completion of flight training in December of that year.

"Fox" Fallon began his naval aviation service flying with RVAH-5 in the RA-5C Vigilante. He made a combat deployment to Southeast Asia during 1969 embarked in USS Ranger *(CVA-61). Transitioning to the A-6 Intruder in 1974 he completed several attack squadron assignments, deploying to the Mediterranean, North Atlantic, and Indian Ocean. He was commanding officer of the Battle "E" award-winning "Fighting Tigers" of Attack Squadron 65 in 1984-85. CAPT Fallon was then selected as the first Atlantic Fleet deputy air wing commander, deploying in* Nimitz *(CVN-68) with Carrier Air Wing 8. He commanded Medium Attack Wing One at NAS Oceana, VA, in 1989-90, and CVW-8 in* Theodore Roosevelt *during a combat deployment to the Arabian Gulf for Operation Desert Storm in 1991.*

Previous tours of duty have included assignments to the staff of Commander Reconnaissance Attack Wing One; aide and flag lieutenant to the Commander, Fleet Air Jacksonville; Commander Operational Test Force; and Commander Naval Air Force, U.S. Atlantic Fleet. Fallon is a graduate of the Naval War College, Newport, RI, and the National War College in Washington, D.C. He holds a Master of Arts Degree in International Studies from Old Dominion University.

Bill served as Deputy J-3, Joint Task Force Southwest Asia in Riyadh, Saudi Arabia, then Deputy Director, Aviation Plans and Requirements Branch of the Air Warfare Division on CNO's staff. His selection to the rank of rear admiral was announced in January 1993, after which he assumed the duties of Chief of Staff for Supreme Allied Commander Atlantic. Later selected for vice admiral, he served in 1998 as Commander, Second Fleet Atlantic.

VADM Fallon has logged more than 1,300 carrier arrested landings and 4,800 flight hours in tactical jet aircraft.

As CAG 8, Fox Fallon led his air wing on missions over Iraq during the Gulf War. He tells of some of these experiences.

"Fox" Fallon, CAG 8, suits up for a mission during CVW-8's pre-deployment strike warfare training at NAS Fallon, Nev., 25 April 1990. Less than one year later, he led his air wing into combat during Desert Storm. USN photo, Randy Duran

"Following our NAS Fallon strike warfare training in April 1990 and work-ups in the Caribbean during the summer, CVW-8 embarked in *Theodore Roosevelt*, CAPT Steve Abbott, commanding, and departed Norfolk 28 December for the Gulf War. 'TR' transited non-stop to the Persian Gulf via the North Atlantic, Med, Suez Canal, Red Sea, and Indian Ocean, arriving at the Strait of Hormuz 19 January. I flew from TR (east of Sicily) to Bahrain with RADM Dave Frost, Commander Carrier Group Eight, in an S-3 Viking on 12 January for a historic prewar conference of key USN/USMC commanders with VADM Stan Arthur, ComUSNavCent, on his flagship. A decision was made to bring TR/CVW-8 straight to the Persian Gulf as soon as possible despite commencement of hostilities 17 January.

TR was just exiting the Red Sea into the Indian Ocean when first strikes on Iraq began. I had all flightcrews and many ship/staff officers to a meeting on the foc's'l the evening of the 16th. I briefed all present that, on the other carriers and airfields around the Middle East, aircrews were preparing to launch the first strikes against Iraq, and that we would join in combat ops as soon as we entered the Persian Gulf. RADM Frost and CAPT Abbott also spoke, and I concluded with a prep talk and directions to prepare for imminent combat. We commenced combat support ops 20 January and began strikes the next day. We operated the entire war from the Gulf and flew the most sorties, dropped most ordnance, etc., of all CVWs in the war. Highlights included destruction of many Iraqi navy vessels and key enemy installations in Kuwait and SE Iraq. Bulk of the strikes were in the area around Basrah, Iraq. We had two combat losses: a VA-36 A-6E with LCDR Barry T. Cook/LT Patrick K. Conner, and a VFA-87 F/A-18 with CVW-8 pilot LT Robert J. Dwyer. (All KIA.)

TR flew mostly the midnight to noon shift during the war, while *Midway* (CV-41) usually did days and *Ranger* (CV-61) nights. *Midway* had three F/A-18 squadrons, while *Ranger* had two F-14 and two A-6, but no F/A-18 or A-7 squadrons.) The first ten days or so were intense with nearly continuous flight ops. We flew only one or two major strikes per 24-hour period, but provided extensive CAP (Combat Air Patrol) and ASUW/SUCAP (Anti-Surface Ship Warfare/Surface CAP) for the Persian Gulf Battle Force. During this time, we played a major role in the destruction of the Iraqi navy, sinking numerous vessels. CVW-8 aircrews played a big part in the battle for Khafji at the end of January. Our E-2Cs, on station in the predawn hours, received calls for help from USMC units on the ground and relayed word to TR for additional strike assets (mostly A-6s), which were launched or diverted to stop Iraqi armor heading south toward Saudi Arabia. Other aircraft participated in the search for an Air Force AC-130 that was shot down in the vicinity, and did some hair-raising low altitude searching for wreckage (subsequently found offshore).

Another key air wing task was the search to locate Silkworm anti-ship missile sites along the shorelines of Iraq and Kuwait. These mobile targets proved elusive, but many sites were destroyed, thanks to some dedicated targeting work by the persistent effort of the aircrews and the ship/air wing intel folks who worked hard to distinguish between decoy and real sites.

CVW-8 fighter crews had a chance to engage the two Iraqi Mirage F-1s which were splashed by a Saudi F-15 in late January. One of our F-14s on CAP station over the Northern Gulf received a vector toward a couple of bogies and actually radar-locked the low-flying Mirages, but lost the lock as they closed for the intercept. AWACS waved off the F-14s as the Saudi F-15 was moving into a better shooting position. An interesting sidenote is the fact that, on the day prior to that engagement, a pair of Iraqi F-1s followed one of our CVW-8 strike groups back out of Iraq, almost to the northern edge of the Gulf before detection by an E-2. A short, unsuccessful chase by a pair of F-14s ensued as the Mirages fled back into Iraq.

In the pre-dawn darkness of 10 February, another interesting episode involved the destruction of a large thermo-electric power plant north of Basrah. This target had been attacked several times in early February but was still functioning. (In fact, the lights were still on at night in Basrah and surrounding areas.) A large strike group was headed toward the target, but various circumstances (weather, aircraft problems, and tanking) reduced strike assets. I went in with two A-6s loaded with two 2,000 lb. LGBs (Laser Guided Bombs) apiece, and we scored direct hits on the plant, knocking it out of action for the duration of the war. (No more lights in Basrah.)

After the ceasefire, TR/CVW-8 remained in the Gulf, flying CAP and recce missions over Iraq for a month, and then moved to the Red Sea where we continued to monitor Iraq's compliance. In mid-April, we were moved into the Eastern Med to support Operation Provide Comfort, the international rescue of the Kurds in Northern Iraq. We flew CAP, armed recce, and photo recce over Iraq, north of the 36th Parallel.

For this one deployment, CVW-8 employed the largest and most powerful air wing ever embarked in a CV. The wing has since been reduced to a standard size."

LCDR Howard L. "Brig" Young was Commander, Enterprise Air Group when his men flew into the middle of the Japanese attack on Pearl Harbor. Young later commanded *Ommaney Bay* (CVE-79) when she was sunk 4 Jan 1945 after being hit by a *kamikazi* during the Philippines campaign. USN photo, courtesy Annie Hunt

(*Left*) Edward Henry "Butch" O'Hare was born in St. Louis, MO, 13 Mar 1914. After attending Western Military Academy, Butch graduated with the Naval Academy Class of 1937. Assigned to USS *New Mexico* (BB-40) from June 1937 to May 1939, he entered flight training at Pensacola with class 127-0 in June 1939. Designated naval aviator 27 May 1940, O'Hare was assigned to VF-3 aboard *Saratoga* (CV-3). After Sara was torpedoed in January 1942, VF-3 transferred to *Lexington* (CV-2). On 20 Feb 1942 in the South Pacific, Butch found himself alone to face eight Japanese Betty bombers after his wingman's guns jammed. Wading into the enemy formation, he shot down five attackers and drove off the other three. This action earned him the Medal of Honor, plus an advancement 30 numbers in permanent rank of LT and promotion to temporary LCDR. Assigned to command VF-3 on 11 May 1942, the squadron became VF-6 on 15 Jul 1943. Butch added another one-and-one-half enemy planes to his credit in action over Marcus Island 31 August and Wake Island 5 October. Named Air Group Six commander 17 Sep 1943, he led two other planes into the first night interception of Japanese planes off Tarawa 26 November. Despite a successful interception, CAG was caught in a crossfire, and his aircraft went into the sea. Exhaustive research into U.S. and Japanese documents by historians John Lundstrom and Steve Ewing for a book on O'Hare refutes the commonly accepted belief of the time that he was accidently shot down by a TBF gunner. Their research indicates that he may have been shot down by a Japanese Betty nose gunner. This theory is not universally accepted by those who were there at the time, and certain truth will probably never be known. RADM Arthur Radford nominated Butch for a second Medal of Honor for breaking up the enemy attack and demonstrating that night interception was possible. However, a posthumous Navy Cross was awarded for O'Hare's last action. USN photo, courtesy PHC Al Labendz, USN(Ret)

Much credit for the U.S. Navy's success at the Battle of Midway was given to LCDR C. Wade McClusky, Commander, Enterprise Air Group. McClusky, a fighter pilot, flew an SBD at Midway and located the Japanese fleet in the early stages of the battle. After leading a successful attack on the Carrier *Kaga*, he was wounded by enemy fighters but was able to return to CV-6. Wade McClusky was awarded the Navy Cross for Midway and retired after the war as an honorary RADM. USN photo, NH 93189

Jimmy Flatley was one of the most respected CAGs of WWII. Born 17 Jun 1906, James H. Flatley II, joined the Naval Reserve as an apprentice seaman 27 Jul 1923. He was posthumously awarded the Distinguished Service Medal.

LCDR Max Leslie relieved Wade McClusky as *Enterprise* CAG in June 1942. During the Battle of Midway, Leslie led his VB-3 strike group against the carrier *Soryu*. Although frustrated by a malfunctioning switch that prematurely released his bomb before reaching his target, the intrepid squadron commander continued his attack by strafing the carrier. Forced to ditch as a result of fuel starvation, Leslie and his radioman-gunner were rescued. Max Leslie earned a Navy Cross for Midway and led Enterprise Air Group during the Battle of the Eastern Solomons. He retired after the war as an honorary RADM. USN photo, courtesy RADM M.F. Leslie, USN(Ret)

CDR William I. "Bill" Martin (center), CO of VT-10, on board *Enterprise* (CV-6) in 1944 with his crewmembers Jerry T. Williams ARM1/c (left), and Wesley R. Hargrove AOM2/c. Williams and Hargrove were later killed during a raid on Saipan when their TBF was shot down. Bill Martin survived and became CAG of Night Air Group 90, where he pioneered carrier night attack aviation. VADM Martin (Ret), resided in Alexandria, VA, until his death in 1997. USN photo, courtesy Joe Houston

Air Group Ten was led from *Enterprise* during the February-July 1944 Pacific battles by CDR William R. "Killer" Kane. Forced to ditch after his F6F was hit by flak 16 Jun 1944, Kane was injured when his head struck the gunsight during the water landing. With his head bandaged, Kane was back flying combat missions three days later. On his second flight of the day he led the *Enterprise* strike group during the Battle of Leyte Gulf. Following the battle and a grueling long trip home in darkness, Kane crashed and disappeared after an attempted waveoff from the Big E. Two days later he was "ransomed" for ice cream from an escorting destroyer. Killer Kane was credited with six enemy aircraft and held a Navy Cross at the time of his death in a flying accident 6 Feb 1954. USN photo 47486, courtesy Barrett Tillman

Quiet and self-effacing, CAPT Albert K. "Bert" Earnest, now retired in Virginia Beach, VA, flew in the first combat mission of the Avenger. Although largely unheralded, Bert was a member of VT-8's TBF contingent that flew against the Japanese from Midway 4 Jun 1942. The only TBF pilot to survive the battle, he brought his aircraft back to Midway with a dead radioman and wounded gunner. Bert Earnest earned two Navy Crosses at Midway, the first of three he would receive for his WWII actions. He later fought at Guadalcanal and with VC-7, flying from *Manila Bay* (CVE-61) before becoming a test pilot at Patuxent River. Bert commanded VA-14A before becoming CAG of ATG-181 and later CO of NAS Oceana. He retired in 1972. USN photo, courtesy CAPT A.K. Earnest, USN(Ret)

Colorful, flambouyant, and loved, describe CDR Paul H. Ramsey (VADM deceased). Ramsey (USNA '27) received his wings in 1929 after serving on the Observation Wing staff in battleships. His varied prewar aviation career led to his command of VF-2, the "Fighting Chiefs," enlisted pilot squadron, which he commanded at the beginning of WWII. During the Battle of Coral Sea, he earned a Navy Cross for shooting down three enemy planes and damaging two others during one mission. Ramsey became the first CAG 11 in November 1942 and became a test pilot the following year. He commanded *Attu* (CVE-102) during 1945 and then skippered VC(N)-1 (1946-49). Following several staff positions, command of *Philippine Sea* (CVA-47) and NATC Patuxent River, Ramsey became DCNO(Air) 13 Mar 1965. Paul Ramsey retired 1 Oct 1966 and lived in Coronado, CA, until his death 15 Feb 1982. He was buried at sea from *Ranger* (CV-61). USN photo 1072227, PH2 William D. Caswell

CDR Bruce S. Weber (right) relieves CDR H. "Monk" Russell 11 Nov 1949 as CAG 17. Weber was awarded a Navy Cross for leading a strike of fighter planes against the battleship Ise at Kure Naval Base, Japan, 28 Jul 1945. Bruce commanded VC-34, VF-34, and VF-98 before becoming commander of Replacement Air Group 98 at Barbers Point during 1944-45. He was also CAG 31 in *Beleau Wood* (CVL-24) during WWII prior to commanding CVG-17. He recalls his last mission of the war: "A major strike was in the air and in sight of Tokyo. We knew the war was about over and had been briefed the strike may be recalled. About 40 miles off the coast, air control called and said to circle and await further orders. There was a lot of chatter and excitement, then I noticed a lot of bombs were exploding on the surface. Most everybody who hadn't dropped figured they had missed the jettison order and did so. Then control said: 'Do not, repeat, do not jettison ordnance!' No one said a word. After about five minutes of sweating it out, the order came through: 'Jettison all ordnance and return home.' You could damn near hear the sighs of the senior CAG to the junior ensign." USN photo, courtesy CAPT B.S. Weber, USN(Ret)

Another holder of three Navy Crosses is CAPT Arthur L. Downing, USN(Ret). Art earned his first Navy Cross as a dive bomber pilot at Tulagi and Coral Sea in May 1942 flying from *Yorktown* (CV-5). His second award was for his actions two months previously at New Guinea, and his third Navy Cross was for his performance as strike group leader during the Battle of Leyte Gulf 26 Oct 1944. During the latter he was CO of VB-14 embarked in *Wasp* (CV-18). He also flew from *Enterprise* (CV-6) in 1942. Art received his wings in 1938 and served with VF-2 in *Saratoga* (CV-3), VF-7 in Wasp (CV-7), and VS-5 in *Yorktown* (CV-5) prior to the war. During the Korean War he commanded CVG-2 flying combat missions from *Boxer* (CV-21). Art Downing is now retired in Sarasota, FL. USN photo

Another of the Navy's colorful CAGs was CDR T. Hugh Winters (CAPT Ret). Holder of two Navy Crosses for his WWII deeds, Winters is credited with eight aerial victories while skipper of VF-19 and later CAG 19 ("99 Mohawk") flying from *Lexington* (CV-16). However, his most cherished achievement is that no aircraft from his air group was ever lost to enemy aircraft while under escort of VF-19. A 1937 Naval Academy graduate, Hugh also served in VB-5 and VF-9 before VF–19. He commanded *Franklin D. Roosevelt* (CVA-42) in 1960-61 and is now retired in Virginia Beach, VA. USN photo, courtesy CAPT T.H. Winters, USN(Ret)

The second of only two Marines to command a U.S. Navy air wing, LCOL Charles H. Ludden was Skipper of VMF(AW)-212 in *Oriskany* (CVA-34) when CAG 16 Jim Stockdale was shot down 9 Sep 1965. Filling in as acting CAG, Ludden led CVW-16 until 4 October when he was relieved by CDR Robert E. Spruit. Ludden was selected as the Marine Corps' 1966 Aviator of the Year for his service and leadership during his Vietnam combat tour in *Oriskany*. COL Ludden was designated naval aviator 15 Jan 1944 and flew combat as a fighter pilot in several squadrons during WWII and Korea. He retired 30 Jun 1974 and lived in Kailua, Hawaii, at the time of his death 10 Sep 1992. USMC photo, courtesy Mrs. Shirley Ludden

Typical of today's recent former Super CAGs is RADM Jay B. "Spook" Yakeley III. Retired in January 1999 from CNO's staff, Spook formerly commanded Carrier Group Threee in *Abraham Lincoln* (CVN-72). Jay, son of a naval aviator, was awarded his wings in May 1967. He is a graduate of the Naval War College and holds a Master's Degree in economics. A former flight instructor (VT-9/VF-124), RADM Yakeley flew F-4 Phantoms with VF-101 and -74. He was the first fleet production acceptance test pilot for the F-14 Tomcat and has flown the F-14 with VF-124, -2, and -114 as XO, then CO. Following a 1983-85 CNO staff tour, he commanded VF-124 until 1987 and then the Naval Fighter Weapons School (Topgun) until 1989. He commanded CVW-14 from 17 Nov 1989 until 10 May 1991, during which time the wing deployed in *Independence* (CV-62) to Operation Desert Shield in the Persian Gulf. CVW-14 entered the Gulf of Oman 6 Aug 1990, and was the first significant force to face Iraqi aggression four days after the invasion of Kuwait. Under then-CAPT Yakeley, the wing was responsible for establishing procedures used by following air wings to counter Iraqi forces during Desert Storm. Returning to Washington in 1991, Jay was Deputy Director of the White House Military Office, serving both Presidents Bush and Clinton. From October 1993 to February 1994, he was Deputy Commander, Joint Task Force-Southwest Asia, in Riyadh, Saudi Arabia. He assumed command of Carrier Group Three in March 1994. During his career, Spook has flown more than 6,000 hours and logged in excess of 1,000 carrier landings, as well as more than 100 combat missions in Vietnam and Iraq.

At this point, although often quoted, it seems appropriate to use an excerpt from James Michener's "The Bridges at Toko-ri," when the carrier division commander asks: "Where did we get such men?"

4

The Machines

"If you don't like helicopters, next time you're in the water, try calling an F-4."

Sign in HS-6 ready room,
USS *Constellation* (CV-64), 1974

CAG Birds

A study of the aircraft flown by CAGs over nearly the past 60 years, is a history of the evolution of carrier aviation in itself. While leading their air groups CAGs have flown virtually every type of tactical and, in more recent years, support aircraft assigned to their units.

Aircraft location records of the late 1920s show planes assigned to the aircraft commanders of both the Battle Fleet and Scouting Fleet. Photographs taken in 1931 have been located indicating a Curtiss O2C-2 Falcon BuNo 8849 and F8C-4 Falcon allocated to Aircraft Squadrons Battle Fleet were assigned for use of the Fighting Wing commander. It is reasonable to assume that other aircraft were assigned to the remaining wing commanders as well. In essence, these airplanes would be the first "CAG birds," as they are known today.

Later, after *Lexington* and *Saratoga* became operational, and the abandonment of the wing commander system, CAG was given his own airplane, which was maintained along with the other aircraft assigned to the carrier's utility unit. The first direct aircraft assignment to the air group commanders that has been found appears on the August 1938 Location of Aircraft list, the first month after the official establishment of the CAG billet. The assignments were: *Lexington*, SB2U-1; *Saratoga* SBC-3; *Ranger* SBU-1; *Yorktown* SBC-3; and *Enterprise* TBD-1. *Yorktown* originally used a BT-1, but it and *Enterprise's* TBD were quickly replaced with SBCs. These were the aircraft types flown by the CAGs until SBD-2s began appearing in 1941.

By 1940, U.S. naval aircraft development was seriously lagging behind that of other nations. This fact was not yet apparent to the Navy's leadership. As related by then-ENS James D. "Jig Dog" Ramage (RADM Ret), the following exchange took place in 1940 on *Enterprise's* flight deck between Secretary of the Navy Frank Knox and CAG LCDR E.C. "Eddy" Ewen. Knox was on board for an official visit, and Ramage was an assigned escort. Knox turned to Ewen and commented on *Enterprise's* obsolete SBCs, F3Fs, and TBDs: "These are a mighty fine bunch of planes." To which the straight-talking CAG Ewen replied: "Mr. Secretary, these planes are not fit to die in."

But new airplanes were on the horizon, such as the Grumman TBF-1, which first appeared in February 1942 with CVG-9 and Ranger Air Group. *Ranger's* CAG, LCDR David B. Basham, flew the TBF during Operation Torch in November 1942. Grumman's F6F-3 Hellcat made its first appearance with CVG-5 and -16 during May 1943, and the Vought F4U-1D Corsair followed with CVG-85 in May 1944. During WWII, CAGs flew either the Hellcat or the Avenger, and later, the Corsair. An interesting deviation from the norm was the lone Curtiss SB2C-3 Helldiver used by CAG 9 for one month when the air group re-formed in San Diego during April 1944. The air group commanders were each assigned one aircraft, of one of these types. However, conversations with former WWII CAGs indicate that many of them flew the type aircraft they preferred. Also, most CAGs flew whatever aircraft was available for the launch regardless of unit assignment, although some CAGs did insist on flying "their" airplane whenever possible.

RADM V.G. Lambert, USN(Ret), CAG 15 during 1948-49 recalls his WWII observations:

> Most of the CAGs flew fighters, but some of them flew TBMs because of longer stay time in the target area, or because they had been VT skippers before fleeting up to CAG. The TBMs were also better communication platforms because of the presence of a radioman and low-frequency radios aboard. We even used CW radios for aircraft-to-ship communications; the TBMs could handle this. Naturally, most CAGs preferred fighters for their offensive/defensive capabilities and greater speed. Also, they could recycle faster if the need existed.

While deployed, only the CVG groups were assigned a CAG aircraft; the CVLGs and CVEGs did not have one. Unaccountably, five CVLGs did have CAG aircraft for short periods during their initial formation stages. By the end of the war, all active CVGs had only one VF type aircraft assigned. This authorization remained the same until 1948.

After 1 Jul 1948, additional aircraft began to be assigned to the CAG unit. However, these were not all "CAG birds," as such. The units were assigned one VF type for the CAG, along with two fighter-photo aircraft and one electronic attack type. CVG-2 was typical with one F4U-5, two F4U-5Ps, and one Douglas AD-1Q Skyraider. Additionally, some CAG units were authorized North American SNJs and later Beech SNBs, for instrument training and logistics flights. In the early 1950s all CAG units were authorized Lockheed TO/TV-2 (T-33B) Shooting Star jet trainers. However, very few of them actually received the aircraft. The first CAG to "own" a jet was CDR Hawley "Monk" Russell, CAG 17, when McDonnell's F2H-1 Banshees reported in to Cecil Field during march 1949. On the West Coast, CDR Harvey P. Lanham, CAG 5, got a Grumman F9F-3 Panther in November at North Island.

Chief of Naval Operations letter Op-55R 285/jk, VV/F40-1 Serial: 167P55 dated 20 Jul 1948 designated the CAG units as composite (VC) squadrons to become effective 1 Sep 1948. The units were numbered by adding a zero to the end of the air group number, thus, CVG-1's CAG unit became VC-10, CVG-11's became VC-110, etc. This system lasted for less than a year, and the VC CAG designation disappeared. Interviews of CAGs of this period produced no recollection of the VC designation ever being used. None of them even recalled the designation, and it appears it was merely another paper exercise by the bureaucracy.

The assignment of aircraft directly to CAG units began to be phased out in mid-1952, although CVG-9 had no aircraft authorization after December 1951. The last two air groups to operate CAG aircraft were CVG-7 on the East Coast and CVG-101 (CVG-12 after February 1953) at Miramar. CVG-7's final assignment was five TV-2s, two JRB-4s, and one SNB-5 on 31 Dec 1954. Air Group 101's was one F9F-8 on 30 Apr 1955. After that date, CAGs began flying aircraft assigned to individual squadrons in the air group, the system still in use today. The new system also introduced the "multi-type CAG," where most air group commanders began flying all or most types of aircraft under their command.

Aircraft Markings

From the very beginning, the aircraft assigned to carrier group commanders bore special markings indicating their status. The first planes carried simple wording, such as "Fighting Wing." Later, after the establishment of the CAG billet, the markings became more elaborate, with wording such as "Air Group Commander USS Enterprise," and a diagonal stripe on the fuselage colored the same as the tail color assigned to each carrier. In 1940, with the advent of tactical camouflage, the colorful markings were removed, leaving only the lettering and the addition of such items as "AGC" or "CAG" on wheel-well doors or engine cowlings.

The origin of "double-nuts," or "00" to identify CAG's airplane is a mystery. The first known use of this marking was by LCDR Jimmy Flatley, CAG 5 in *Yorktown*. A photograph dated 6 May 1943 shows Flatley standing on the wing of his F6F-3 BuNo 04872 with 00 on its fuselage. This use of 00 is also an enigma because during WWII 00 was Navy terminology for the "boss," and usually was reserved for the admiral. In radio terminology, CAG was "99," and these numbers were used by some WWII CAGs on their airplanes instead of 00. Over the radio, Flately was referred to as "99 Sniper." Dave McCampbell, CAG 15, had no numbers on either of his three "Minsi" Hellcats, only AGC on the wheel-well doors and vertical stabilizer of the first Minsi, and CAG on the following two planes. His callsign was "Rebel 99."

Following the war, the 00 marking was generally the identifier of choice by most CAGs, although some use a single 0. CNO Aviation Circular Letter (ACL) No. 156-46 of 7 Nov 1946 directed that carrier squadrons would use a three-digit numbering system on their aircraft, and that the air group commander's planes *may* be marked 100, 200, 300, or 400, according to the type of aircraft. However, the ACL did not direct these numbers to be used, and apparently CAGs chose the system they wanted until 1948. ACL 69-48 of 4 Aug 1948 directed that the numbers 0-99 *shall* be used to identify aircraft assigned to the VC CAG division. After the VC designation was terminated in 1949, ACL 43-50 dated 19 Jul 1950 redirected the 0-99 system to continue in use for CAG aircraft.

As the CAGs began losing their individualy assigned airplanes in the mid-'50s, the 0-99 numbering system became outmoded and was gradually replaced over the next few years with 100, 200, etc. Having said all that, CAGs, being the independent individualists they are, have used just about whatever they pleased for their individual markings, ranging from 0 to 000, and CDR James Bond Stockdale's classic 007.

The heyday of colorful CAG markings began in the late '50s and continued until the advent of the drab tactical paint schemes introduced in the '80s. OpNav Instruction 03561.2A of 24 Jun 1956 directed the use of specific colors and numbering for each squadron in a carrier air group as follows: (100) insignia red, (200) orange yellow, (300) light blue, (400) international orange, (500) light green, and (600) black. Later instructions added maroon for VC dets, and after they were abandoned, for dedicated squadrons or dets; (700) insignia blue, along with maroon for the 800 and 900 squadrons, were also added. The introduction of these colors precipitated the tradition of CAG's airplane having a design that incorporated the color of each squadron in his air group. Soon, the squadrons began competing with each other to produce the most colorful CAG-bird design—and some of the results were spectacular. A related note contained in the 1 Oct 1938 BuAer Newsletter is of interest:

> Carrier Group Commanders' planes lack that certain "esprit de corps" present in squadron planes which is visibly indicated by the squadron insignia on the fuselage. In the near future the SARATOGA Group Command plane will roll out of the VS-3 hangar with insignia on its fuselage that will no doubt cause many a pair of eyes to look twice—and hard. It is a combination of the insignia of the SARATOGA squadrons. On the back of an "animal," whose forward part is Torpedo Three's firey dragon and whose after part is Scouting Three's sturdy pointer, stands Fighting Three's Felix poised with a bomb in his hand and jauntily wearing Bombing Three's high hat.

Unfortunately, no record or photo of this airplane has been located, and it is unknown if it ever actually existed. The idea does establish the amalgamation of squadron markings on CAG's airplane that followed.

During 1976 the Naval Air Systems Command began a Tactical Paint Scheme Program that would sound a virtual death knell (at least for nearly 20 years) for the colorful CAG markings. Over the next few years, several new camouflage schemes were put into use on carrier aircraft, replacing the gray/white scheme in use from 1956 to 1977, and by the late '80s several directives had been published abolishing the use of any color on tactical aircraft. CAG's airplane once again lacked *esprit de corps*, along with those of his air wing. Although the Navy completely shifted to the drab tactical paint schemes by the late '80s, there remained a few daring independent CAGs and squadron COs who defied the no-color ban. Although subdued, ways were found by many aviators to retain some individuality on their CAG birds. By 1992 some color had begun creeping back onto the carriers. (CVW-14 became notorious for painting its CAG birds in resplendent colors while on cruise, then painting them out before returning to the wrath of AirPac.)

The subject of aircraft markings and its effect on unit morale was addressed in a 1986 Navy Times interview by CDR Fredrick G. "Wigs" Ludwig, then-CO of VF-31 and later CAG 14. Defending his red-tailed "Tomcatters" F-14s Wigs stated: "(When I was an instructor at the Naval Fighter Weapons School) I made it a point that the first thing that I was going to do as a CO was to get rid of the black noses and red tails. But after a month I realized what that paint scheme does for our morale, *esprit de corps*, and identity. I did a 180-degree change in my thinking."

Finally, by 1994, the powers-that-be recognized the *esprit* value and authorized the use of limited colorful markings on one aircraft of each squadron, either the CAG bird or the CO's aircraft. However, there shall always be a CAG somewhere willing to "explore the outer limits of the envelope."

The photos in this section depict in much better detail the spirit of carrier aviation and the markings of the CAGs than the author could ever possibly write.

Prewar - 1949

Color profiles by Steve Ferguson

The earliest photos located of wing commander aircraft are of Curtiss Falcons, O2C-2 BuNo 8849 and an F8C-4. These planes were flown in 1931 by LCDRs Ralph E. Davison and Gerry Bogan during their Fighting Wing Commander tours with the Aircraft Battle Force at NAS San Diego. Photos courtesy Dave Ostrowski

Yorktown's **(CV-5) CAG, LCDR Miles R. Browning, briefly used a Northrop BT-1 BuNo 0633. This photo taken 3 Aug 1938 shows one of the first specialized paint schemes for a CAG aircraft. WM. F. Yeager/ Steve Ferguson**

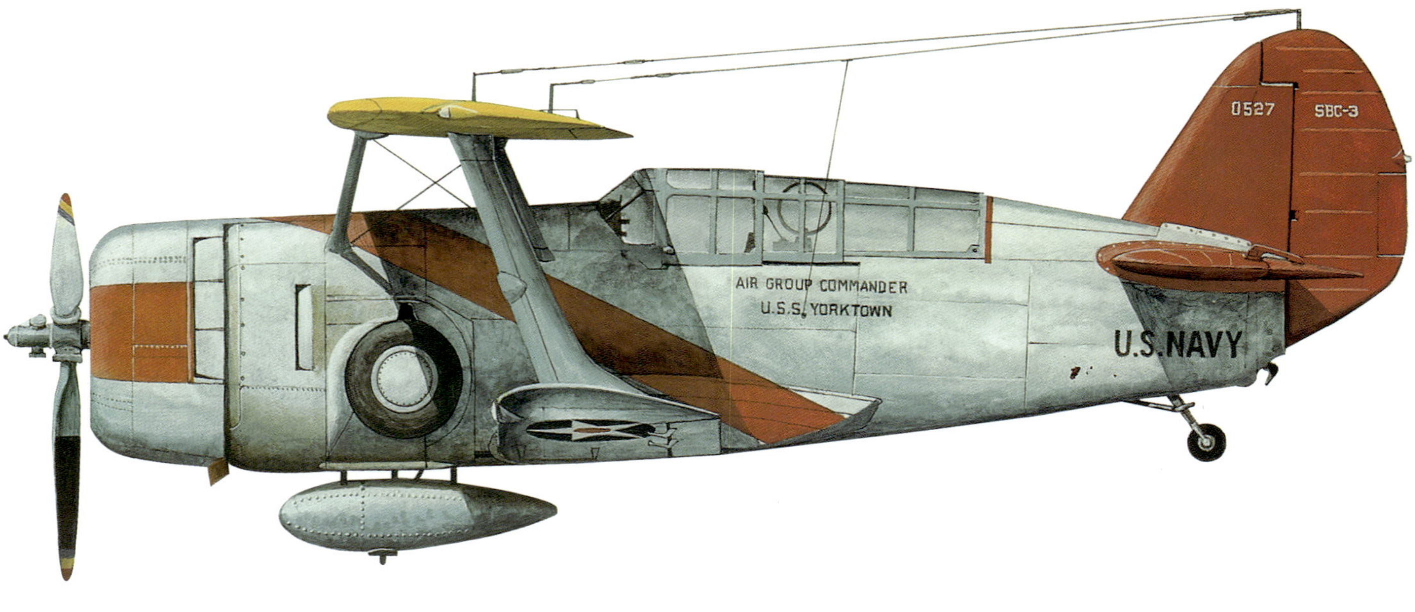

By the time of the September 1938 Cleveland Air Races, Browning was flying a Curtiss SBC-3 Helldiver BuNo 0527 (*see historical photo next page*). Donald F. Kauer, courtesy Wm. T. Larkins/Steve Ferguson

Enterprise's (CV-6) CAG also flew a Curtiss Helldiver during the late '30s. SBC-4 BuNo 1295 was on a cross-country flight during this undated photo. Shertzer, courtesy Pete Bowers/Steve Ferguson

October 1938 photo of SBC-4 BuNo 1295 about to launch from *Enterprise* in an exercise with *Yorktown*. USN photo

Enterprise CAG LCDR Eddy Ewen flew SecNav Frank Knox out to CV-6 off Pearl Harbor 8 Sep 1940 in BuNo 1295. Replying to Knox's admiration of his obsolete airplanes, Ewen replied, "Mr. Secretary, these planes are not fit to die in" USN photo, courtesy Bob Cressman

Arriving in the fleet during December 1937, the Vought SB2U Vindicator was used by five carriers for its CAG bird. SB2U-1 BuNo 0773 was flown by *Ranger* CAG, LCDR Oscar A. Weller, at the time of this 29 May 1940 photo. James C. Fahey, courtesy Wm. T. Larkins/Steve Ferguson

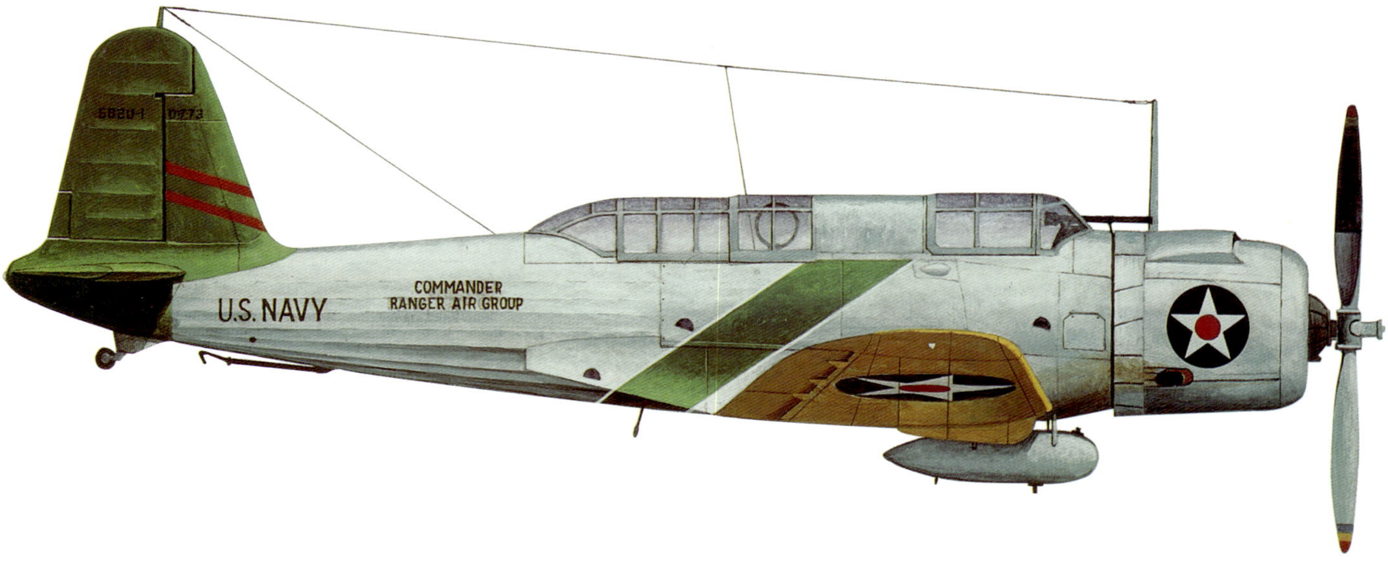

SB2U-2 BuNo 1379 wearing *Wasp's* (CV-7) black tail and CAG stripe. This Vindicator was flown by LCDRs Grover B.H. Hall II and Charles W. Crawford during its July 1939-July 1941 CAG tour. Howard Levy, courtesy Wm. T. Larkins

Douglas' outstanding SBD 2 Dauntless (profile previous page) began fleet service with VB and VS-6 on board *Enterprise* (CV-6) during December 1940 with the -2 model. All -1s went to the Marine Corps. LCDR Howard L. "Brigham" Young is shown in BuNo 2162 during a late 1941 flight. USN photo, courtesy Harry Gann/Steve Ferguson

The earliest known use of a special numerical designation for CAG aircraft is CDR Harry Felt's SBD-3 BuNo 03213 "Queen Bee," used during the Battle of the Eastern Solomon's 24 Aug 1942. A white 0 is visible on the cowling, along with a Japanese flag under the Saratoga (CV-2) CAG's canopy rail, symbolizing Felt's hit on the carrier *Ryujo*, contributing to her sinking. Photo courtesy ADM H.D. Felt/Barrett Tillman

Earlier 29 Jul 1941 photo shows Brig Young with Under Secretary of the Navy, James V. Forrestal, in the rear seat of BuNo 2126 saying farewell to RADM Wm. F. Halsey, following a visit to *Enterprise*. Originally, the SBD CAG markings were in white, as seen here. USN photo, courtesy M.H. Law

LCDR David B. Overfield, CAG Ranger Air Group (insert) flew a VT-4 TBF-1 during Operation Torch, the November 1942 invasion of North Africa. CAG Overfield's markings were rather subdued, consisting only of black "Cs" on the top of the vertical stabilizer and in the wing-fold well. USN photos, courtesy Bob Cressman

Above: The first known use of "double nuts," the 00 modex, was by CAG 5 CDR Jimmy Flatley, in early 1943.

Left: Flatley leads his air group during a work-up mission July 1943. Flatley flew his personal aircraft for the 31 Aug 1943 attack on Marcus Island that introduced the Hellcat into combat. Note the additional fuel tanks carried to allow CAG more loiter time over the target. USN photo/Steve Ferguson

Medal of Honor winner CDR Dave McCampbell, CAG 15, and the Navy's all-time top ace, was nearly as hard on his airplanes as he was on the enemy. "Rebel 99" wore out two "Minsis" before finishing his tour in Minsi III while bagging 34 enemy aircraft. McCampbell established two other records: the most kills in a single combat tour and nine kills during one sortie. (Above left) CAG holds in position for launch from *Essex* (CV-9) as strike leader during July 1944 in F6F-3 "The Minsi," the first of the series. This Minsi was shot up by ground fire during the 19 May 1944 Marcus raid and jettisoned over the side. It was the only Minsi to carry "00" markings. (Center) F6F-5 "Minsi II" undergoes maintenance on board *Essex* 30 Jul 1944. CAG was infamous for overboosting his engine in the heat of combat. (Above right) For the Philippine campaign, Dave flew F6F-5 BuNo 70143 as "Minsi III." Shown here, he taxies forward after making his 21st kill, 21 Oct 1944. The Minsis were names for an old girlfriend, Mary Blatz. USN photos/Steve Ferguson

AGC
The Minsi

CAG
Minsi II

CAG
Minsi III

CVG-16's CAG bird is in trouble, as it misses the arresting gear but picks up a wire of the barrier during landing aboard *Bon Homme Richard* (CV-31) 16 Jan 1945. CDR Ray Davis' F6F-5 (profile opposite, top) carries CAG's "99" radio call number, vice 00. USN photo, courtesy Richard M. Hill

F6F-5N of VF(N)-53 wears LCDR Vincent F. McCormack's 00 CAG markings as it is respotted after recovery on board *Saratoga* (CV-3) 2 Feb 1945. USN photo The profile on the following page shows the aircraft after the addition of geometric pattern markings for *Saratoga's* air group. Geometric symbols were added in Feb 1945 to all Pacific Fleet carrier aircraft.

Opposite, bottom: CVLG-27's VF-27 Hellcats wore a menacing sharkmouth on its F6F-5s.The air group flew from *Princeton,*(CVL-23) under CAG LCDR Frederic A. Bardshar, during the 1944 Philippines Campaign. Steve Ferguson

Another example of the 99 marking was CDR Phillip H. Torrey, Jr.'s, CAG 9 F6F-5 on board *Lexington* (CV-16) during February 1945. CAG Torrey was killed 16 February over Japan while flying a Hellcat. USN photo, courtesy Richard M. Hill (See color profile p. 60)

Vought's F4U-4 Corsair's first appearance with the fast carriers was in late 1944 following the 1943 ban on Corsairs as a result of the poor performance of VF-17's F4U-1s on board *Bunker Hill* (CV-17). Two Marine squadrons, VMF-124 and -213, were assigned to *Essex* (CV-9) to help counter the new *Kamikaze* threat. Here, CDR Allen M. Shipp's CVG-89 CAG F4U-4 taxis out of the gear after landing aboard *Antietam* (CV-36) 14 Mar 1945 during her shakedown cruise. USN photo, courtesy CDR Pete Clayton, USN(Ret)

99

00

CDR Hugh T. Winters earned two Navy Crosses for his combat actions during WWII in North Africa and as CAG 19 while flying F6F-5 Hellcats in the Pacific. Steve Ferguson

Another variation of CAG markings was CAG 12, CDR Ralph A. Embree's use of a single digit 0 on his F6F-5. Embree was killed 12 February over Okinawa.

CDR Bernard M. "Smoke" Strean, leads his CVBG-75 over the Caribbean 6 May 1946. The air group was established 1 Jun 1945, then in the postwar mass confusion of re-designations, became CVBG-3 on 15 Nov 1946, then CVG-4 on 1 Sep 1948, before standing down 8 Jun 1950. Steve Ferguson

CVG-153 was another short lived postwar air group, in active service for less than five years, 26 Mar 1945-1 Dec 1949. It was commanded by CDRs Richard A. Teel, John M. "Jocko" DeVane, and Vlademar G. Lambert. CAG 153's F6F-5 Hellcat is shown in early markings before the red bar was added to the national insignia 14 Jan 1947. Steve Ferguson

First known use of "CAG" in large fuselage lettering, vice smaller lettering on wheel doors or tail surface, was that of CDR J.T."Tommy" Blackburn, CAG 74 during mid-1945. Tommy won fame early in WWII with VF-17 land-based in the Solomons, scoring eleven victories in F4U-1s. Courtesy CAPT Tom Blackburn, USN(Ret)

The large CAG fuselage lettering became popular with many air groups at the end of the war as seen on CDR Lou Bauer's F6F-5 (profile on p. 161) near Atlantic City, NJ, in late 1945. Lou used the same markings for his follow-on F8F-1 the next year. USN photo (See profile p. 161)

CDR R.B. "Bob" Dosé, CAG 13 (profile center p. 162), reverted to the smaller tail lettering on his F4U-4 during late 1946. Courtesy CAPT R.G. Dosé, USN(Ret)

Possibly the first post-war use of specialized paint schemes for a CAG aircraft was CVG-19's Grumman F8F-1 Bearcat (profile bottom p. 162) in 1946. CAG 19 at the time was CDR Albert P. "Scooter" Coffin. Courtesy CAPT Hugh J. Tate, USN(Ret)

Changing tailcodes in 1947, Scooter's Bearcat also added a spectacular extension of the red bar in the national insignia that went into effect 14 January. Wm. T. Larkins, courtesy Pete Bowers/ Steve Ferguson

CAG 1, CDR Frederick E. Bakutis, is about to lose his CVAG-1 CAG bird during a landing accident on board *Tarawa* (CV-40) 4 Sep 1947. This F8F-1 BuNo 95324 is an early example of the formal use of three digit markings using double zero for CAG. USN photo, Galloway

Variation on a theme was the addition of "CAG 13" on the cowling of CVAG-13's F8F-2, flown by CDR John J. Lynch from *Tarawa* during a 23 Oct 1947 CarQual period. Roland J. Sorensen

This F8F-2, BuNo 121701, was flown by CVG-15 CAG CDR V.G. Lambert at NAS Alameda during 1949. Wm. T. Larkins

CDR V.G. Lambert stands on the wing of his Bearcat 10 Mar 1949. USN, courtesy RADM V.G. Lambert, USN(Ret)

CDR Wm. H. "Bill" Leonard, CAG 17, brought his Douglas AD-4Q BuNo 124056 Skyraider into Wilmington, NC, for a precautionary landing because of a serious oil leak. As a result of the 1948-49 VC CAG unit designation, most CAG birds reverted to 00 as the standard marking for the late '40s and early '50s. Paul J. McDaniel, courtesy Jim Sullivan.

CAG, CDR J.D. "Jack" Blitch, flew this F4U-5 during his tour with CVBG-1, later CVG-2. The Corsair is about to deck launch from *Wright* (CVL-49) during a November 1948 CQ period. CDR R.E. Bennett, USN(Ret)

Another example of the late '40s-early '50s CAG 00 standard markings was CDR James W. O'Grady's CVAG-11 F8F-1 on board *Valley Forge* (CV-45) 27 Apr 1945. Author's coll.

First known example of the multi-colored CAG markings were on this F4U-4 assigned to CDR Turner F. Caldwell, Jr., CAG 4, launching from *Midway* (CVB-41) in 1948. Apparently, Caldwell skipped 0 and began his numbering with 1. Author's coll.

1950 - 1959

CDR R.H. Burns, CAG 4 during 1950, also eschewed the use of zero on his Vought F4U-5 during CVG-4's Med cruise in *Midway* (CVB-41). His Corsair was another example of the early multi-color markings. Courtesy CAPT R.A. Burns, USN(Ret)/Steve Ferguson

CAG 4, CDR John B. Sweeney's McDonnell F2H-2 Banshee is spotted on the deckedge elevator on board *Coral Sea* (CVB-43) while the air group was conducting a July 1952 carqual period. CVG-4 was working up for its 1953 Korean War deployment in *Lake Champlain* (CVA-39), due to be recommissioned in September 1952. USN photo, AF3 Stan Grocki.

CDR Syd Bottomley (3rd from right, standing), who earned a Navy Cross flying SBDs at the Battle of Midway, was CAG 6 on board *Franklin D. Roosevelt* (CVB-41) during her 1951 Med cruise. His Vought F4U-4 BuNo 82164 (profile center p. 173) also wore the multi-color tail markings popular on the Corsair. USN, courtesy CAPT H.S. Bottomley, USN(Ret)/Steve Ferguson

A unique variation of the multi-color CAG markings was this Grumman F9F-5 Panther of CAG 11, CDR Erwin B. Schwab, during a CQ period on board *Boxer* (CV-21) 25 Feb 1953. USN photo/Steve Ferguson

CDR Roland W. "Bud" Schumann, CAG 7, also used another numbering variation, "01," on his F9F-5 during deployment to the Med in *Bennington* (CVA-20) 6 Nov 1958. CAPT R.S. "Beaver" Blake, USNR(Ret)

Left: Another WWII SBD Navy Cross winner, CDR James D. "Jig Dog" Ramage, flew this McDonnell F2H-3 "Big Banjo" Banshee from *Oriskany* (CVA-34) during his 1953 CAG 19 tour. Author's coll.

Below: F9F-5 BuNo 125082 was resplendent in CDR Richard W. Parker's CAG 2 markings. Air Group 2 deployed to WestPac in *Essex* (CVA-9) during 1954-55. Steve Ferguson

Not officially a CAG, CDR Michael J. Hanley, Jr., chose to use only the 00 markings on his ATG-181 Grumman F9F-6 BuNo 131010 Cougar during their 1954-55 deployment to the Med in *Randolph* (CVA-15). Otherwise, the aircraft is in standard VF-21 markings. CDR Ed Foy, USN(Ret)

CDR Pierre N. Charbonnet's North American FJ-3 Fury CVG-8 CAG bird is parked on the ramp at NAS Oceana with its multi-colored nose c. 1956. USN photo/Steve Ferguson

CDR John E. Lacouture had the great fortune to command three air groups, CVG-4 (acting), CVG-6, and RCVG-4. His CVG-6 FJ-3 is spotted aft on board *Lake Champlain* (CVA-39) in 1956. Tom Keenan

Intrepid's (CVA-11) catapult crew guides CDR Pierre N. Charbonnet's CVG-8 Grumman F9F Cougar onto cat two for launch 20 Apr 1956. USN photo, courtesy Dick Starinchak

A change in the basic paint scheme from Sea Blue to gray and white on carrier aircraft came about in 1956, as seen on CAG CDR Charles N. "Tex" Conaster's CVG-12 FJ-3M during it 1956 WestPac deployment in *Lexington* (CVA-16). Ray Sinwell

A late Air Task Group Commander, CDR David S. "Davey" Crockett, ATG-1, did use the multi-color CAG markings. His FJ-4B was at NAS Oakland 15 Sep 1958. Steve Brown, courtesy Steve Ginter

ATG-4's CAG bird leads a formation of FJ-4Bs over Honolulu 26 Jan 1958 during pre-deployment work-ups for their WestPac tour in *Hornet* (CVA-12). CDR Ronald L. Johns led ATG-4 at the time. *USN photo*

Entering fleet service with VA-72 in September 1956, McDonnell Douglas' magnificent A4D Skyhawk was quickly adopted by many CAGs. CAG 3, CDR Richard H. Mills', A4D-1 BuNo 139970, assigned to VA-34, reflects the system of each squadron dedicating one aircraft to CAG and applying the 00 suffix to their squadron number. Photo was taken during CVG-3's 1958 Med cruise in *Saratoga* (CV-60). Jack Anderson, courtesy Duane A. Kasulka

Showing a mix of 00 and X00 markings in use at the same time, CVG-7's VF-84 FJ-3M CAG bird was used by CDR Armistead B. "Chick" Smith flying from *Randolph* during September 1958. Author's coll.

Even the "cats and dogs" got into the CAG act. Here VAH-9's Douglas A3D-1 BuNo 139864 Skywarrior carries CAG 3 Richard Mills' colors during landing aboard *Sara* in September 1959. Author's coll.

The McDonnell F3H-2 "Screamin' Demon" (not very screamin', more demon) night fighter came aboard the carriers with VF-14 in 1956. VF-41's CAG bird F3H-2 BuNo 143479, assigned to CAG 7 CDR Richard C. Reinhardt, was at MCAAS Yuma for Fleet Air Gunnery Units' Top Gun competition 3 Dec 1959. Wm. L. Swisher

During the late '40s through the early '50s, while not deployed, CAGs were given a variety of support aircraft to fly for proficiency and logistics missions, such as CAG 7's Beechcraft SNB-5 Navigator (never used, always called the Beech or Bouncing Bitch). Author's coll.

CAG 1, CDR W.F. "Bush" Bringle, flew this F4U-5 Corsair during 1950 while CVG-1 was assigned to the newly completed *Oriskany* (CV-34). Steve Ferguson

Prior to flying the FJ Fury, Tex Conatser flew the F9F-8 Cougar as CAG 12. CVG-12 became the replacement training air group at NAS Miramar in April 1958. Steve Ferguson

CDR J.W. Boyum's CVG-2 North American FJ-3M Fury BuNo 136155, at rest during 1962 at NAS Miramar. The FJ-2/3/4 series served with fleet squadrons during 1954-62. Steve Ferguson

CAG 2, CDR John R. Bowen II's FJ-4B assigned to VA-63 makes the 72,000 landing on board *Midway* (CVA-41), 21 Jan 1959. USN photo, courtesy CAPT Wm. F. Chapman, USN(Ret)/Steve Ferguson

A rarity in CAG aircraft was the short-legged Douglas F4D-1 Skyray, that was in fleet service during 1956-64. However, for only a short period of that time did they serve aboard carriers. CDR Lawrence Heyworth Jr.'s CVG-8 Skyray is seen in the colors of VF-102. Steve Ferguson

VA-75 Sunday Punchers' Douglas AD-6 BuNo 139802 was adorned with CDR David S. Stears' CAG 7 colors during 1962. Steve Ferguson

1960 - 1969

The 1960s were the true "golden era" of naval aviation's spectacular (sometimes downright flamboyant) CAG markings. The beautiful innovative color schemes would last for 20 years, and carrier aviation's *esprit* was never higher. Above, (l-r) CDRs William H. Shawcross, CAG 16; Carl Yeagle, CO VA-165; and Robert P. Coogan, CO VA-162 and former acting CAG 16, admire the "Boomers" multi-colored tail markings during April 1961 at NAS Cecil Field, FL. USN photos, courtesy RADM W.H. Shawcross, USN(Ret)/Steve Ferguson

VA-55 ordnancemen accomplish last-minute safety checks on CDR Walter W. Alldredge's (CAG 5) Douglas A-4 Skyhawk prior to launch on a training mission c. 1963. McDonnell Douglas photo, courtesy Harry Gann

During the early '60s, 00 was still the popular choice for numbering CAG's aircraft, as seen on CDR Dickson W. Alderton's (CAG 19) Vought F-8E Crusader on board *Bon Homme Richard* (CVA-31) in July 1963. McDonnell Douglas photo, courtesy Fred W. Roos

CAG 7, CDR Harry E. Gerhard, preferred the use of 000 for his McDonnell Douglas F-4B Phantom II during VF-41's 1965 Med deployment. McDonnell Douglas photo, courtesy Fred W. Roos

VA-195's Dambusters have upheld a tradition of outstanding paint schemes for its CAG birds for more than 30 years, beginning with this A-4C 147819 flown by CAG 19, CDR Billy Phillips, seen 7 Jun 1966 out of NAS Lemoore, CA. USN photo, courtesy Don Montgomery

With the advent of the Vietnam War, mission markings began showing up on the CAG birds. Interestingly, no one seemed to be particularly worried about CAG markings possibly making their aircraft special targets. CDR Ken Burrows', CAG 10 A-4C is on NAS Cecil Field's ramp following a Vietnam combat tour on board *Intrepid* (CVS-11). McDonnell Douglas photo, by Paul Quapp, courtesy Harry Gann

Air Wing 19 was a leader with colorful CAG birds, as with this splendid example of CDR Billy Phillips' F-8E 150909 at NAS Miramar, CA, 16 Jul 1966. VF-194's 400 modex was used instead of the more common at the time 00. Doug Olson

Continuing CVW-19's colorful tradition, VA-52 dressed up its McDonnell Douglas A-1H 134569 Skyraider for CAG Billy Phillips for a 1967 Vietnam deployment in *Ticonderoga* (CVA-14). The use of 300 for a Skyraider squadron was unusual, as they normally were the 500 squadron of each air wing. Neil Schneider, courtesy Richard M. Hill

VF-53's CAG bird F-8E 150323 wore "double nuts" on its nose for CDR Jack L. Snyder, CAG 5, at Miramar 17 Sep 1966. Bud deDonato

CAG 5, CDR William H. Harris', A-4E Skyhawk has more than 60 combat mission markers painted on its nose at NAS Lemoore in April 1967. Doug Olson

A mid-50s requirement for a long-range night attack, medium attack carrier bomber resulted in the Grumman A-6 Intruder series, which began arriving at VA-42 and 75 on the East Coast in February 1963, just in time for the Vietnam War. For nearly 30 years Grumman was noted for its Navy fighters, but outdid itself with the superb Intruder. CAG 9, CDR Alexander R. Burts' A-6A undergoes final safety checks prior to launch from NAS North Island, CA, in January 1969. Doug Olson

Another new attack aircraft joined the fleet during the Vietnam War when Vought's capable A-7A Corsair II reported to VA-147 in early 1967. VA-97's A-7A 153252, flown by CAG 14, CDR Kenneth E. Enney, is at NAF Washington, DC, 7 Jun 1969. Joe Handleman

Another variation of the 00 marking was this VF-162 F-8E 149203 with eyeballs, flown by CDR Jim Stockdale, CAG 16, in 1966. Photo courtesy CAPT Cal Swanson, USN(Ret)

1970 - 1979

By 1970, VF-194's F-8J 150683 CAG bird was somewhat subdued using only multi-colored lightning bolts representative of the squadron nickname, The Lightnings. CDR Douglas F. Mow was CAG 19 at the time of this October 1970 photo during CVW-19's Vietnam deployment in *Oriskany* (CVA-34). Don Priest, courtesy Steve Ginter

CAG 9 CDR Lowell F. "Gus" Eggerts' A-7E 156833 is on a combat mission over North Vietnam in May 1972 with a VA-165 A-6A Intruder. CDR Thomas C. Watson, Jr., relieved Eggert 20 May. USN photo, courtesy Bob Carlisle

Possibly the most flamboyant squadron markings were worn by VF-51's Screaming Eagles during CVW-15's 1971-72 Vietnam combat cruise. The squadron's CAG F-4B BuNo 150456 had multi-colored feathers on the "Supersonic Eagle." However, irreverent fighter pilots from other squadrons more often referred to it as a "supersonic can opener." CVW-15 had two CAGs for the cruise, CDR Thomas E. Dunlop was killed over North Vietnam 6 Apr 1972 in an A-7E, upon which CDR Roger E. "Blinky" Sheets assumed command. CDR Rick Webb, USN(Ret), courtesy Tom Gates.

Possibly the most famous of all CAG birds was Gus Eggert's F-4J 155800, shown here launching from *Constsellation* (CVA-64). The Fighting Falcons' VF-96 Phantom II "Showtime 100" was flown 10 May 1972 by LT Randy Cunningham and LTJG Willie Driscoll during the largest air battle of the Vietnam War. The Fighting Falcons' aircrew downed their third, fourth, and fifth MiGs to become the Navy's only aces of the war. They were shot down the same day by a SAM and rescued unharmed. CAPT David P. Erickson, USN

CDR George A. Aicheson, CAG 15, personalized his F-4B by using his initials for the modex number. VF-51 had also subdued their colors by the time of this 23 Nov 1973 photo at NAS Miramar. Bob Lawson

The Intruder became a virtual bombing workhorse in Vietnam. It had the ability to carry a larger bomb load than the WWII B-17 Flying Fortress. CDR Richard D. "Deke" Bordone's CAG 3 VA-75 A-6A is on a joint USAF, USN, USMC mission over Vietnam during December 1972. Photo courtesy Larry Davis

VAW-115's Liberty Bells were resplendent in 1973 with CVW-5's Grumman E-2B Hawkeye CAG bird. CDR Marion H. Isaaks was the air wing commander in 1973 until relieved 3 August by CDR John B. Stuart. USN photo

VF-21 Freelancers "Sundown 200" F-4J 158378 carrying the colors of CAG 2 CDR Roger A Massey, Jr., during a training flight out of Miramar 22 Jan 1974. CDR Jan C. Jacobs, USNR(Ret)

VA-147's Argonauts added "CAG" to the vertical stabilizer of A-7E 156834, along with the multi-colored rudder for CDR James M. Seely, CAG 9. CVW-9 deployed to WestPac and the Indian Ocean during 1974 in *Constellation* (CV-64) to become the first U.S. Navy carrier presence in the Persian Gulf since 1948. Bob Lawson

CVW-1 took the first East Coast Grumman F-14A Tomcat squadrons to the Med during 1975-76. VF-32 Swordsmen CAG bird lands aboard *John F. Kennedy* (CV-67) during the cruise. CDR Gerald L. Riendeau was relieved early in the deployment by CDR Dwight D. Timm as CAG 1. USN photo, courtesy Bob Carlisle

Another Vietnam MiG killer, CDR Lowell L. "Moose" Myers, CAG 17, temporarily lost the use of this VA-83 A-7E 157480 after it took the barricade on board *Forrestal* (CV-59) in May 1974. CDR Pete Clayton, USN(Ret)

CAG 6's VA-15 Lions A-7E wore a beautiful blue and white tail scheme with multi-colored stars during this 1974 weapons deployment to MCAS Yuma, AZ. CVW-6 was commanded by CDRs Guy Cane (a MiG killer) and Henri B. Chase during that year. USN photo

Carrier-based ASW squadrons began integrating into the attack carrier air wings in the early '70s. VS-83's Grumman S-2G Tracker, assigned to CVW-11 for deployment in *Kitty Hawk* (CV-63), carried CDR Forrest S. "Tooter" Teague's CAG markings during this April 1975 photo at NAS North Island. Teague was better known as an effusive VF-51 skipper and MiG killer during 1971-72. Bob Lawson

The highly-capable Lockheed S-3A Viking began replacing the S-2s in the fleet during 1974. VS-21 made the first deployment with the new ASW aircraft to the Med in 1975 with CAG 1 CDR Gerald L. Riendeau embarked in *Kennedy.* Bob Lawson

Not to be outdone by their tactical squadron contemporaries, VFP-63 dressed up its Det 2 RF-8G 146890 for CAG 6, CDR Henri B. Chase, during its 1975 Med cruise in *Franklin D. Roosevelt* (CVA-42). USN photo

CDR Hugh Lynch was CAG 9 at the time of this 1975 photo of VA-165 Boomers A-6A Intruders near NAS Fallon, NV. Under the new air group designation system instituted during WWII, CVG-9 became the first numbered air group, vice the prewar system of naming air groups after the ships to which they were assigned. WWII's CVG-9 was disestablished 15 Oct 1945. A new CVG-9 was established 26 Mar 1952 and became CVW-9 on 20 Dec 1963. The Navy abandoned the "establish" "disestablish" system for its naval aviation units 19 Mar 1998. Units are now activated and deactivated while keeping the number in perpetuity as one constant unit. USN photo.

Marine Corps presence in Navy air wings began becoming more predominant during the mid-70s. VMFA-451Warlords' F-4J 153775, bears CAG 17 CDR Dean E. Cramer's colors as it taxis at NAS Miramar in April 1975. Bob Lawson

In honor of the Nation's 1976 Bicentennial celebration, many Navy squadrons had one aircraft adorned with special bicentennial markings. VF-161 chose to use CAG 5, CDR John L. "Black Jack" Finley's F-4N 151433 for this purpose at NAS Atsugi, Japan. Shinichi Ohtaki

The last CAG 19, CDR Robert C. Coffey, took his air wing aboard *Roosevelt* (CV-42) for CVW-19's final deployment during October 1976-April 1977. VA-215's A-7B 154381 CAG bird is about to launch during workups on board *Oriskany* (CV-34). CDR Pete Clayton, USN(Ret)

VF-111's Sundowners have a long tradition of spectacular markings, which they continued with Bob Coffey's Phantom F-4B 153012 during CVW-19's '76-'77 deployment in *FDR*. CDR Pete Clayton, USN(Ret)

By the time of CVW-19's FDR Med cruise, CAG's VF-51 F-4N 152326 had returned to the traditional 100 nose number. Courtesy LTJG Tom Kilcline, Jr.

During 1977 the Navy began experimenting with new camouflage schemes for its tactical aircraft. VF-1's CAG 14 aircraft flown by CDR Roger Box, was selected to test the unsuccessful Ferris scheme, seen here at NAS North Island 14 May 1977. Roy Lock

CDR Charles L. "Chuck" Tinker's VF-201 CAG 20 aircraft carries a "1000" modex number during an August 1977 weapons deployment to NAS Fallon. The markings honor Chuck's recent 1,000th trap. Doug Olson

CAG markings on helos are a rare occurrence, such as these seen on HS-2's SH-3D 154122. CDR David N. Rogers was CAG 14 at the time of this March 1979 photo at North Island. Doug Olson

A first in naval aviation history occurred during 1979-80 when two Marine squadrons deployed as sole fighter cover for CVW-14 in *Coral Sea* (CV-43). VMFA-531's F-4N 152323, with CAG CDR David N. Rogers' colors, is in burner for launch during August 1979. CDR Pete Clayton, USN(Ret)

VMFA-323's Death Rattlers' was the sister Marine squadron for CVW-19's 1979 *Coral Sea* deployment to the Med. F-4N 150480 is over Southern California 7 Nov 1979 in Dave Rogers' markings. Harry Gann

CAG, CDR H. David Alexander, led his CVWR-30 during its 1979 weapons deployment to NAS Fallon, NV. Reserve air wings are commanded by regular Navy CAGs. VA-305 A-73 154554 over Fallon 4 Sep 1979. Bob Lawson

1980 - 1989

CDR Luther F. Schreifer's CAG 3 markings are worn by VF-31's F-4J 155530 on board *Saratoga* (CV-60) during a January 1960 CarQual period. Schreifer was the first ASW pilot to command an integrated air wing. Bob Lawson

Most of CVW-3's squadrons of the early '80s exhibited great pride in their squadron and CAG markings as seen on this VA-105 Gunlslingers A-7E 159269 over Pinecastle Target in Florida. USN Photo

CAG 3's VA-75 Sunday Punchers' A-6E Intruder lands aboard *Saratoga* 12 Jan 1980 in CVW-3 colors. Bob Lawson

A-7E Corsair II of VA-97 wore special markings, along with CDR Vincent J. Huth, CAG 14's, colors for the Iran hostage raid from *Coral Sea* (CV-43) 25 Apr 1980. USN photo

Another rare sight is CAG markings on an EA-6B Prowler, such as VAQ-137's 158807 on board *Ranger* (CV-61) during September 1980. CVW-2 was led by CDR John M. Waples, one of the Navy's leading aviators in carrier landings. Because of the multiple use of 6XX series numbers in current air wings, EA-6B squadron numbers begin with 620. CDR Pete Clayton, USN(Ret)

VF-84's F-14A 160386 had only the CVW-8 logo on its tail to specialize it as CDR Fred. L. "Bad Fred" Lewis' aircraft on this 19 Jan 1981 flight over Fallon. Bad Fred later became one of the Navy's first two "Super CAGs" when he again took command of CVW-8. Bob Lawson

A unique marking was CDR Dave Layton's Reserve CAG 20 colors on his VA-205 A-7B at Fallon, in July 1981. Mike Grove

VA-196 Main Battery's painters perfomed a max effort on A-6E 154170 for CDR Richard A. Wilson, CAG 14 on board *Coral Sea* during October 1981. CDR Pete Clayton, USN(Ret)

Accompanying the decade of the '80s was a drastic change in Navy aircraft markings. Several new camouflage schemes were evaluated (all various shades of dull gray) to replace the gull-gray/white scheme that had been standard since 1956. Additionally, high visibility markings were virtually banned, seemingly sounding a death knell for CAG markings. One concession was to allow one aircraft in the squadron to have a special paint scheme, as seen here on VF-143's F-14A CAG 7 aircraft in flight 19 May 1982 with the CO's aircraft AG 101. Bob Lawson

CAG 14. CAPT John L. "Swede" Zerr, wore his squadron's colors on his helmet during the dull dark '80s. Bob Lawson

By the mid '80s, the Phantom was fast disappearing from the fleet, being replaced by F-14s and F/A-18s. VF-151's F-4J 158348 still had its full colors during 1984. CAG 5 was led by CDR Timothy R. Beard during 1984-85. USN photo

The McDonnell Douglas F/A-18A Hornet began fleet service in 1980, replacing the last Phantom squadrons. VFA-137 Kestrels pilots show off the versatility of the Hornet during this flight near Fallon c. 1987. CDR Paul C. "Paco" Campbell was CVW-10's only CAG during its short-lived rebirth during 1986-88. USN photo

VF-111 subdued their markings somewhat during 1986 on its CAG bird F-14A 160660 seen here escorting a Soviet Tu 95 Bear long-range bomber over the Western Pacific. CAPT Ronald J. "Zap" Zlatoper, the first West Coast "Super CAG," commanded CVW-15 that year during deployment in Carl Vinson (CVN-70). USN photo

VF-1 Wolfpack, the Navy's first Tomcat squadron, seemingly never went for multi-colored markings on its CAG aircraft, opting instead for a modified paint scheme. CAPT Christopher T. "Boomer" Wilson was CAG 2 during this 1988 photo of F-14A 152597. USN photo

CAG 2 Boomer Wilson (right) celebrated his 1,000th trap with a 1000 nose number on his VF-1 Tomcat 23 Jul 1989 while deployed in *Ranger* (CV-61). LCDR Dan "Darth" Cain, CVW-2 operations officer, made his 900th landing on the same flight. USN photo

Although their markings were subdued, carrier aircrews still maintained their camaraderie and *esprit* throughout the gloomy days of the late '80s. VA-94's A-7E carries only a minimal red shrike and "CAG" lettering on CAPT Thomas J. Ford's CAG 11 bird. USN photo

Another short-lived air wing was the last CVW-13 (1 Mar 1984-1 Jan 1991), but it was around long enough to participate in the 1986 raid against Libya. VMFA-451 Warlords was one of the squadrons that participated in the operation. CAPT Warren B. "Butch" Christie's F/A-18A is at Fallon 12 Dec 1988. Doug Olson

Another variation of the subdued schemes is seen on this VA-37 Bulls A-7E 160617 landing at NAS Cecil Field 6 Jan 1989. CAG 6 was CAPT Patrick C. "Maztini" Havert. Bruce Trombecky

VF-33's Starfighter F-14As had only CAG Johnson's aircraft wearing any color when this 2 Jun 1989 photo was taken. Bob Lawson

CAPT Jay L. Johnson, CAG 1 during this April 1989 photo, went on to become Chief of Naval Operations, the Navy's top job. VFA-82 F/A-18Cs carry full weapons loads over Fallon's training range. AB 300 163467 wears CAG's colors. McDonnell Douglas photo

CVW-14 remained irrepressible fighting for its black-tailed aircraft and colorful CAG markings. VF-21 F-14A launches from Miramar during June 1989 in CAPT William S. "Bud" Orr's marking. Ray Rivard

CAPT Jay L. "Spook" Yakely carried on CVW-14's black tail tradition during his air wing's 1990 deployment to the Persian Gulf for Operation Desert Shield against Iraq. VFA-113's F/A-18A CAG bird in over *Constellation* (CV-64) during her 1989 WestPac deployment under Bud Orr. LCDR Dave Kennedy

1990 - 1999

The last decade of the millennium began with a dearth of colorful markings on U.S. Navy aircraft. Another example of the subdued CAG birds was VAQ-141's EA-6B AJ 620 wearing only small red stripes on the base of the rudder. CAG at the time of CVW-8's January 1990 Fallon weapons deployment was CAPT John F. "Archie" Manning, Jr. Bob Lawson

Contrasting paint schemes of VF-32 Swordsmen's F-14A CAG bird and the rest of the squadron is seen in this May 1990 photo as CVW-3 participated in Operation Desert Shield. CAG was CAPT A. Hardin White, Jr. LCDR Dave Parsons, USN(Ret)

CAPT Jay "Rabbit" Campbell's VAW-116 E-2C had a colorful temporary water-based camouflage scheme during the squadron's det to NAF El Centro, CA, in mid-1990. USN photo

Another rare helo "bubba" CAG bird was HS-8's SH-3D 149713 carrying CAT 14 CAPT Patrick D. "Münt" Moneymaker's colors at NAS North Island 22 Dec 1990. Bob Lawson

VA-72 specialized its A-7E CAG 3 bird during the Gulf War of early 1991. Thirty-three "camel" bombing mission symbols are on the nose of CAPT Hugh D. "Ranger" Connell II's Corsair returning to *John F. Kennedy* (CV-67) after a mission. This deployment would be the last hurrah for the A-7, which was phased out of tactical service in 1991. CAPT John R. Leenhouts, USN

After the advent of Super CAGs and Deputy CAGs, it became common practice to place CAG's name and title on the left side of an aircraft and those of DCAG on the opposite side. CAPT Jay B. "Spook" Yakely's CVW-14 CAG birds (above and below) reflected this usage, along with the proud colorful markings of early 1991. Doug Olson, Mick Roth

VS-38's Griffins were still flying the older S-3A model of the Viking from *Ranger* (CV-61) during Desert Storm. Thus, Jay Campbell's CAG bird still retained the old gray/white basic paint scheme and full-color tail markings. CDR Pete Clayton, USN(Ret)

The Tigers of VA-65 adhered to the "one colorful airplane" dictate during CVW-8's Desert Storm deployment in *Theodore Roosevelt* (CVN-71). CAG 8, CAPT William J. "Fox" Fallon, flew A-6E 161675 for the last combat deployment of Grumman's Intruder. LCDR Rick Morgan, USN(Ret)

In various modified forms, VF-111's Sundowners remained faithful to their "setting sun" markings. In early 1992 their F-14A was back to a full-color tail for CAG 15, CAPT Frederic G. "Luds" Ludwig. CDR Tom Twomey, USN

Certainly the most artistic of all CAG birds was VFA-195 Dambusters' F/A-18C 163758 of the mid-'90s. "Chippie 400" is landing at NAF Atsugi 9 Mar 1995 during CAPT Brian M. "Friar" Calhoun's tour as CAG 5. Shinichi Ohtaki

Air Wing Five revived the colorful spirit with its CAG birds during the '90s. VAQ-136 Gauntlets' EA-6B 161883 flies past Japan's famed Mt. Fuji c. 1995 in Brian Calhoun's colors. USN photo, courtesy LT Michael Arthurs

VS-21's S-3B also poses near Mt. Fuji during a 1995 photoex. USN photo, courtesy LT Michael Arthurs

An all-black tail became popular with several Tomcat squadrons after the "one airplane colored" edict. VF-154 Black Knights F-14A 161621 in CAPT Richard J. Mauldin's markings during 1996. CDR Tom Twomey, USN

CVW-14 had to tone down its black tails upon returning stateside, but still retained colorful CAG markings, as seen on VF-11 Red Rippers' F-14D and VFA-113 Stingers' F/A-18C as *Carl Vinson* (CVN-70) steams beneath the Golden Gate for Fleet Week in October 1995. Both aircraft were launched inside San Francisco Bay to the awe of the onshore spectators. CAG 14 was CAPT Mark Fitzgerald. George Hall

The last two A-6E Intruder squadrons went out in style. VA-75 Sunday Punchers (foreground) was the first squadron to deploy the Intruder and take it into combat. VA-196 Main Battery flew the Intruder from NAS Whidbey Island for nearly 31 years. Both squadrons were disestablished at Whidbey 28 Feb 1997, ending the career of Grumman's great medium attack bomber. AA 500 164382 was CAG 17 CAPT James M. Zortman's aircraft, and NK 500 159579 CAG 14 CAPT Hamlin B. "Ham" Tallent's. Both aircraft were on their squadron's and the A-6's last deployment. They are over the North Arabian Gulf 26 Sep 1996. USN photo, courtesy LT Pat Etienne

The rarest of all modern CAG birds is this ES-3A of VQ-5 Det B NK 727. The ES-3A is an S-3A Viking modified for carrier airborne electronic warfare missions. CAPT Tom Kilcline's aircraft is over *George Washington* (CVN-72) during a mid-1998 mission. USN photo

VAW-113's E-2C 164353 bearing CAG 14 CAPT Thomas J. "Killer" Kilcline's colors lands aboard *Carl Vinson* (CVN-70) during 1997 in "stateside markings." LT Ian "Chunx" Anderson

The most colorful of all Prowlers was EA-6B 161115 of VAQ-139 seen landing aboard Vinson in 1997. LT Ian "Chunx" Anderson

VF-31 Tomcatters F-14D claws for altitude bearing Killer Kilcline's CVG-14 markings. VF-31 traces its "Felix Cat" heritage back to the prewar VF-1B established 1 Jul 1935. LT Ian "Chunx" Anderson

Lts Craig "Hitter" Larson and Dave "Fur" Gergstrom show off VF-14's CAG bird painted to honor the Tophatters' 80-year history of continuous operation, the Navy's oldest squadron. The F-14A is over Dare County, VA, Military Operations Area 22 Mar 1999. CAG 8 was CAPT Dale E. Lyle. CDR Tom Twomey, USN

CAG LIST BY NAME

Last Name	First Name		Rank	AirWing	Command Dates						Comments
Adams	John	F.	LCDR	CVEG-29	1944	JAN	05	1944	JAN	21	
Ady Jr.	Howard	P.	CDR	CVG-101	1951	OCT	25	1953	MAY	15	CVG-14 4 FEB 53
Aitcheson	George	A.	CDR	CVW-15	1973	FEB		1974	MAR	15	
Albertson	William	H.	CDR	CVW-2	1971	JUL	02	1972	MAY	03	
Albertson	William	H.	CDR	CVW-14	1970	JAN	02	1970	JAN	26	ACTING
Aldern	Donald	D. 'Dean'	CDR	CVW-16	1969	OCT		1969	DEC	12	
Aldern	Donald	D. 'Dean'	CDR	CVW-19	1969	DEC	12	1970	JUN	29	MIA over Laos in A-7A
Alderton	Dickson	W.	CDR	CVG-19	1963	MAR	22	1964	MAR		CVW-19 20 DEC 63
Alexander	H.	David	CAPT	CVWR-30	1978	NOV	11	1980	MAY	31	
Alldredge	Walter	W.	CDR	CVG-5	1963	JAN	18	1964	JAN	10	CVW-5 20 DEC 63
Allen Jr.	Lloyd	E.	CAPT	CVW-2	1984	DEC	27	1985	OCT	28	
Allen	Richard	C. 'Sweetpea'	CDR	CVW-6	1979	MAY	04	1980	JUN	06	1st NFO CAG
Allen	William	B.	CDR	CVG-10	1957	MAR	10	1957	MAR	25	ACTING
Altmeyer	John	M.	CAPT	CVWR-30	1971	NOV	21	1973	JUN	24	
Anderson	William	D.	CDR	CVG-9	1942	MAR	01	1942	APR	10	1st CAG
Anderson	Robert	L.	MAJ	MCVG-10							ONLY CAG?/UNKNOWN EXISTENCE DATES
Andrews	Edward	K. 'Hunyak'	CAPT	CVW-6	1982	NOV	01	1984	JUL	19	
Armstrong Jr.	Clarence	E. 'Skip'	CDR	CVW-6	1975	OCT	14	1977	JAN	26	
Arnold	H.	Duff	CDR	CVW-11	1968	JAN	13	1968	NOV	02	or 14 JAN 68?
Arnold	Jackson	D.	CDR	CVG-2	1944	JUN	01	1944	DEC	01	
Askey	Charles	B.	CAPT	CVWR-20	1996	MAR	26	1997	JUL	12	
Atherton	Raymond		CDR	CVW-10	1969	AUG	29	1969	NOV	28	LAST CAG
Ault	Frank	W.	CDR	CVG-10	1960	MAR	18	1961	MAY	22	
Ault	William	B.	CDR	LEXINGTON	1941	AUG		1942	MAY	08	LAST CAG/KIA at Coral Sea in SBD-3
Avery	Howard	M.	CDR	CVG(N)-91	1944	OCT	05	1946	JUN	21	1st CAG/ONLY?
Axtell Jr.	George	C.	MAJ	MCVG-16							ONLY CAG?/UNKNOWN EXISTENCE DATES
Badger	Heber	J.	CDR	ATG-202	1955			1957			1st CAG
Bagdanovich	Michael	P. 'Seabags'	CDR	CVG-17	1943	JAN	01	1944	JAN	08	1st CAG
Bagwell	Ralph	M.	CDR	CVG-5	1956	JUL	06	1957	DEC	13	
Bagwell	Wallace	B.	CDR	CVW-8	1966	NOV	10	1967	OCT	05	
Baird	Don	W.	CAPT	CVW-2	1982	APR	07	1983	AUG	11	
Baker	Steven	H.	CAPT	CVW-7	1992	JUL	29	1994	JAN	14	
Bakutis	Frederick	E.	CDR	CVAG-1	1947	JUN		1948	JUL		
Bakutis	Frederick	E.	CDR	CVAG-5	1948	JUL	26	1949	FEB	04	CVG-5 1 SEP 48
Bakutis	Frederick	E.	LCDR	CVG-20	1943	NOV	10	1943	DEC	03	ACTING
Baldwin	Oa	Fredrick 'Fred'	CDR	CVW-2	1977	APR	29	1978	JUL	08	1ST NAME Oa!
Baldwin	Robert	B.	CDR	CVG-16	1963	FEB	09	1964	FEB	03	CVW-16 20 DEC 63
Ball	Ronald	F.	CDR	CVW-6	1977	JAN	26	1978	MAY		
Ballintine	John	J.	LCDR	TrpBmbWing	1932	JUL	01				AS OF DATE
Bandy	Jack	I.	LCDR	CVEG-28	1943	AUG	01	1943	SEP	30	ACTING
Bangs	Louis	L. 'Bangor'	CDR	CVG-9	1956	SEP		1958	MAY		
Bardshar	Frederic	A.	LCDR	CVLG-27	1944	JUN	19	1945	OCT	26	REFORMED 02 JAN 45/LAST CAG
Barker	Raymond	H.	CDR	CVSG-57	1968	AUG	16	1969	JUL	09	
Barksdale	David	A.	CDR	CVSG-52	1965	AUG	09	1966	JUN	15	
Barleon Jr.	John	S.	CDR	CVG-98	1946	OCT	01	1946	NOV	15	CVAG-21 15 NOV 46
Barleon Jr.	John	S.	CDR	CVAG-21	1946	NOV	15	1947	AUG	05	LAST CAG
Barnett	Marvin	E.	CDR	CVG-3	1956	OCT		1957	SEP	04	
Barr	Max	D.	CDR	CVW-1	1967	MAR		1968	JUN	27	
Barrow	John	C.	CDR	CVW-3	1967	JUL	04	1968	NOV	14	
Barth Jr.	Joseph	J.	CDR	CVW-21	1971	JUN	18	1972	AUG	19	
Bascom	Wade	R. 'Bob'	CDR	CVG-15	1963	JUL	01	1964	JUL	08	CVW-15 20 DEC 63
Bass	Harry	B.	LT	CVEG-29	1944	JAN	21	1944	APR	10	REFORMED 1 MAR/CVLG 01 MAR 44
Bates	Fred	L.	LT	CVEG-37	1943	JAN	02	1943	AUG		1st CAG
Bauer	Louis	H.	CDR	CVG-3	1945	AUG	08	1947	JUL	30	CVAG-3 15 NOV 46
Beakley	Wallace	M.	LCDR	WASP	1942	JAN		1942	SEP		LAST CAG
Beard	Timothy	R.	CDR	CVW-5	1984	AUG	23	1986	APR	11	
Beaty III	William	E. 'Jesse'	CDR	CVW-6	1986	JAN	23	1987	MAY	29	
Beauchamp	Ernest	M.	CDR	CVG-21	1955	JUL	01	1957	FEB	05	1st CAG
Beebe	Marshall	U.	CDR	CVG-5	1950	DEC	12	1952	APR	30	
Beebe	Marshall	U.	LCDR	CVG-17	1944	APR	20	1944	APR	24	ACTING
Bellinger	Richard	M. 'Billy'	CDR	CVW-16	1966	OCT	26	1966	DEC	19	ACTING
Berns Jr.	Max	A.	CDR	CVG-7	1950	APR	01	1950	MAY	01	ACTING?
Berrey	Samuel	B.	CDR	CVG-9	1953	AUG		1953	AUG		ACTING/ASSUMED PERIOD
Bert	George	W.	LT	CVEG-38	1945	FEB		1945	FEB		ACTING
Best	James	B. 'Red'	CAPT	CVW-1	1980	APR	07	1981	JUL	10	
Bien	Lyle	G. 'Ho-Chi'	CAPT	CVW-15	1987	MAY	08	1987	SEP	28	
Bien	Lyle	G. 'Ho-Chi'	CAPT	CVW-15	1989	JUN	02	1991	JUL	01	

Billings	Randell	K.	CDR	CVW-14	1969	MAR	28	1970	JAN	02	KIFA over Luzon in RA-5C
Bitting	James	T.	CDR	CVSG-53	1961	APR	07	1961	OCT	12	
Black	Joseph	D.	LCDR	CVLG-30	1943	APR	01	1943	DEC	21	1st CAG
Blackburn	J.	Thomas	CDR	CVBG-74	1945	MAY	10	1946	AUG		1st CAG
Blattman	Walter	C.	CDR	CVG-5	1958	DEC	10	1958	DEC	24	ACTING
Blick	Robert	E.	LCDR	SARATOGA	1941	APR		1941	JUL		
Blitch	John	D. 'Jack'	CDR	CVBG-1	1948	AUG	02	1950	JAN	01	CVG-2 1 SEP 48
Blitch	John	D. 'Jack'	CDR	CVG-14	1944	NOV	02	1945	JAN	01	
Blouin	Clarence	A.	CDR	ATG-201	1955	JUN	01	1956	APR	16	1st CAG?/ASSUMED PERIOD
Bloyer	Stanley	F.	CAPT	CVW-6	1991	JAN	04	1992	APR	01	LAST CAG
Bogan	Gerald	F.	LCDR	FitWing	1928	JAN	01				AS OF DATE
Bogan	Gerald	F.	LCDR	FitWing	1931	APR	01				AS OF DATE
Bogusch	Harry	R.	LCDR	TrpBmBWing	1928	OCT	01				AS OF DATE
Boice	Grant		CDR	CVG-3	1961	APR	20	1962	JUL	18	
Bordone	Richard	P. 'Deke'	CDR	CVW-3	1972	FEB	04	1973	FEB	28	or 2 FEB 72?
Bottomley Jr.	H.	Sydney	CDR	CVG-6	1950	FEB		1951	MAY	19	
Bottomley Jr.	H.	Sydney	LCDR	CVG-13	1949	AUG		1949	SEP	30	ACTING
Bouffard	Edward	N.	CDR	CVSG-56	1967	JUL	12	1968	JUL	25	
Bowen II	John	R.	CDR	CVG-2	1957	MAY		1959	MAR	18	
Bowen	Llewellyn	D. 'Lew'	CDR	RCVSG-51	1965	SEP	24	1966	MAY		
Bowen	Llewellyn	D. 'Lew'	CDR	CVSG-53	1964	NOV	06	1965	SEP	10	
Bowman	Michael	L.	CAPT	CVW-1	1990	SEP	05	1991	NOV	13	
Bowman	Michael	L.	CAPT	CVW-5	1986	APR	11	1987	NOV	07	
Bowman	Michael	L.	CAPT	CVW-13	1990	JUL	26	1990	AUG	21	
Box	Roger	E.	CDR	CVW-14	1976	FEB	28	1977	APR	21	
Boyd Jr.	Alfred	I.	CDR	CVAG-7	1947	JUN	21	1948	DEC	10	CVG-7 1 SEP 48
Boyd Jr.	Alfred	I.	LCDR	CVLG-38	1944	MAY	21	1944	NOV	28	REFORMED 14 JUN 44/CVEG 15 AUG 44
Boyd	Edward	A.	CDR	CVSG-55	1964	OCT	20	1965	SEP	29	
Boyd	Richard	C.	CDR	CVW-6	1967	APR	08	1968	FEB	21	
Boydstun	Howard	J.	CDR	CVG-21	1959	DEC	11	1961	JUN	16	
Boyum	John	H.	CDR	CVG-2	1956	FEB		1957	MAY		ASSUMED PERIOD
Bracken	Leonard	A.	CDR	CVSGR-80	1974	FEB	23	1975	JUN	21	
Bradbury	Thomas	B.	LCDR	CVLG-51	1943	SEP	22				ASSUMED 1st CAG
Bradt	Douglas	J. 'Buckeye'	CAPT	CVWR-30	1986	FEB	01	1987	AUG	30	
Brehm	William	W.	CDR	CVG-101	1950	AUG	21	1951	OCT	25	1st CAG
Bremner	Bruce	B.	CAPT	CVW-17	1980	JUL	10	1981	NOV	12	
Briggs	Steven	R.	CAPT	CVW-5	1979	MAY	31	1980	OCT	20	
Bringle	William	F. 'Bush'	CDR	CVAG-1	1948	JUL		1950	JUN		CVG-1 1 SEP 48
Bringle	William	F. 'Bush'	CDR	CVG-17	1945	OCT		1946	SEP		
Brodsky	Robert	G. 'Bubba'	CAPT	CVW-9	1989	JUN	09	1990	JUL	13	
Brodsky	Robert	G. 'Bubba'	CAPT	CVW-17	1985	AUG	23	1986	OCT	23	
Brooks	Richard	S.	CDR	CVSG-53	1963	DEC	10	1964	NOV	06	
Brow	Harold	J.	LCDR	TrpBmbWing	1933	JUL	01				AS OF DATE
Brown	Charles	R. 'Cat'	LCDR	SARATOGA	1940	MAY		1941	APR		
Brown	Donald	C.	CAPT	CVW-2	1991	JUL	26	1993	FEB	25	
Brown	F.	Taylor	CDR	CVW-9	1965	JAN	11	1966	JAN	04	
Brown	Floyd	H.	CDR	CVSG-60	1966	MAY	12	1967	JUN	09	
Brown	Gaylord	B.	CDR	CVG-7	1952	JAN	01	1953	FEB		
Brown	Gaylord	B.	LCDR	CVLG-25	1944	JUN	12	1944	AUG	28	
Brown	Gordon	J.	CDR	ATG-202							WHEN? LAST CAG?
Brown	Gordon	J.	CDR	CVG-4	1958	MAR		1958	JUL	03	RCVG-4 1 APR 58
Brown	Ian	F.	CDR	RCVSG-50	1964	JUL	06	1965	JUL	30	
Brown Jr.	Samuel	R.	CDR	CVG-14	1945	FEB	16	1945	DEC		
Brown III	Thomas	F.	CDR	CVW-19	1972	JUN	05	1973	JUN	28	
Browning	Miles	R.	LCDR	FitWing	1934	OCT	01				AS OF DATE
Browning	Miles	R.	LCDR	YORKTOWN	1938	JUL	01	1939	JUN	26	1st Formal CAG
Bruce	Donald	E.	CDR	CVG-19	1951	JUL	27	1951	SEP	14	ACTING
Bruning	Robert	M.	CDR	CVG-1	1953	MAR		1954	APR		
Brush	Frederick	J.	CDR	CVG-81	1944	MAR	01	1945	MAY	13	1st CAG
Bryan	William	C.	CDR	CVG-1	1957	JUN	20	1957	DEC	21	
Bryan	William	C.	CDR	CVG-8	1956	SEP		1957	MAY		
Bucchi	Torey	M. 'Smiles'	CAPT	CVW-3	1992	DEC	31	1994	APR	15	
Burbage	Charles	L.	CDR	CVG-19	1955	JUL		1957	SEP	20	
Burin	James	M.	CAPT	CVW-5	1989	SEP	09	1991	JUN	14	
Burns	Richard	H.	CDR	CVG-4	1949	OCT	18	1950	JUN	08	LAST CAG
Burns	Richard	H.	CDR	CVG-8	1948	SEP	27	1949	OCT	18	ONLY CAG
Burns	Richard	H.	CDR	CVG-152	1945	MAR	19	1945	SEP	21	ONLY CAG?
Burroughs Jr.	Sherman	E. 'Evie'	CDR	SARATOGA	1942	OCT		1943	MAR		
Burrows	Kenneth	A.	CDR	CVW-10	1966	JUN	23	1968	JAN	12	
Bursik	Vlada	D.	CDR	CVSG-57	1961	JAN	03	1962	JUN	01	1st CAG

Burt	Alexander	R.	CDR	CVW-9	1968	AUG	02	1969	NOV	13	
Butler	Charles	A.	CDR	RCVSG-51	1969	MAR	14	1969	DEC	18	
Butts	W.	Spencer	CDR	CVG-20	1943	OCT	15	1943	OCT	16	1st CAG
Byng	Weston	H.	CDR	CVW-11	1975	JUN	03	1976	DEC	23	
Cady	Joseph		CDR	CVSG-52	1960	JUN	01	1961	AUG	14	1st CAG
Caine	Paul	E.	CDR	CVSGR-80	1975	JUN	21	1976	JUL	01	LAST CAG
Caldwell	Howard	H.	CDR	CVG-12	1943	JAN	09	1944	JAN	01	1st CAG/1 JAN 44 APPROX.
Caldwell Jr.	Turner	F.	CDR	CVBG-3	1947	NOV	19	1949	FEB		CVG-4 1 SEP 48
Caldwell Jr.	Turner	F.	CDR	CVLG(N)-41	1944	AUG	28	1945	FEB	25	ONLY CAG?
Calhoun	Brian	M. 'Friar'	CAPT	CVW-5	1994	JUL	13	1996	JUL	09	
Calhoun Jr.	Otis	V.	MAJ	MCVG-8							ONLY CAG?/UNKNOWN EXISTENCE DATES
Califf	Toxey	H.	CDR	CVSGR-70	1970	DEC		1972	APR	15	
Callaghan Jr.	William	M.	CDR	CVSG-54	1967	JUN	10	1968	JUN	13	
Callahan	Earle	R.	CDR	RCVSG-50	1970	JUL	10	1971	FEB	19	LAST CAG
Calland	Robert	W.	CDR	CVG-11	1957	APR	12	1958	NOV	26	
Campbell	Grafton	B. 'Soupy'	CDR	CVG-6	1948	JUN		1950	FEB		CVG-6 1 SEP 48
Campbell	Jay	A. 'Rabbit'	CAPT	CVW-2	1990	FEB	21	1991	JUL	26	
Campbell	Norman	D.	CAPT	CVWR-30	1975	NOV	23	1977	MAY	28	
Campbell	Norwood	A. 'Soupy'	CDR	SARATOGA	1943	MAR		1943	SEP		CVG-3 25 SEP 43
Campbell	Paul	C. 'Paco'	CDR	CVW-10	1986	NOV	07	1988	JUN	01	ONLY CAG
Campbell	William	R.	LCOL	MCVG-2							ONLY CAG?/UNKNOWN EXISTENCE DATES
Campman	John	H.	LCDR	FitWing	1930	JUL	01				AS OF DATE
Campman	John	H.	LCDR	LitBmbWing	1929	OCT	01				AS OF DATE/LAST
Canaan	Gerald	C.	CDR	CVSG-59	1969	MAY	29	1970	JUN	01	
Cane	Guy		CDR	CVW-6	1973	JUN	14	1974	JUL	26	
Canepa	L.	Robert 'Bullet'	CAPT	CVW-5	1983	APR	27	1984	AUG	23	
Canepa	L.	Robert 'Bullet'	CAPT	CVW-11	1987	MAY	29	1989	MAR	03	
Cannon	James	D.	CDR	CVWR-20	1990	JUN	16	1993	FEB	20	
Capalbo	Joseph	J 'Condo'	CAPT	CVW-8	1995	DEC	14	1997	JUL	01	NOTE: NO MIDDLE NAME
Carey Jr.	Henry	A.	LT	CVG(N)-52	1945	FEB	12	1945			ACTING?
Carius	Robert	W.	CDR	CVSG-55	1968	JUL	10	1968	SEP	27	LAST CAG
Carlson Jr.	Olof	Magnus	CDR	CVW-19	1970	JUN	29	1970	JUL	10	ACTING
Carmichael	E.	Inman 'Hoagy'	CDR	CVW-15	1974	MAR	15	1975	MAY	19	
Carmody	Martin	D. 'Red'	CDR	CVG-8	1957	AUG	01	1958	JUL	09	
Carr Jr.	Donald	E.	CDR	CVG-15	1957			1958	MAY	23	PRIOR TO JUL 57
Carroll	David	L.	CAPT	CVW-5	1987	NOV	07	1989	SEP	09	
Carroll	David	L.	CAPT	CVW-11	1984	MAY	18	1985	OCT	23	
Carter	George	H. 'Spanky'	CDR	CVG-17	1951	FEB		1952	JUN	13	
Carter Jr.	John	T.	CAPT	RCVSG-50	1968	JUN	18	1969	JUL	07	
Carter	Rodney	B.	CDR	CVW-16	1966	SEP	30	1966	OCT	26	Killed by Oriskany fire
Carver	Lamar	P. 'Pete'	LCDR	RANGER	1943	MAR	15	1943	JUL	15	
Casey	Vincent	F.	CDR	CVAG-5	1947	JUL	01	1948	JUL	26	
Casey	Vincent	F.	LCDR	CVLG-21	1943	SEP	27	1945	JAN	16	REFORMED
Cash	Paul	D.	CAPT	CVW-1	1991	NOV	13	1993	MAY	04	
Cash	Paul	D.	CAPT	CVW-13	1990	AUG	21	1991	JAN	16	LAST CAG
Cash Jr.	Roy		CAPT	CVW-14	1982	JUN	10	1983	SEP	09	
Cates	Charles	W.	CAPT	CVW-16	1968	FEB	05	1969	AUG	18	KIFA at Okinawa in A-7B
Cebrowski	Arthur	K.	CAPT	CVW-8	1982	DEC	13	1984	AUG	22	
Cellar	Charles	J.	CDR	CVW-2	1972	MAY	03	1973	JUN	11	
Chaffee	George	B.	LCDR	CVLG-23	1942	NOV	16	1943	APR	24	1st CAG
Chalbeck	John	A.	CDR	CVW-10	1968	JAN	12	1969	MAR	26	
Champlin	Norman	D.	CDR	CVSG-56	1960	MAY	26	1961	AUG	22	1st CAG
Chapman	Austin	E.	CAPT	CVW-9	1982	DEC	03	1984	JUN	22	
Chapman	Austin	E.	CAPT	CVW-17	1986	OCT	23	1988	OCT	09	
Charbonnet Jr.	Pierre	N.	CDR	CVG-8	1955	MAY		1956	SEP		
Charbonnet Jr.	Pierre	N.	LT	CVEG-24	1945	APR	10	1945	SEP	25	LAST CAG
Charles	Nelson	R.	CDR	CVSG-58	1960	JUN	06	1961	JUL	15	1st CAG
Chase	Gordon	P.	CDR	CVG-102	1952	JAN	17	1953	JUN	30	CVG-12 4 FEB 53
Chase	Henri	B.	CDR	CVW-6	1974	JUL	26	1975	OCT	14	
Chase	Robert	P.	LT	CVEG-60	1945	JAN	02	1945	APR	19	REFORMED/ACTING
Chatham	W.	Lewis	CAPT	CVW-5	1974	JUL	23	1975	OCT	24	
Cheney	Ralph	E.	LCDR	CVG-10	1952	MAY	01	1952	MAY	30	ACTING
Chick	Lewis	W.	CDR	ATG-182	1955	JUL	01	1957			1st CAG?
Christiansen	John	S. 'Big Coolie'	CAPT	RCVW-4	1965	JUN	28	1967	JUL	14	
Christiansen	John	S. 'Big Coolie'	CDR	CVG-6	1963	MAR	07	1964	JUL	04	CVW-6 20 DEC 63
Christie Jr.	Warren	B. 'Butch'	CAPT	CVW-13	1988	NOV	18	1990	JUL	26	
Christiensen Jr.		Ernest	E.	CAPT	CVW-2		1983		AUG	11	1984 DEC 27
Clancy Jr.	Albert	H. 'Hat'	LCDR	CVLG-47	1945	MAR	20	1945	SEP	21	LAST CAG
Clark	Carlton	H.	CDR	CVG-6	1953			1954	OCT	22	22 OCT 54 APPROX.
Clark	Douglas	A.	CDR	CVG-10	1955	FEB	10	1957	MAR	10	

Clark	Douglas	A.	LCDR	CVLG-30	1945	FEB	23	1945	APR	15	ACTING
Clark	Douglas	A.	LCDR	CVLG-30	1945	APR	15	1945	SEP	12	LAST CAG
Clark	Sam	E.	CDR	CVG-4	1950	SEP	01	1951	OCT		1st CAG
Clark	Sam	E.	CDR	CVG-18	1946	APR	01	1946	APR	19	ACTING
Clifton	Joseph	C.'Jumpin' Joe'	CDR	CVG-12	1944	JAN	01	1944	JUL	20	01 JAN 44 APPROX.
Coffey	Robert	C.	CDR	CVW-19	1975	NOV	28	1977	JUN	30	LAST CAG
Coffin	Albert	P. 'Scoffer'	CDR	CVG-19	1945	JAN	24	1946	NOV	25	REFORMED 20 JAN 45
Cogswell	George	W.	CDR	CVSG-58	1963	JUN	26	1964	MAY	21	
Coleman	Thaddeus	T.	CDR	CVG-11	1961	MAR	30	1962	JAN	31	
Coleman	Wilson	McC.	CDR	CVG-18	1944	NOV	03	1944	NOV	11	ACTING
Coleman	Wilson	McC.	CDR	CVG-18	1944	NOV	11	1946	APR	01	
Colenda	Herbert	F.	CDR	RCVSG-51	1969	DEC	18	1970	JUN	30	LAST CAG
Collins	Vincent	W.	CDR	CVSG-55	1961	DEC	11	1962	DEC	10	
Coln	Royce	W.	LCOL	MCVG-3							ONLY CAG?/UNKNOWN EXISTENCE DATES
Colvin	Robert	D.	CDR	CVSG-56	1968	JUL	25	1969	MAY	20	
Combs	Robert	E.	CDR	CVSG-54	1970	JUN	23	1971	JUL	13	
Combs	Thomas	S.	LCDR	ENTERPRISE	1937	JUL	01	1938	JUN		Senior Squadron CO
Conatser	Charles	N. 'Tex'	CDR	CVG-12	1954	OCT	29	1956	APR	06	
Conn	John	P.	LCDR	CVG-84	1944	MAY	01	1944	MAY	05	ACTING
Connell II	Hugh	D. 'Ranger'	CAPT	CVW-3	1991	JUN	24	1992	DEC	31	
Conner Jr.	Andrew	B.	CDR	CVG-6	1959	JUN	28	1960	MAY	26	
Conrad	Robert	W.	LCDR	CVLG-34	1945	APR	01	1945	DEC	05	ONLY CAG?
Coogan	Robert	P.	CDR	CVG-16	1960	SEP	01	1961	FEB	13	ACTING
Cook Jr.	Harry	E.	CDR	CVAG-19	1946	NOV	25	1948	SEP	01	
Cooke Jr.	Emmett	M.	CDR	CVW-5	1964	JAN	10	1965	JAN	12	
Cooley Jr.	Samuel	M.	CDR	CVSG-52	1963	JUL	29	1964	JUL	14	
Coon Jr.	Ralph	H.	CAPT	CVW-7	1994	JAN	14	1995	APR	05	
Coonan	John	J.	CAPT	CVW-1	1983	OCT		1985	MAR		
Cooper	Damon	W. 'Hutch'	CDR	ATG-3	1955	MAR	05	1956	JUN	23	1st CAG
Copeland Jr.	William	W.	CDR	CVW-1	1986	JUL	02	1988	FEB	05	
Cormier	Richard	L. 'Zeke'	CDR	CVG-11	1958	NOV	26	1959	NOV	12	
Cotten	Junius	L.	LCDR	TrpBmbWing	1931	APR	01				AS OF DATE
Cottingham	Claybrook	B.	CDR	CVAG-9				1948	SEP	01	CVG-9 1 SEP 48
Cottingham	Claybrook	B.	CDR	CVG-9	1948	SEP	01	1949	DEC	01	LAST CAG?
Cottingham	Claybrook	B.	LCDR	CVG-99	1944	AUG	14	1945	JAN	09	1st CAG
Counihan	John	L.	CDR	CVG-80	1944	JUN		1944	OCT	16	
Cousins	Ralph	W.	CDR	CVG-11	1945	APR	21	1945	JUN	10	ACTING
Cousins	Ralph	W.	CDR	CVG-11	1945	JUN	10	1946	FEB	14	CVAG-11 15 NOV 46
Coward IV	Asbury	'Sandy'	CAPT	CVW-17	1988	OCT	09	1990	JUL	18	
Crabill Jr.	Charles	H. 'Fritz'	CDR	ATG-1	1951	SEP	28	1952	JUL	03	1st CAG
Cramer	Dean	E.	CDR	CVW-17	1975	SEP	09	1976	DEC	03	
Craven	Phillip	R.	CAPT	RCVW-12	1968	OCT	11	1969	JUN	27	
Craven	Phillip	R.	CAPT	CVW-19	1967	JUL	11	1968	SEP	27	
Crawford	Charles	W.	LCDR	WASP	1941	APR		1942	JAN		
Crawford	Roderick	P.	CAPT	CVSGR-70	1973	OCT	27	1975	MAY	10	
Crenshaw Jr.	Lewis	W.	CAPT	CVW-1	1994	MAY	20	1995	JUL	23	
Crim Jr.	George	N.	CAPT	CVW-17	1991	NOV	27	1993	AUG	26	
Crinkley	Andrew		LCDR	FitWing	1936	JUL	01				AS OF DATE/LAST
Crist	Marion	E.	LCDR	SARATOGA	1937	JUL	01	1937	OCT	01	Senior Squadron CO
Crockett	David	S. 'Davey'	CDR	ATG-1	1957	JAN	03	1959	FEB	28	LAST CAG
Crommelin	Charles	L.	LCDR	CVG-5	1943	FEB	15	1943	APR	07	ACTING
Crommelin	Charles	L.	CDR	CVG-5	1943	SEP		1943	NOV	21	WIA over Mille in F6F-3
Crommelin	Charles	L.	CDR	CVG-12	1944	JUL	20	1945	MAR	13	REFORMED/DETACHED TO TF-58 STAFF
Crommelin	Quentin	C.	CDR	CVG-17	1955	OCT	07	1958	MAR	10	
Crosby	Derrill	P.	CDR	CVSG-55	1963	SEP	06	1964	OCT	20	
Cruise	Edgar	A.	LCDR	SARATOGA	1941	JUL		1941	DEC	29	
Cruse	Carl	M.	CDR	CVG-19	1962	MAR	07	1963	MAR	22	
Cummings	Arthur	H. 'Jim'	CDR	CVW-7	1964	JUL	17	1965	MAY	15	
Cunninghame	Craig	T	CAPT	CVW-1	1997	DEC	31				NOTE: NO MIDDLE NAME
Curtis	John	O.	LCDR	CVLG-24	1943			1943	SEP	06	KIFA off CVL-24
Cyr	Richard	F.	CDR	CVG-11	1959	NOV	12	1961	MAR	30	
Dahlston	Vincent	A.	CDR	CVG-8	1954	MAR	14	1954	APR		ACTING/APR 54 ASSUMED
Dale	Roland	H. 'Brute'	CDR	CVG-17	1944	JAN	08	1944	MAR	10	
Dale	Roland	H. 'Brute'	CAPT	CVG-12	1958	MAR	10	1959	JUL	03	RCVG-12 1 APR 58
Dale	Roland	H. 'Brute'	CDR	CVLG-24	1943	OCT		1943	DEC	24	
Daniels III	James	G.	CDR	ATG-2	1951	NOV	30	1953	FEB	06	ATG-2 30 NOV 51
Dankworth Jr.	Edwin	G.	CDR	CVG-3	1960	MAR		1961	APR	20	
Dankworth	Theodore	P. 'Big Daddy'	CDR	CVG-21	1961	JUN	16	1962	JUN	21	
Darcy	Robert	T.	CDR	RCVSG-51	1967	MAR	31	1968	APR	05	
Davidson	James	J.	CDR	CVG-14	1957	AUG	09	1958	AUG	29	

Davis	Arthur	C.	LCDR	LitBmbWing	1928	JUL	01	1929	MAY		
Davis	Donald	C.	CDR	CVG-5	1962	FEB	28	1963	JAN	18	
Davis	Ray		CDR	CVG-16	1944	AUG	24	1945	NOV	06	REFORMED/LAST CAG
Davis	Robert	E. 'Sundance'	CAPT	CVW-17	1994	JUL	31	1996	JAN	26	
Davison	Ralph	E.	LCDR	FitWing	1931	JAN	01				AS OF DATE
Day	Barton	E.	LCDR	CVEG-37	1943	AUG		1944	MAY	23	
De Lorenzi	Robert	M.	CDR	CVW-15	1965	JUN	25	1965	JUL	10	ACTING
Dean Jr.	William	A.	CDR	CVG-17	1946	SEP		1947	SEP		CVBG-5 15 NOV 46
Dean Jr.	William	A.	CDR	CVG-97	1944	NOV	01				ASSUMED 1st CAG
Dearth	Randolph	S.	CAPT	CVW-2	1997	JAN	10	1998			
Deaver Jr.	William	N. 'Billy D'	CAPT	CVW-1	1993	MAY	19	1994	MAY	20	
Decker	Arthur	T.	CDR	CVAG-3	1947	JUL	30	1949	JAN		CVG-3 1 SEP 48
Demmler	Charles	F.	CDR	CVG-2	1963	DEC	12	1964	NOV		CVW-2 20 DEC 63
Denniston Jr.	Radcliffe		LCDR	CVG-11	1944	NOV	01	1944	NOV	13	ACTING/KIA over Manila Bay/TBM-1C
Denton Jr.	William	D.	CDR	CVG-19	1951	SEP	14	1952	DEC	10	REFORMED
DeVane	John	M. 'Jocko'	CDR	CVG-153	1946	OCT	16	1948	JUL	07	CVAG-15 16 NOV 46
Dew	Irwin	L.	LCDR	CVG-15	1943	SEP	01	1943	SEP	25	ACTING
Dewenter Jr.	Jack	R.	CDR	CVW-17	1966	DEC	22	1967	DEC	19	1st CAG
Dibble	Edgar	J.	CDR	RCVSG-50	1961	JUN	05	1961	SEP	11	ACTING
Dillon	Wallace	M.	LCDR	FitWing	1933	JUL	01				AS OF DATE
Dimon Jr.	Charles	G.	CDR	CVW-8	1970	JUN	06	1970	JUN	20	ACTING
Dixon	John	C.	CDR	CVW-8	1967	OCT	05	1968	OCT	16	
Dobbin	John	F.	LCOL	MCVG-1							ONLY CAG?/UNKNOWN EXISTENCE DATES
Donaldson Jr.	James	C.	CDR	CVG-1	1959	JUN	30	1960	SEP	16	
Dose'	Robert	G.	CAPT	RCVG-4	1958	JUL	03	1960	JAN	22	
Dose'	Robert	G.	CDR	CVG-81	1946	MAR	13	1947	MAY		CVAG-13 15 NOV 46
Dose'	Robert	G.	CDR	CVG-14	1945	DEC	14	1946	MAR		
Downing	Arthur	L.	CDR	CVG-2	1951	AUG	30	1952	NOV	05	
Doyle	Austin	K. 'Artie'	LCDR	SARATOGA	1939	JUN		1940	MAY		
Drane	William	M.	CDR	CVG-15	1943	SEP	25	1944	FEB	09	1st CAG
Duckworth	Herbert	S. 'Duckie'	LCDR	LEXINGTON	1941	MAY		1941	AUG		
Duff	Byron	L.	CAPT	CVW-9	1987	OCT	21	1989	JUN	09	
Duff	Byron	L.	CDR	CVW-13	1985	NOV	29	1987	JAN	16	
Duke	Paul	D.	LCDR	CVG-8	1945	JAN	01	1945	JAN	25	REFORMED/ACTING
Dumas	Glenn	I.	CDR	CVSG-52	1961	AUG	14	1962	JUL	05	
Duncan	David	O.	LT	CVG-93	1945	JUN	07	1945	JUN	18	ACTING
Duncan	Donald	B.	LCDR	ScoutWing	1935	JUL	01				AS OF DATE
Duncan	George	C. 'Duke'	CDR	CVG-5	1953	JUL	31	1954	NOV	30	
Dunlop	Thomas	E.	CDR	CVW-15	1971	JUN	02	1972	APR	06	KIA over NVN in A-7E
Dunn	Robert	F.	CDR	CVW-7	1970	JUL	23	1971	JAN	10	
Durgin	Calvin	T.	LT	CVG-150	1945	JAN	26	1945	MAR	01	ACTING
Earnest	Albert	K. 'Bert'	CDR	ATG-181	1957	AUG	12	1958	AUG	15	LAST CAG
Eason	William	R.	CDR	CVG-9	1963	MAR	02	1964	FEB	17	CVW-9 20 DEC 63
Eder	Willard	E.	LCDR	CVLG-29	1944	APR	10	1945	JUN	10	
Edge	Donald	B.	CDR	CVW-14	1964	MAR	24	1965	MAR	05	
Edmands	Allan	C.	LCDR	CVEG-60	1943	JUL	15	1943	DEC	29	1st CAG
Edmondson	Robert	F.	LCDR	CVG-101	1950	AUG	01	1950	AUG	21	ACTING
Edney	Leon	A. 'Bud'	CDR	CVW-2	1974	OCT	30	1976	JAN	08	
Edwards	David	R.	CDR	CVW-15	1980	MAR	28	1981	JUL	22	
Eggert	Lowell	F. 'Gus'	CDR	CVW-9	1971	JAN	29	1972	MAY	20	
Eisenhart	George	N.	LCDR	CVLG-32	1944	AUG	24	1944	DEC	09	
Eisenhart	George	N.	LCDR	CVLG-32	1944	DEC	09	1945	AUG	22	REFORMED
Ekelund Jr.	Kenneth	O.	CDR	CVSG-54	1965	JUN	03	1966	JUN	03	
Elder	Robert	M.	CDR	CVG-12	1956	APR	06	1957	NOV	08	
Elliott	Orville	G. 'Tex'	CAPT	CVWR-20	1973	FEB	10	1974	AUG	10	
Ellis Jr.	Robert	L.	CAPT	CVW-7	1984	MAY	04	1985	AUG	04	
Ellis	William	E.	CDR	CVG-18	1943	AUG	30	1944	NOV	03	1st CAG
Ellis	William	H.	CDR	CVW-21	1965	JUN	03	1966	JUN	03	
Ellison	David	J.	CDR	CVW-5	1969	NOV	24	1970	DEC	10	
Embree	Ralph	A.	CDR	CVG-12	1945	MAR	13	1945	APR	08	ACTING
Embree	Ralph	A.	CDR	CVG-12	1945	APR	08	1945	MAY	12	KIA over Okinawa in F6F-5
Emerson	Mark	T.	CAPT	CVW-9	1997	DEC	05				
Emig	Alvin	F.	CDR	RCVSG-51	1963	OCT	04	1964	JUL	24	
Emig	Alvin	F.	CDR	CVSG-55	1962	DEC	10	1963	SEP	06	
Emrick	Paul	E.	LCDR	CVG-9	1943	OCT	24	1943	DEC	01	ACTING
Emrick	Paul	E.	LCDR	CVG-9	1943	DEC	01	1943	DEC	22	
Engen	Donald	D.	CDR	CVG-11	1962	JAN	31	1963	JAN	03	
English Jr.	Elbert	H.	CDR	ATG-201	1956	JUN		1958	NOV	28	REFORMED/ASSUMED PERIOD/LAST?
Enney	Kenneth	E.	CDR	CVW-14	1968	FEB	05	1969	MAR	28	
Epes Jr.	Horace	H.	CDR	CVG-15	1954	MAY	01	1955	JAN		

Erickson	Oscar	W.	LCDR	ScoutWing	1931	JUL	01				AS OF DATE
Erickson	Oscar	W.	LCDR	TrpBmBWing	1928	JAN		1928	OCT		
Eslick	Mark		LCDR	CVEG-28	1943	SEP	30	1944	FEB	11	REFORMED/CVLG 20 JAN 44
Ethridge	Walker		CDR	CVLG-47	1944	MAY	15	1945	MAR	20	1st CAG/KIA over TF-58 in F6F-5
Evans	Delmar	H.	CDR	CVSG-60	1967	JUN	09	1968	JUL	02	
Evans	Jack	R.	CDR	RCVSG-51	1964	JUL	24	1965	JUL	23	
Ewen	Edward	C.	LCDR	ENTERPRISE	1940	MAY		1941	APR		
Ewing	Kent	W.	CAPT	CVW-17	1984	MAR	23	1985	AUG	23	
Fair	John	W.	CDR	CVG-1	1957	DEC	21	1959	JUN	30	
Fairlamb Jr.	George	R.	LCDR	TrpBmbWing	1934	JUL	01				AS OF DATE
Falconer	Alastair	S.	CDR	CVWR-30	1970	MAY	02	1971	NOV	21	1st CAG
Fallon	William	J. 'Fox'	CAPT	CVW-8	1990	MAR	16	1991	AUG	12	or 8 MAR 90?
Farrington	Robert	F.	CDR	CVG-15	1951	APR	05	1952	MAY	04	1st CAG
Feeks	Edmund	E.	CDR	CVWR-20	1971	JUL	17	1973	FEB	10	
Feher	Melvin	A.	CDR	CVW-1	1964	NOV	28	1966	JAN	19	
Feightner	Edward	L. 'Whitey'	CDR	CVG-10	1959	MAR	20	1960	MAR	18	
Feilbach	Harvey	O.	LCDR	CVEG-60	1943	DEC	29	1945	JAN	02	
Felt	Harry	D.	CDR	SARATOGA	1941	DEC	29	1942	OCT		
Fenwick	Joseph	E.	CDR	CVSG-55	1965	SEP	29	1966	OCT	03	
Ferguson	Robert	E.	CDR	CVW-5	1967	MAY	10	1967	AUG	31	ACTING
Ferguson	Robert	H.	CDR	CVW-8	1979	APR	10	1980	JUN	12	
Ferguson	Robert	E.	CDR	CVW-17	1967	DEC	19	1969	JAN	18	
Ferris	James		CAPT	RCVG-4	1963	JUN	20	1965	MAY	27	RCVW-4 20 DEC 63
Ferris	James		CDR	CVG-8	1960	DEC	01	1961	NOV	04	
Fetterman Jr.	John	H. 'Jack'	CAPT	CVW-8	1975	JUL	26	1976	AUG	10	
Fidel	John	A.	LCDR	CVG-11	1945	APR	09	1945	APR	21	REFORM/ACTING
Finley	John	L. 'Black Jack'	CAPT	CVW-5	1975	OCT	24	1976	OCT	20	
Finney	James	H. 'Jay'	CAPT	CVW-9	1984	JUN	22	1986	FEB	28	
Finney	James	H. 'Jay'	CAPT	CVW-15	1987	SEP	28	1989	JUN	02	
Fischer	Edwin	A.	CDR	CVG-8	1952	JUN	25	1952	AUG	11	ACTING
Fitzgerald	Mark	P. 'Lobster'	CAPT	CVW-14	1994	JUN	09	1995	DEC	15	
Flatley II	James	H.	CDR	CVG-5	1943	APR	07	1943	SEP		1st CAG
Flatley III	James	H.	CDR	CVW-7	1975	FEB	07	1976	MAR	05	
Flight Jr.	John	W.	CAPT	CVSGR-70	1975	MAY	10	1976	JUN	30	LAST CAG
Flinn Jr.	George	H.	LCDR	CVG-10	1944	SEP	15	1944	SEP	19	ACTING
Flint Jr.	Lawrence	E.	CAPT	RCVG-12	1963	JAN	18	1964	JUL	31	RCVW-12 20 DEC 63
Flower	Roger	P.	CAPT	CVW-5	1980	OCT	20	1982	JAN	20	
Fly	William	E.	CDR	CVG-4	1955	DEC		1958	MAR		ASSUMED PERIOD
Flynn Jr.	Samuel	C.	CDR	CVW-2	1976	JAN	08	1977	APR	29	
Foley Jr.	S.	Robert		CDR	CVW-11		1968		NOV	02	1969 OCT 02
Fonvielle Jr.	Charles	D.	CDR	ATG-1	1955	JUN		1957	JAN	03	REFORMED/BETWEEN 9-27 JUNE
Fonvielle Jr.	Charles	D.	LT	CVEG-38	1945	FEB	15	1945	APR	29	ACTING
Fonvielle Jr.	Charles	D.	LCDR	CVEG-38	1945	APR	29	1946	JAN	31	LAST CAG
Foote	Everett	W. 'Hoot'	CDR	CVW-8	1974	AUG	08	1975	JUL	26	
Ford	Arlo		LT	CVG-99	1944	JUL	15	1944	AUG	14	ACTING
Ford	Thomas	J.	CAPT	CVW-11	1989	MAR	03	1990	JUL	19	
Forkner	Levern	E. 'Lee'	LT	CVLG-45	1945	MAR	18	1945	MAY	22	ACTING
Fowler Jr.	Richard	E.	CDR	CVG-1	1962	DEC	19	1963	NOV	29	
Fowler	Richard	L.	CDR	CVAG-11	1946	FEB	14	1948	JUL	01	
Fox	Albert	D.	CDR	CVSG-59	1968	JUL	30	1969	MAY	29	
Franger	Marvin	J.	CDR	CVG-19	1959	OCT	01	1960	JUN	01	
Franson Jr.	William	E.	CDR	CVWR-20	1986	MAR	08	1987	SEP	13	
Fraser	Robert	B.	MAJ	MCVG-7				1945	JUN	18	A CAG?/KIFA near Santa Barbara
Frazier	Walter	J.	CDR	CVG-14	1953	MAY	15	1954	SEP	03	
French	Lawrence	C.	CDR	CVEG-2	1948	APR		1948	SEP	01	ASSUMED LAST CAG
Fretwell	Uncas	L.	CDR	CVG-19	1958	SEP	05	1959	OCT	01	
Frick	Joseph	F.	CDR	CVW-14	1972	MAY	26	1973	JUL	11	
Friedman	Malcom	C.	LCDR	CVG-15	1952	MAY	04	1952	MAY	21	ACTING
Fritz	Charles	W.	CDR	RCVW-4	1965	MAY	27	1965	JUN	28	ACTING?
Funk	Harold	N.	CDR	CVG-102	1950	AUG	23	1952	JAN	17	1st CAG
Funk	Harold	N.	LCDR	CVEG-26	1943	NOV	15	1945	JAN		REFORMED/1st CAG?
Fuoss	Ralph	A.	CDR	CVG-5	1945	AUG	11	1947	JUL	01	CVAG-5 15 NOV 46
Fuoss	Ralph	A.	LCDR	CVLG-23	1944	MAY	12	1944	JUN	25	
Furlong Jr.	George	M. 'Skip'	CDR	CVW-14	1973	JUL	11	1974	NOV	21	
Furney	Maynard	M. 'Bud'	LCDR	CVLG-28	1945	MAR	08	1945	NOV	06	REFORM/LAST CAG
Gabriel Jr.	Daniel	W.	CAPT	CVW-11	1992	JAN	31	1993	SEP	09	
Gaines	Richard	K.	CDR	CVG-10	1942	APR	16	1943	APR	07	1st CAG
Gallatin	Robert	E.	CAPT	RCVW-12	1966	JUL	22	1968	OCT	11	
Gallatin	Robert	E.	CDR	CVG-14	1963	APR	13	1964	MAR	24	CVW-14 20 DEC 63
Gambrill	N.	Jackson	CDR	CVW-11	1963	DEC	20	1964	DEC	17	CVW-11 20 DEC 63

Gammon	James	M.	CDR	CVSG-53	1966	JUL	01	1967	JUL	07	
Gay Jr.	Donald		CDR	CVG-3	1952	JUL		1953	OCT		
Gay	John	P. 'Phil'	CAPT	CVW-13	1984	MAR	29	1985	NOV	29	1st CAG
Geiger	Anson	D.	CDR	CVG-21	1963	JUN	22	1964	APR	29	CVW-21 20 DEC 63
Geis	Lawrence	R. 'Larry'	LCDR	CVG-3	1949	JAN		1949	FEB	14	ACTING
Geis	Lawrence	R. 'Larry'	CDR	CVG-4	1951	OCT		1952	OCT		
Gerhard	Harry	E.	CDR	CVW-7	1965	MAY	15	1966	JUN	10	
Gibson	Charles	E.	CDR	CVLG-1	1947						
Gilfry	Mason	C.	CDR	CVW-21	1973	JUL	09	1974	JUN	26	
Gilkeson	Fillmore	B.	CDR	CVEG-33	1944	MAY	15	1945	NOV	19	ONLY CAG
Gill	James	E. 'Ed'	CDR	CVWR-20	1982	JUN	26	1983	JUN	28	
Gillcrist	Paul	T. 'Punchy'	CDR	CVW-3	1971	JAN	28	1972	FEB	04	
Gillespie	Dennis	M. 'Dizzy'	CAPT	CVW-11	1995	MAR	10	1996	MAY	14	
Gillespie	George	S.	LCDR	ScoutWing	1929	JUL	01				AS OF DATE
Gillooly	John	F.	CDR	CVSG-56	1963	JUN	17	1964	OCT	30	
Girard	Jean	L.	CDR	CVSG-59	1965	SEP	06	1966	SEP	09	
Giuliani	Leonard	E. 'Lex'	CDR	CVW-19	1974	JUL	25	1975	NOV	28	
Gladman	Dennis	St. Clair	CAPT	CVW-11	1985	OCT	23	1987	MAY	29	
Glindeman Jr.	Henry	P. 'Hank'	CDR	CVW-15	1964	JUL	08	1965	APR	29	
Godman	Robert		CDR	CVG-5	1960	OCT	17	1962	FEB	28	
Goldenstein	Gordon	R. 'Goldie'	CAPT	CVWR-20	1979	MAY	26	1980	DEC	19	
Gormley	Robert	H.	CDR	CVW-7	1966	JUN	10	1967	MAY	16	
Gould	Richard	K.	LCDR	CVG-5	1945	JUN	05	1945	AUG	11	ACTING
Graning	Leonard	G. 'Tiny'	CDR	CVG-14	1962	MAR	24	1963	APR	13	
Grant	Mark	A.	LCDR	CVLG-25	1942	FEB	15	1943	OCT	06	1st CAG
Grant	Vernon	F.	LCDR	ScoutWing	1936	JUL	01				AS OF DATE/LAST
Gray Jr.	James	S.	CDR	CVG-3	1945	MAY	07	1945	AUG	08	REFORMED
Gray	Paul	N.	CDR	ATG-3	1956	AUG		1958	APR	01	REFORMED/ASSUMED PERIOD/LAST
Gray III	Walter	S.	CDR	CVW-16	1969	DEC	12	1970	AUG	07	ASSUMED ACTING(VA-122 CO)
Greer	Howard	E.	CDR	CVG-10	1961	SEP	11	1962	SEP	05	
Greer	Marshall	R.	LCDR	ScoutWing	1933	JUL	01				AS OF DATE
Greiwe	William	H.	CDR	CVW-21	1972	AUG	19	1973	JUL	09	
Griffin	Charles	D.	LCDR	CVG-9	1942	JUN	19	1943	AUG	13	
Griffin	Thomas	H.	CDR	CVSG-54	1964	JUN		1965	JUN	03	
Griffin Jr.	Virgil	C. 'Squash'	LCDR	ScoutWing	1928	OCT	01				AS OF DATE
Grimm	William	F.	CDR	CVSG-53	1968	JUL	18	1969	MAY	09	
Grimmell Jr.	Howard	L.	LT	CVLG-23	1945	JUL	15	1945	AUG	19	LAST CAG
Guinn	Dick	H. 'Tex'	CDR	CVG-6	1956	AUG	10	1957	NOV		10 AUG 56 APPROX.
Gureck	William	A.	CDR	CVW-19	1968	SEP	27	1969	DEC	12	
Hacker	Daniel	M.	CAPT	CVW-2	1993	FEB	25	1994	JUL	13	
Hagen	Thomas	L. 'Hook'	CAPT	CVW-3	1997	SEP	18	1999	APR	01	
Hall II	Grover	B.H.	LCDR	WASP	1939	JUL	01	1941	APR		1st CAG
Halland	Herman	E.	LCDR	TrpBmbWing	1935	OCT	01				AS OF DATE
Hamberg	Harold	A.	CDR	CVSG-57	1964	SEP	18	1965	AUG	03	
Hamilton	Charles	B.	CDR	CVSG-53	1962	NOV	28	1963	DEC	10	
Hamilton	James	B.	CAPT	CVWR-30	1980	MAY	31	1981	NOV	21	
Hamilton	Weldon	L.	LCDR	CVG-11	1943	FEB	08	1943	JUN	08	KIFA en route Sydney in R4D
Hammett	Wayne	E.	CDR	CVG-15	1962	JUL	27	1963	JUL	01	
Hampton	Isaiah	M. 'Ike'	CDR	CVG-2	1944	DEC	01	1945	NOV	06	REFORMED/LAST CAG
Hanley Jr.	Michael	J.	CDR	ATG-181	1953	NOV	01	1955	JUL		1st CAG
Harden	Hubert	B. 'El Groupo'	CDR	CVG-1	1944	OCT	01	1945	OCT	25	REFORMED/ASSUMED LAST CAG
Hardison	Osborne	B.	LCDR	LitBmbWing	1929	JUL	01				AS OF DATE
Hardisty	Huntington		CDR	CVW-11	1971	JUL	23	1972	MAY	08	
Harkins	John	A.	CDR	RCVSG-50	1961	SEP	11	1963	FEB	08	
Harlow	Robert	J.	CDR	CVSG-57	1967	JUN	29	1968	AUG	16	
Harmer	Richard	E. 'Chick'	CDR	CVBG-74	1946	AUG		1947	MAY		CVBG-1 15 NOV 46
Harrell	Robert	H.	LCDR	TrpBmbWing	1936	JUL	01				AS OF DATE/LAST
Harris	Leroy	E.	CAPT	RCVG-12	1959	JUL	31	1959	OCT	30	ACTING
Harris Jr.	Robert	D.	CDR	CVW-15	1969	MAY	29	1970	JUL	10	
Harris	Thomas	D.	CDR	CVG-9	1952	MAR	26	1953	AUG		1st CAG
Harris	William	H.	CDR	CVW-2	1965	DEC	28	1967	MAR	31	
Harrison	Harry	W. 'Stinkey'	LCDR	CVG-6	1944	APR	10	1944	JUN	13	ACTING
Hart	Gordon	M.	CDR	CVG-9	1961	DEC	20	1962	MAR	27	
Hart	James	M.	CAPT	CVW-17	1999	MAR	12				
Hartley	Gordon	E.	CDR	CVG-14	1959	AUG	12	1960	DEC	28	
Haskell	William	C.	CDR	CVSG-54	1968	JUN	13	1969	JUN	26	
Hatcher	Martin	T.	CDR	CVG-3	1949	FEB	14	1950	SEP		
Hathaway	Robert	R.	CAPT	CVWR-20	1993	FEB	20	1994	DEC	10	
Hathorn	Vincent	L.	CDR	CVG-8	1951	APR	09	1952	JUN	25	1st CAG
Hattersley	Julian		CDR	CVSG-53	1965	SEP	10	1966	JUL	01	

Hauert	Patrick	C. 'Maztini'	CAPT	CVW-6	1987	MAY	29	1989	APR	18	
Haugh	Edward	M.	CDR	RCVSG-50	1963	FEB	08	1964	JUN	02	
Hawkins	A.	Ray	CDR	CVG-1	1960	SEP	16	1961	OCT	09	
Hayward	Thomas	B.	CDR	CVW-10	1965	JUL	03	1966	JUN	23	
Hazelrig	John	P.	CAPT	CVWR-30	1989	APR	15	1992	FEB	02	
Hazelton Jr.	Dewitt	W.	CDR	RCVSG-51	1961	JUN	09	1962	SEP	28	
Hazelton Jr.	Dewitt	W.	CDR	CVSG-55	1960	SEP	01	1960	DEC	16	ACTING
Heap	George	L.	CDR	CVG-82	1944	APR	01	1945	AUG	20	1st CAG
Hedrick	James	G.	CDR	CVG-9	1953	AUG		1955	JAN		ASSUMED PERIOD
Hedrick	Roger	R. 'Rog'	LCDR	CVG-84	1944	MAY	05	1944	MAY	29	ACTING
Hedrick	Roger	R. 'Rog'	LCDR	CVG-84	1945	MAR	24	1945	APR	27	ACTING
Hedrick	Roger	R. 'Rog'	LCDR	CVG-84	1945	APR	27	1945	JUL	12	
Heimgartner	Kenneth	F.	CAPT	CVW-5	1993	JAN	22	1994	JUL	13	
Heishman	Jack	C.	CDR	CVG-8	1957	MAY		1957	AUG	01	ACTING?
Helsel	Kenneth	D.	CDR	CVSG-60	1960	MAY	02	1961	JUN	08	1st CAG
Henderson	James	R.	CDR	CVSG-54	1966	JUN	03	1967	JUN	10	
Hendrickson Jr.	Dean	M. 'Milo'	CAPT	CVW-17	1990	JUL	18	1991	NOV	27	
Herbig	Henry	F.	CDR	CVSG-56	1970	MAY	13	1971	MAY	22	
Hernandez	Diego	E.	CDR	CVW-7	1974	FEB	01	1975	FEB	07	
Herrick	E.	Cady	LCDR	CVEG-38	1945	FEB		1945	FEB	15	ACTING
Hershey Jr.	Merle	M.	CDR	ATG-182	1957						
Hershey Jr.	Merle	M.	CDR	CVG-8	1958	JUL	09	1959	NOV	30	
Hessel	Edward	W.	CDR	CVG-7	1948	DEC	10	1950	APR	01	
Heyworth Jr.	Lawrence		CDR	CVG-8	1959	NOV	30	1960	DEC	01	
Hickey	Robert	F.	LCDR	RANGER	1940	JUN	01	1941	APR		
Hickey Jr.	Robert	P. 'Burner Bob'	CDR	CVW-11	1982	NOV	06	1984	MAY	18	
Highsmith	Frederick	L.	CDR	CVSG-52	1966	JUN	15	1967	JUN	30	
Hill	Allen	E. 'Boot'	CAPT	CVW-2	1969	JUN	12	1970	MAY	16	
Hill	John	S.	CDR	CVG-1	1955	AUG		1957	JUN	20	
Hill	John	S.	CDR	CVG-7	1951	DEC	01	1952	JAN	01	ACTING
Hill	Raymond	E.	LCDR	CVG-84	1945	JUL	22	1945	JUL	31	ACTING
Hilton	John	J.	LCDR	CVLG-29	1945	JUN	10	1945	SEP	10	REFORM 07 JUN 45/LAST CAG
Hindman	Stanley	E.	LCDR	CVEG-37	1944	MAY	23	1944			
Hizer	Devon	M.	CDR	ATG-1	1954	JAN		1955	APR	11	REFORMED/ASSUMED PERIOD
Hlavka	Stanford	H.	CAPT	CVW-15	1994	JUN	17	1995	MAR	31	LAST CAG/DISEST DATE
Hodges	William	M.	LCDR	CVEG-60	1945	MAY	15	1945	NOV	19	LAST CAG?
Hodgson	Gordon	S.	CDR	CVG-19	1961	MAR	17	1962	MAR	07	
Hoffman	David	L. 'Hawk'	CAPT	CVW-8	1981	SEP	19	1982	DEC	13	or 9 SEP 81?
Holbrook	James	L.	CDR	CVG-6	1961	APR	26	1962	JUN	22	
Holder	Billy	D.	CDR	CVG-2	1961	DEC	20	1962			
Holley Jr.	Edward	B.	CDR	CVG-2	1959	MAR	18	1960	APR	08	
Holley Jr.	Edward	B.	CAPT	RCVW-12	1964	JUL	31	1966	JUL	22	
Hollingsworth	George		LCDR	CVLG-27	1943	NOV	08	1944	APR	26	
Holt	Robert	E.	CAPT	CVWR-20	1970	APR	01	1971	JUL	17	1st CAG
Hope Jr.	Herbert	A.	CDR	CVW-8	1971	JAN	13	1972	FEB	10	
Hopper	Richard	S.	CDR	CVW-6	1970	SEP	02	1971	JUL	28	
Horne	M.	Kent	CAPT	CVWR-20	1997	JUL	12	1999	JAN	30	
Houck	Herbert	N.	LCDR	CVG-9	1945	FEB	16	1945	OCT	15	LAST CAG
House	William	H.	CDR	CVG-11	1955	JAN		1957	APR	12	
House	William	H.	LCDR	CVG-14	1945	JAN	01	1945	FEB	16	REFORMED/ACTING
Houser	Robert	E. 'Stout'	CDR	CVW-3	1985	DEC	17	1987	JUL	24	
Howard	Philip	G. 'Cycle'	CAPT	CVW-17	1993	AUG	26	1994	JUL	31	
Howerton	Charles	C. 'Sunshine'	CDR	CVG-13	1943	NOV	01	1944	AUG	08	1st CAG
Howland	John	B.	CDR	CVG-4	1954	FEB		1955	DEC		DEC 54 ASSUMED
Hubbard Jr.	Samuel	W.	CDR	CVW-19	1971	MAY	28	1972	JUN	05	
Huff	Alvin	F.	CDR	CVSG-59	1964	OCT	02	1965	SEP	06	
Hughes	David	L.	CDR	CVSG-59	1967	JUL	14	1968	JUL	30	
Hulme	John		CDR	CVG-11	1943	JUN		1943	NOV		
Hulme	John		LCDR	CVLG-21	1943	MAY	15	1943	JUN	25	1st CAG
Humphrey	Ernest	W.	CDR	CVG-6	1951	MAY	19	1952	JUL	08	
Hunt	Robert	F.	CDR	CVG-14	1960	DEC	28	1962	MAR	24	
Huth	Vincent	J.	CAPT	CVW-14	1979	DEC	22	1981	APR	21	
Hyland	John	J.	CDR	CVG-10	1944	SEP	19	1945	NOV	23	LAST CAG
Iarrobino	Charles	A.	CDR	CVG-7	1954	MAR		1956	JAN		
Ingerslew	Donald	B.	LCDR	CVG-6	1944	FEB	16	1944	APR	10	
Ingerslew	Donald	B.	CDR	CVAG-19	1948	SEP	01	1949	OCT	10	
Irelan	Dennis	W.	CAPT	CVW-14	1986	OCT	11	1988	MAY	31	
Isaacks	Marion	H.	CAPT	CVW-5	1972	NOV	15	1973	AUG	03	
Issitt	Donald	K.	CDR	CVG-14	1958	AUG	29	1959	AUG	12	
Jack	Gary	M.	CAPT	CVW-8	1994	JUN	03	1995	DEC	14	

Jackson	Andrew	McB.	CDR	CVG-8	1943	JUN	01	1944	MAY		1st CAG
Jackson	Henry	S.	LCDR	CVLG-48	1944	JUN	15	1945	JAN	02	ONLY CAG?
Jacoby	Raymond	B. 'Jake'	LCDR	CVG-11	1943	NOV		1943			ACTING CAG?
James	Jack	M.	CDR	CVG-6	1958	JAN	21	1959	JUN	28	
Jan	Kolin	M.	CAPT	CVW-7	1996	JUN	04	1997	SEP	04	
Jaskot	Richard	D.	CAPT	CVW-7	1999	FEB	26				
Jenkins	Thomas	H.	LCDR	CVLG-22	1943	DEC	16	1945	FEB		
Jensen	Wayne	L.	CAPT	CVSG-59	1970	JUN	01	1971	MAY	24	
Jensen	William	G.	CDR	CVSG-59	1966	SEP	09	1967	JUL	14	
Jernigan Jr.	William	A.	CDR	CVG-9	1959	JUN	12	1960	SEP	02	
Jewell Jr.	Robert	W. 'Louie'	CDR	CVW-8	1973	JAN	31	1974	AUG	08	
Jewett	Charlton	B.	CAPT	CVW-17	1997	JUL	01	1999	MAR	12	
Johns	Clifford	M.	CDR	CVW-3	1975	FEB	10	1976	MAY	17	
Johns	Ronald	L.	CDR	ATG-4	1957	NOV		1959	JAN	19	REFORMED/ASSUMED PERIOD/LAST?
Johnson	Cecil	V.	CDR	CVG-5	1952	APR	30	1953	JUL	31	
Johnson	Gregory	G. 'Grog'	CAPT	CVW-3	1994	APR	15	1995	MAR	25	
Johnson	J.	Michael	CAPT	CVW-8	1993	MAY	20	1994	JUN	03	
Johnson	Jay	L. 'Bird'	CAPT	CVW-1	1985	MAR		1986	JUL	02	
Johnson	Jay	L. 'Bird'	CAPT	CVW-1	1988	FEB	05	1989	JUL	08	
Johnson	Robert	F. 'Buzz'	CAPT	CVW-14	1985	MAY	08	1986	OCT	11	
Johnson	Roy	L.	CDR	CVG-2	1943	JUN	01	1944	JUN	01	1st CAG
Johnson	Theodore	L.	CDR	CVG-1	1963	NOV	29	1964	NOV	28	CVW-1 20 DEC 63
Jones	Donald	S.	CDR	CVSG-53	1971	JUN	09	1972	JUN	06	
Jones	Keith	S. 'Casey'	CDR	CVW-3	1978	OCT	17	1980	JAN	04	
Joy	James	A.	CDR	CVW-8	1972	FEB	10	1973	JAN	31	
Jung	Karl	E.	CDR	CVG-19	1943	AUG	18	1943	SEP	22	ACTING
Jung	Karl	E.	CDR	CVG-19	1943	SEP	22	1944	SEP	17	1st CAG
Kane	William	R. 'Killer'	CDR	CVG-10	1944	FEB	12	1944	JUL	17	
Keating	Timothy	J. 'Timbo'	CAPT	CVW-9	1993	JUL	26	1994	OCT	26	
Keats	Edgar	S.	CDR	CVG-5	1949	FEB	04	1949	SEP	30	
Keely	Leroy	B.	CAPT	CVWR-30	1974	NOV	17	1975	NOV	23	
Keighley	William	H.	CDR	CVLG-1	1948	JUL	09	1948	NOV	20	ASSUMED LAST CAG
Kelly	J.	Curtis	CDR	CVG-1	1950	JUN		1951	OCT		
Kelly	J.	Curtis	CDR	CVG-13	1949	SEP	30	1949	NOV	30	LAST CAG
Kempf	Cecil	J.	CDR	CVSG-60	1968	JUL	02	1968	OCT	01	LAST CAG
Kemple Jr.	Morris	M.	CAPT	CVWR-20	1983	JUN	28	1984	SEP	04	
Kendall	David	C.	CAPT	CVWR-30	1994	FEB	26	1994	DEC	31	LAST CAG
Kennedy	John	E.	CDR	ATG-4	1955	MAR	30	1956	SEP	13	1st CAG
Kennedy	Ronald	W.	CDR	CVSG-59	1971	MAY	24	1972	MAY	12	
Ketcham	Dixwell	'Dixie'	LCDR	SARATOGA	1938	JUL	01	1939	MAY	01	1st Formal CAG
Kibbe	Richard	L.	CDR	CVG-13	1944	AUG	08	1945	JAN	01	
Kibbe	Richard	L.	CDR	CVG-13	1945	FEB	01	1945	OCT	20	LAST CAG
Kiehl	Richard	L.	CAPT	CVWR-20	1974	AUG	10	1976	FEB	28	
Kilcline Jr.	Thomas	J.	CAPT	CVW-14	1997	MAR	19	1998	OCT	01	
Kinnear II	George	E.R. 'Gus'	CDR	CVW-2	1967	MAR	31	1968	JUN	28	
Kirkconnell	William	B.	CDR	CVW-5	1976	OCT	20	1978	JAN	21	
Kirkpatrick	Raleigh	C.	CDR	CVEG-50	1944	OCT	01	1945	OCT	29	REFORMED/CVEG 01 OCT 44/LAST?
Kirn	Lewis	J.	CDR	CVG-20	1943	OCT	16	1943	NOV	10	ACTING
Klahn	Dale	C.	CDR	CVG-6	1960	MAY	26	1961	APR	26	
Klapka	Edward	J.	CDR	RCVSG-51	1968	APR	05	1969	MAR	14	
Kline	Raymond	P. 'Rip'	LCDR	CVG-8	1948	SEP	15	1948	SEP	27	ACTING
Kline	Raymond	P. 'Rip'	CDR	CVG-10	1952	MAY	30	1954	JAN	18	1st CAG
Klinsmann	G.	Otto	CDR	CVG-4	1944	MAY	06	1945	JAN	15	KIA over Pescadores Is. in F6F-5
Kneip	John	B.	LCDR	FitWing	1930	JAN	01				AS OF DATE
Knight	James	T.	CAPT	CVW-11	1997	OCT	09	1999	MAR	27	
Knight	Page		CDR	CVG-151	1945	MAR	09	1945	OCT	06	ONLY CAG?
Knudson	Irving	H. 'Knut'	CDR	CVSG-54	1963	AUG		1964	JUN		
Knutson	Danny	C.	CAPT	CVW-2	1998						
Knutson	Wilbert	D. 'Bill'	CDR	CVW-7	1971	JAN	10	1971	DEC	18	
Kober Jr.	H.	Patrick	CDR	CVW-3	1977	AUG	27	1978	OCT	17	
Koch	Ferdinand	B. 'Fred'	CDR	CVW-10	1964	AUG	06	1965	JUL	03	
Kodak	Preston	F.	LT	CVG-151	1945	FEB	12	1945	MAR	09	ACTING
Koen	George		CDR	CVG-21	1962	JUN	21	1963	JUN	22	
Kohn Jr.	Edwin	R.	CDR	CVW-1	1973	JUN	23	1974	JUN	26	
Kolakowski	Roman		CDR	CVG-5	1959	JAN	14	1959	NOV	13	
Konard	Edmond	G.	LCDR	CVLG-32	1943	JUN	01	1944	MAY	12	1st CAG
Konrad	Edmond	G.	CDR	CVG-17	1944	MAY	13	1945	JUN	29	
Koval	Francis	P.	CDR	CVSG-53	1967	JUL	07	1968	JUL	18	
Krebsbach	Eugene	W.	LT	CVG-84	1945	JUL	12	1945	JUL	22	REFORMED/ACTING
Kreutz	Arthur	R.	CDR	RCVSG-50	1966	AUG	10	1967	SEP	08	

Kroeger	Edwin	J.	CDR	CVG-6	1952	AUG	04	1953			
Kunkle	Steven	A. 'Knuckle'	CAPT	CVW-2	1995	SEP	07	1997	JAN	10	
LaCouture	John	E. 'Tiger One'	LCDR	CVG-4	1945	MAY	18	1945	JUL	04	ACTING
LaCouture	John	E. 'Tiger One'	CAPT	RCVG-4	1960	JAN	22	1960	APR	28	
LaCouture	John	E. 'Tiger One'	CDR	CVG-6	1954	OCT	22	1956	AUG	10	APPROX. DATES
LaFlair	Keith	R. 'Floo'	CAPT	CVWR-30	1992	FEB	02	1994	FEB	26	
Lair	James	A.	CAPT	CVW-6	1981	AUG	06	1982	NOV	01	
Lake	Julian	S.	CDR	CVG-8	1962	NOV	05	1963	NOV	27	
Lamade	John	D. 'Lucky'	CDR	CVG-7	1944	JAN	03	1945	FEB		1st CAG
Lambert	Valdemar	G.	CDR	CVAG-15	1948	JUL	07	1949	DEC	01	CVG-15 on 1 SEP 48/LAST CAG
Landon	James	B.	CDR	CVSG-59	1960	MAR	25	1961	MAY	12	1st CAG
Lang	Harold	F.	CDR	CVG-7	1960	MAR		1961	JUL		
Langdon	Stewart	D.	CAPT	CVW-5	1978	JAN	21	1979	MAY	31	
Langston	Arthur	N. 'Bud'	CAPT	CVW-5	1991	JUN	14	1993	JAN	22	
Lanham	Harvey	Peter	CDR	CVG-5	1949	SEP	30	1950	DEC	12	
Lanham	Harvey	Peter	CDR	CVG-81	1945	MAY	18	1946	MAR	13	REFORMED 13 MAY
Larsen	Bertrand	O.	CDR	CVW-19	1965	DEC	18	1966	MAR	10	ACTING
Latendresse	Thomas	B.	CAPT	CVW-2	1987	JAN	21	1988	AUG	19	
Latendresse	Thomas	B.	CDR	CVWR-30	1983	MAY	21	1984	JUL	07	
Laubach	Luther	W.S.	CDR	CVSG-52	1964	JUL	14	1965	AUG	09	
Lawler	Frederick	W.	CDR	CVW-1	1972	MAY	24	1973	JUN	23	
Lawler	Joseph	T.	CDR	CVG-11	1949	DEC	28	1950	JUL	01	ACTING?
Lawrence	Hamilton	'Ham'	LCOL	MCVG-9	1945	MAY	05	1946	FEB		ONLY CAG?/UNKNOWN EXISTENE DATES
Laws	Walter	T.	CDR	CVW-11	1964	DEC	17	1965	DEC	31	
Layton	David	R.	CAPT	CVWR-20	1980	DEC	19	1982	JUN	26	
Lee	Bobby	C.	CAPT	CVW-9	1977	DEC	16	1979	FEB	22	
Lee	James	A. 'Buckey'	LCDR	CVG-6	1943	APR	15	1943	OCT	11	1st CAG
Lee	Kent	L.	CDR	CVG-6	1962	JUN	22	1963	MAR	07	
Leeds	Rene	W. 'Sam'	CAPT	CVW-11	1976	DEC	23	1977	DEC	10	
Lenox	Glen	W.	CDR	CVW-14	1971	MAY	10	1972	MAY	26	
Leonard	William	N.	CDR	CVG-17	1950	NOV	02	1951	FEB		Relieved after 25 FEB 51
Leslie	Maxwell	F.	LCDR	ENTERPRISE	1942	JUN	19	1942	SEP		LAST CAG
Less	Anthony	A.	CDR	CVW-9	1976	AUG	09	1977	DEC	16	
Leue	David	E.	CDR	CVW-7	1969	APR	23	1970	JUL	23	
Lewis	Frederick	L. 'Bad Fred'	CDR	CVW-8	1980	JUN	12	1981	SEP	19	or 9 JUN 80?
Lewis	Frederick	L. 'Bad Fred'	CAPT	CVW-8	1986	JAN	08	1987	JUL	31	1ST 'SUPER CAG'
Lewis Jr.	Willis	I.	CDR	CVW-6	1972	JUL	03	1973	JUN	14	
Linder	James	B.	CDR	CVW-15	1967	FEB	03	1968	MAY	24	
Lindner	Robert	H.	LCDR	CVLG-30	1944	JUN	25	1945	FEB	23	REFORM/KIA off Chichi Jima in F6F-5
Lineberger	Perton	H. 'Bud'	CDR	CVW-17	1978	MAR	30	1979	JUN	01	
Link Jr.	Everett	M.	LCDR	CVLG-24	1943	DEC	24	1944	JUL		CVEG 15 AUG 44
Little	George	A.	LCDR	CVG-93	1944	DEC	15	1944	DEC	28	ACTING
Lockwood	Warren	H.	CDR	RCVSG-50	1967	SEP	08	1968	JUN	18	
Longino Jr.	James	C.	LCDR	CVEG-40	1945	APR	05	1945	NOV	19	LAST CAG?
Lonnquest	Theodore	C.	LCDR	ScoutWing	1933	JAN	01				AS OF DATE
Loux	Raymond	E.	CAPT	CVW-1	1970	MAY	01	1971	MAY		
Lovelace	Paul	C.	CDR	CVG-2	1952	NOV	05	1954	MAR	12	
Lovelace	Robert	H.	CDR	CVW-8	1968	OCT	16	1969	SEP	12	
Lovelady	David	E.	CDR	CVWR-20	1984	SEP	08	1986	MAR	08	
Lowe Jr.	John	T. 'Jigger'	CDR	CVG-3	1944	DEC	16	1945	FEB	23	ACTING
Lowe Jr.	John	T. 'Jigger'	CDR	CVG-3	1945	FEB	23	1945	APR	01	
Ludden	Charles	H.	LCOL	CVW-16	1965	SEP	09	1965	OCT	04	ACTING
Ludwig	Frederic	G. 'Wigs'	CAPT	CVW-15	1991	JUL	01	1992	DEC	10	
Luecke	John	M. 'Ice Man'	CAPT	CVW-7	1989	APR	14	1991	JAN	18	
Luker	George	R.	CDR	CVG-86	1944	JUN	15	1945	NOV	21	ONLY CAG
Lyle	Dale	E.		CAPT	CVW-8	1999			JAN	15	
Lynch	Hugh	F.	CDR	CVW-9	1975	JUL	03	1976	AUG	09	
Lynch	John	J.	CDR	CVAG-13	1947	MAY		1948	AUG	10	
Lyons	Kenneth	H.	CDR	CVSG-56	1964	OCT	30	1965	OCT	11	
Mackey	William	A.	CDR	CVG-9	1962	MAR	27	1963	MAR	02	
MacKnight	Harding	C.	CDR	CVG-7	1962	JUL	21	1963	JUL	31	
Macomber	Brainard	T.	LCDR	CVLG-21	1945	JAN	16	1945	NOV	05	REFORMED 02 JAN 45/LAST CAG
Macon	Benjamin	H.	CDR	CVSG-56	1965	OCT	11	1966	NOV	16	
Macri	Guiseppi	'Gus'	CDR	ATG-2	1956	JAN		1957	SEP	18	REFORMED/ASSUMED PERIOD
Madden	Walter	F. 'Wally'	CDR	CVG-3	1950	SEP		1951	OCT		KOREAN TOUR CAG
Mahler III	George	H.	CDR	CVG-12	1957	NOV	08	1958	MAR	10	
Mahler III	George	H.	CDR	CVG-15	1958	MAY	23	1959	JAN		
Maltby Jr.	Arthur	L.	CDR	CVG-19X	1951	JUN	11	1951	OCT	03	ACTING
Mandarich	Stevan		LCDR	CVEG-35	1943	JUL	15	1944	MAR	15	1st CAG
Mandarich	Stevan		CDR	CVG-94	1944	NOV	15	1945	NOV	07	ONLY CAG?

Mandeville Jr.	Robert	C.	CAPT	CVW-3	1968	NOV	14	1969	DEC	13	
Manning Jr.	John	F. 'Archie'	CAPT	CVW-8	1988	DEC	09	1990	MAR	16	
March	Daniel	P.	CDR	CVW-1	1978	DEC	11	1980	APR	07	
Marr	Harold	L.	CDR	CVW-1	1968	JUN	27	1969	JUN	17	
Marshall	Jack	L.	CDR	CVSGR-80	1972	AUG	20	1974	FEB	23	
Marshall	David	E.	MAJ	MCVG-13							ONLY CAG?/UNKNOWN EXISTENCE DATES
Martin	William	I.	CDR	CVG(N)-90	1944	SEP	22	1945			1st CAG
Masek	William		LCDR	TrpBmbWing	1930	OCT	01				AS OF DATE
Massa	Emiddo		CDR	CVSGR-80	1969	APR	01	1971	JAN	23	1st CAG
Massey	Forsyth	'Hi'	CDR	CVLG-24	1942	DEC	31	1943			1st CAG
Massey Jr.	Roger	A.	CAPT	CVW-2	1973	JUN	11	1974	OCT	30	
Matheny	James	T.	CAPT	CVW-1	1981	JUL	10	1982	JUL		
Matters	Alfred	R.	CDR	CVLG-50	1943			1944	APR		1st CAG?/DOWNED OVER TRUK 29/30 APR
Matthews Jr.	H.	Spencer	CDR	CVG-2	1962			1963	DEC	12	
Matton	Michael	N.	CAPT	CVW-7	1985	AUG	04	1987	FEB	27	
Mattson	Royce	R.	CAPT	CVWR-20	1987	SEP	13	1989	MAR	18	
Mauldin	Richard	J.	CAPT	CVW-5	1996	JUL	09	1998	MAY	30	
Mayer	George	E.	CAPT	CVW-7	1997	SEP	04	1999	FEB	26	
Mazach	John	J.	CAPT	CVW-3	1982	JUL	25	1984	APR	21	
McArthur Jr.	James	D.	CAPT	CVW-11	1993	SEP	09	1995	MAR	10	
McCabe	Michael	J. 'Wizzard'	CAPT	CVW-14	1992	DEC	04	1994	JUN	09	
McCamish	Michael	J.	CAPT	CVW-15	1992	DEC	10	1994	JUN	17	
McCampbell	David		CDR	CVG-15	1944	FEB	09	1945	JAN	10	
McCanna Jr.	Marvin	G. 'Red Dog'	CDR	CVW-17	1971	APR	03	1972	APR	07	
McCarthy Jr.	Paul	F.	CDR	CVW-9	1973	MAY	24	1974	JUL	14	
McClain III	Kirby	L.	CDR	CVSG-52	1969	APR	18	1969	DEC	15	LAST CAG
McClelland	Thomas	L.	CAPT	CVW-9	1986	FEB	28	1987	OCT	21	
McClusky Jr.	C.	Wade	LCDR	ENTERPRISE	1942	MAR	21	1942	JUN	19	
McCormack	Vincent	F.	LCDR	CVLG(N)-42	1944	AUG	25	1945	JAN	02	1st CAG/ONLY?
McCormack	Vincent	F.	LCDR	CVG(N)-53	1945	JAN	02	1945			1st CAG (to MAR 45?)
McCormack	Vincent	F.	CDR	CVEG(N)-63	1945	JUN	20	1945	DEC	11	1st CAG/ONLY?
McCrimmon	Douglas	R.	CAPT	CVW-11	1973	APR	13	1974	MAY	01	
McCrory	Donald	L.	CDR	CVW-7	1982	MAR	31	1983	JUN	25	
McCuddin	Leo	B.	CDR	CVG-3	1958	OCT	23	1960	MAR		
McCue	Hartsel	F.	CDR	ATG-182							RELIEF? WHEN? A CAG?
McCullough	Robert	J.	LCDR	CVG(N)-90	1944	AUG	25	1944	SEP	22	ACTING
McCutchan	George	T.	CDR	CVG-11	1943			1944	AUG	24	KIFA during training
McDonald	Harvey	B. 'H-MAC'	CAPT	CVW-14	1998	OCT	01				
McDonald	Wesley	L.	CDR	CVW-15	1965	JUL	10	1966	JUN	25	
McElraft	Ronald	D.	CAPT	CVW-7	1995	APR	05	1996	JUN	04	
McHarg	Richard	O.	CAPT	CVW-5	1998	MAY	30				
McKay	James	H.	LCDR	LEXINGTON	1939	JUN	01	1939	DEC	06	
McKellar Jr.	Edwin	D.	CDR	CVW-8	1970	JUN	20	1971	JAN	13	
McKenzie II	James	A.	CDR	CVW-11	1972	MAY	08	1973	APR	13	
McKnight	Jesse	E.	CAPT	CVW-2	1970	MAY	16	1971	JUL	02	
McLane	Robert	L.	CAPT	CVWR-20	1994	DEC	10	1996	MAR	26	
McLaughlin Jr.		William	H.	CDR	CVSG-59	1963			APR	30	1964 OCT 02
McNair	William	D.	CDR	CVSG-55	1960	DEC	16	1961	DEC	11	1st CAG
McNeely	Henry	E.	CDR	CVG-7	1950	MAY	01	1951	DEC	01	
McWhinney	John	L. 'Mac'	CDR	CVW-9	1981	SEP	25	1982	DEC	03	
Mehle	Roger	W.	LCDR	CVLG-28	1944	MAR	16	1945	FEB		
Merrick	Richard	C.	CDR	CVG-19	1950	JUL	07	1951	MAY	18	KIA over Korea in AD-4Q
Merrick	Richard	C.	LCDR	CVLG-28	1944	FEB	11	1944	MAR	16	
Metsger	Alfred	B. 'Abe'	CDR	CVG-80	1944	FEB	01	1944	JUN		1st CAG
Metz	Fredrick	J.	CDR	CVW-17	1976	DEC	03	1978	MAR	30	
Metzger III	Lewis	W.	CDR	RCVSG-51	1960	JUN	30	1961	JUN	09	1ST CAG
Metzger	Robert	L.	CDR	CVSG-60	1965	JUL	16	1966	MAY	12	
Meyer	William	F.	CDR	CVW-7	1973	JAN	26	1974	FEB	01	or 20 JAN 73?
Meyers	Frederick	P. 'Boom Boom'	CDR	CVW-15	1978	NOV	22	1980	MAR	28	
Michaelis	Frederick	H.	CDR	CVG-11	1954	FEB	01	1955	JAN		
Miller	D.	Bruce	CDR	CVW-5	1967	AUG	31	1968	AUG	24	
Miller	Dana	F.	CAPT	CVWR-20	1989	MAR	18	1990	JUN	16	
Miller	Gerald	E.	CDR	CVG-17	1958	MAR	10	1958	SEP	15	LAST CAG
Miller	Henry	L. 'Slag One'	CDR	CVG-6	1944	JUN	13	1945	OCT	29	LAST CAG
Miller	Henry	L. 'Hank'	LCDR	CVLG-23	1943	APR	24	1944	MAY	12	
Miller	Richard	A.	CDR	CVSG-53	1969	MAY	09	1970	JUN	12	
Miller	Robert	N.	CDR	CVG-3	1954	OCT		1956	OCT		
Miller	Robert	N.	LCDR	CVLG-32	1945	OCT	14	1945	NOV	13	LAST CAG
Millington	William	A.	LCOL	CVG-4	1945	JAN	15	1945	FEB		ACTING
Mills	Richard	H.	CDR	CVG-3	1957	SEP	04	1958	OCT	23	

Milner	Robert	M.	CDR	CVG-20	1945	APR	16	1946	NOV	15	REFORMED/CVAG-9 15 NOV 46
Milner	Robert	M.	CDR	CVAG-9	1946	NOV	15				
Mitchell	John	R. C.	CDR	CVW-9	1969	NOV	13	1971	JAN	29	
Mitchell Jr.	William	T.	CDR	CVSG-55	1966	OCT	03	1967	JUN	07	
Mitchie	Donald		CDR	CVG-5	1958	DEC	24	1959	JAN	14	ACTING
Mixon	Riley	D.	CAPT	CVWR-30	1977	MAY	28	1978	NOV	11	
Moebus	Lucien	A.	LCDR	LEXINGTON	1939	DEC	06	1940	MAY		
Moffitt	Mathew	G. 'Mofat'	CAPT	CVW-3	1996	APR	12	1997	SEP	18	
Monahan	Alfred	E.	CDR	RCVSG-51	1962	SEP	28	1963	OCT	04	
Moneymaker	Patrick	D. 'Munt'	CAPT	CVW-14	1991	MAY	10	1992	DEC	04	
Monger	A.	Jack	CDR	CVW-21	1966	JUN	03	1967	OCT	06	
Mongilardi Jr.	Peter	J.	CDR	CVW-15	1965	APR	29	1965	JUN	25	KIA over NVN in A-4C
Montgomery	Alfred	E.	LCDR	TrpBmBWing	1929	OCT	01				AS OF DATE
Moore Jr.	Charles	W.	CAPT	CVW-8	1991	AUG	12	1993	MAY	20	
Moore Jr.	Charles	L.	CDR	CVLG-51				1944			DURING CVL-30 1944 CRUISE
Moore	Edward	P.	LCDR	LEXINGTON	1940	MAY		1941	MAY		
Moore Jr.	Frederick	T.	LCDR	CVEG-35	1944	MAR	15	1944	DEC	01	
Moore Jr.	Horace	N. 'Whitey'	CDR	CVSG-58	1962	JUN	02	1963	JUN	26	
Moore	Robert	E.	CDR	CVW-2	1964	NOV		1965	DEC	28	
Moore Jr.	Robert	H.	CDR	CVG-5	1959	NOV	13	1960	OCT	17	
Moorer	Joseph	P.	CDR	CVG-10	1962	SEP	05	1963	AUG	05	
Moranville	Kendall	E.	CDR	CVW-7	1971	DEC	18	1973	JAN	26	
Moreau	Ronald	F. 'Moon'	CDR	CVW-8	1976	AUG	10	1978	JAN	31	
Morgan	Alfred	D.	LCDR	CVG-99	1945	JAN	09	1945	JUL	14	
Morin	James	B.	CAPT	RCVW-4	1969	JUN	20	1970	JUN	30	LAST CAG
Morin	James	B.	CDR	CVW-7	1968	JUN	12	1969	APR	23	
Morton	Albert	O.	CDR	CVG-10	1957	MAR	25	1959	MAR	20	
Morton	William	B.	CDR	CVG-1	1954	APR		1955	AUG		
Mouton	Edison	E.	CDR	CVSG-57	1965	AUG	03	1966	JUL	08	
Mow	Douglas	F.	CDR	CVW-19	1970	JUL	10	1971	MAY	28	
Mueller	Richard	C.	CDR	CVG-19X	1951	OCT	03	1951	NOV	30	ACTING
Mueller	Richard	C.	LCDR	CVBG-74	1945	MAY	01	1945	MAY	10	ACTING
Munsinger	Melvin	D.	CAPT	CVW-9	1979	FEB	22	1980	JUL	04	
Munson	Arthur	H.	CDR	CVSG-59	1962	MAY	14	1963	APR	30	
Murphy	Daniel	J.	CDR	CVSG-57	1962	JUN	01	1963	AUG	22	
Murto Jr.	Thomas	V.	LCOL	MCVG-7							ONLY CAG?/UNKNOWN EXISTENCE DATES
Musick	Kenneth	F. 'Mr. 5x5'	CDR	CVG-92	1945	JAN		1945	DEC	18	ONLY CAG
Myers	Carroll	E. 'CE'	CAPT	CVW-5	1971	DEC	10	1972	NOV	15	
Myers	Lowell	R. 'Moose'	CDR	CVW-17	1973	MAR	24	1974	JUN	22	
Nance	James	W.	CDR	CVG-8	1963	NOV	27	1964	NOV	30	CVW-8 20 DEC 63
Nearman	Leonard	M.	CDR	CVG-7	1963	JUL	31	1964	JUL	17	CVW-7 20 DEC 63
Needham	William	R. 'Buzz'	CDR	CVW-7	1979	DEC	14	1981	MAR	31	
Neilson	Thomas	L.	CDR	CVW-6	1964	JUL	04	1965	MAR	19	
Nelson Jr.	Robert	W. 'Brick'	CAPT	CVW-3	1999	APR	01				
Netherland	Roger	M. 'Dutch'	CDR	CVW-5	1966	NOV	21	1967	MAY	10	KIA over NVN in A-4C
Newell	James	H.	CDR	CVG-18	1946	APR	19	1947	JUN	21	CVAG-7 15 NOV 46
Newell	James	H.	CDR	CVG-84	1945	JUL	31	1945	OCT	08	LAST CAG
Newhall	Albert	W.	CDR	CVG-14	1954	SEP	03	1956	FEB	20	
Newman	Roscoe	L.	CDR	CVG-10	1943	JUL	15	1944	FEB	12	
Newman	William	E.	CAPT	CVW-9	1980	JUL	04	1981	SEP	25	
Nichols	David	C	CAPT	CVW-2	1994	JUL	13	1995	SEP	07	
Nicholson Jr.	John	L. 'Nick'	CDR	CVW-16	1970	AUG	07	1971	JUN	30	ACTING(VA-122 CO)/LAST CAG
Nickerson	Richard	H. 'Nick'	CDR	CVSG-56	1966	NOV	16	1967	JUL	12	
Nicol	George	B.	CDR	CVG-93	1944	DEC	28	1945	JUN	07	1st CAG
Nielson	John	L.	LCDR	CVG-152	1945	MAR	05	1945	MAR	19	ACTING
Norman	Geoffrey	P.	CDR	CVG-82	1945	AUG	20	1947	JUN	20	CVAG-17 15 NOV 46
Norris	Jerry	D.	CAPT	CVW-3	1987	JUL	24	1989	JUN	09	
O'Brien	Frank	H.	CDR	CVG-21	1958	DEC	22	1959	DEC	11	
O'Connor	Francis	E.	CDR	CVW-10	1969	MAR	26	1969	AUG	29	
O'Grady	James	W.	CDR	CVAG-11	1948	JUL	01	1949	JUL	01	CVG-11 1 SEP 48
O'Hara	Jack	F.	CDR	CVW-11	1970	JAN	21	1971	JUL	23	
O'Hare	Edward	H. 'Butch'	LCDR	CVG-6	1943	MAR	15	1943	APR	15	ACTING
O'Hare	Edward	H. 'Butch'	LCDR	CVG-6	1943	OCT	11	1943	NOV	25	KIA off Gilberts in F6F-3
O'Neil	Warren	H.	CDR	CVG-11	1963	JAN	03	1963	DEC	20	
O'Neill	John	T. 'Tex'	CDR	CVG-3	1953	OCT		1954	OCT		
O'Shea	George	A.	CDR	CVSG-53	1962	OCT	31	1962	NOV	28	ACTING
Oberg	Owen	H.	CDR	CVW-8	1965	NOV	21	1966	NOV	10	
Oechslin	Raymond	E.	CDR	CVW-6	1966	APR	08	1967	APR	08	
Ofstie	Ralph	A.	LCDR	FitWing	1934	JUL	01				AS OF DATE
Olson	Donald	K.	CDR	CVG-19	1954	MAY	28	1955	JUL		

Onstott	Jacob	W. 'Jake'	CDR	CVG-11	1951	MAY	22	1952	SEP	29	or 23 MAY 51?
Orr Jr.	William	S. 'Bud'	CAPT	CVW-14	1988	MAY	31	1989	NOV	17	
Orrik Jr.	Frederick	J.	CDR	CVW-16	1969	AUG	18	1969	OCT		ACTING
Ortega	Joseph	J. 'Jay'	CDR	CVSG-53	1972	JUN	06	1973	JUN	01	LAST CAG
Osterhoudt	Raymond	S. 'Gus'	CDR	ATG-2	1957	OCT	04	1958	APR	01	LAST CAG
Ostrom	Charles	H.	LCDR	CVLG-21	1943	JUN	25	1943	SEP	27	
Ostrom	Rexford	J.	LCDR	CVEG-24	1944	JUL		1945	APR	10	KIA near Okinawa in F6F-5
Ottinger	George	M. 'Bunky'	CDR	CVG-84	1944	MAY	29	1945	MAR	24	1st CAG/KIA over Okinawa in F6F-5
Outlaw	Edward	C. 'Eddie'	CDR	CVBG-5	1947	SEP		1948	JUN		
Outlaw	Edward	C. 'Eddie'	LCDR	CVLG-32	1944	MAY	12	1944	AUG	24	
Overend	Edmund	F.	MAJ	MCVG-11							ONLY CAG?/UNKNOWN EXISTENCE DATES
Overfield	David	B.	CDR	RANGER	1942	AUG	30	1942	DEC	15	
Owens	Darrel	D.	CDR	CVW-14	1977	APR	21	1978	JUL	10	
Owens Jr.	Robert	G.	LCOL	MCVG-6	1945	MAR	25				ONLY CAG?/UNKNOWN EXISTENCE DATES
Paddock	Merlin		LT	CVLG-23	1944	JUN	25	1944	JUL	01	ACTING?
Paddock	Merlin		LCDR	CVLG-23	1945	FEB	17	1945	APR	16	ACTING/KIA over Japan in F6F-5
Page	Arthur	M.	CDR	CVW-7	1976	MAR	05	1977	JUN	03	
Page Jr.	Louis	C.	CDR	CVW-2	1968	JUN	28	1969	JUN	12	
Page Jr.	Louis	C.	CAPT	RCVW-12	1969	JUN	27	1970	JUN	01	LAST CAG
Palmer	Fitzhugh	L.	CDR	CVG-150	1945	MAR	01	1945	NOV	02	ONLY CAG?
Palmer Jr.	Frederick	F.	CDR	CVW-14	1966	MAR	08	1967	JAN	23	
Palmer	Gary	H.	CAPT	CVW-21	1970	MAY	14	1971	JUN	18	
Palus	George	R.	LCDR	CVEG-26	1945	JUL	08	1945	NOV	13	LAST CAG
Pardee	William	M.	CDR	CVSG-53	1960	MAR	25	1961	APR	07	1st CAG
Park	David	V.	CAPT	CVW-1	1993	MAY	04	1993	MAY	19	
Parker	Edelen	A.	CDR	CVG-9	1955	JAN		1956	SEP		JAN 55 ASSUMED
Parker	Edwin	B.	CDR	CVG-5	1944	JUN	25	1945	MAY	08	
Parker	Edwin	B.	CDR	CVG-5	1945	MAY	09	1945	JUN	05	
Parker	Richard	W.	CDR	CVG-2	1954	MAR	12	1955	JUN	03	
Parks	John	E.	CDR	CVG-15	1952	MAY	22	1953	OCT		
Parrett	Gaylord	S.	LCDR	CVG-13	1945	JAN	01	1945	FEB	01	REFORMED/ACTING
Pattie	Donald	A.	LT	CVLG-23	1945	APR	16	1945	APR	21	ASSUMED ACTING
Paul	Paul	M. 'Paco'	CDR	ATG-2	1954	SEP		1955	NOV	23	REFORMED/ASSUMED PERIOD
Paul	Paul	M. 'Paco'	LCDR	CVEG-25	1945	MAY	05	1945	SEP	20	LAST CAG
Pawka	Edward	J.	CDR	CVG-12	1945	MAY	12	1945	SEP	17	LAST CAG
Payne	Douglas	W.	CDR	CVSG-59	1972	MAY	12	1973	JUN	30	LAST CAG
Peck	Paul	A.	CDR	CVW-9	1967	JUL	13	1968	AUG	02	
Pedersen	Dan	A.	CDR	CVW-15	1975	MAY	19	1976	AUG	07	
Pederson	Oscar	'Swede'	LCDR	YORKTOWN	1942	APR	25	1942	JUN		LAST CAG
Perry	Frank	C.	CDR	CVG-10	1954	JAN	18	1955	FEB	10	
Peters	James	M.	CDR	CVG-1	1943	MAY	01	1944	OCT	01	ASSUMED 1st CAG
Peters	James	McC.	LCDR	CVLG-22	1942	SEP	30	1943	DEC	16	1st CAG
Peterson	Dale	K.	CDR	CVG(N)-90	1945			1946	JUN	21	LAST CAG
Petrea Jr.	Howard	A.	CAPT	CVW-9	1990	JUL	13	1992	FEB	27	
Phares	Everett	L.	CDR	CVG-87	1944	JUL	01	1945	NOV	02	ONLY CAG?
Phillips	Billy		CAPT	RCVW-4	1967	JUL	14	1969	JUN	20	
Phillips	Billy		CDR	CVW-19	1966	MAR	10	1967	JUL	11	
Phillips II	Brewster		LT	CVEG-25	1945	APR	08	1945	MAY	05	ACTING
Phillips Jr.	John	L.	LCDR	CVG-6	1943	NOV	26	1944	FEB	16	KIA over Truk in F6F-3
Phillips	Richard	W.	CDR	CVG-2	1951	JUL	01	1951	AUG	30	ACTING?
Pierce	Edwin	L.	LCDR	CVEG-36	1944	MAY	15	1944	JUN		ACTING
Pitz	Marcellus	T.	CDR	CVSG-60	1964	JUN	19	1965	JUL	16	
Pokorny Jr.	William	A. 'Reef'	CAPT	CVW-11	1999	MAR	27				
Poor	Richard	L.	CDR	CVG-15	1945	JAN	10	1945	OCT	30	REFORMED/LAST CAG
Potosnak	Joseph	E.	CDR	CVW-19	1973	JUN	28	1974	JUL	25	
Potter	Richard	W. 'Duke'	LCDR	CVW-1	1983	AUG		1983	OCT		ACTING
Powell	J.	Richard	CAPT	CVW-3	1969	DEC	13	1971	JAN	28	
Powell Jr.	Lucian	C.	CDR	ATG-2	1953	APR		1954	JUL	12	REFORMED/ASSUMED PERIOD
Powell	Ward	F.	CDR	CVG-19	1949	OCT	10	1950	JUL	07	
Powers	Danny	J.	CAPT	CVW-3	1984	APR	21	1985	DEC	17	
Presley	Jack	C.	CAPT	CVW-1	1976	AUG	20	1977	SEP	23	
Preston	Benjamin	G.	CDR	CVG-7	1954	MAR		1954	MAR		ACTING
Preston	John	P.	CDR	CVG-7	1945	MAR		1946	JUL	08	REFORMED/LAST CAG
Price	George	D.	LCDR	ScoutWing	1932	JAN	01				AS OF DATE
Price	George	D.	LCDR	ScoutWing	1933	APR	01				AS OF DATE
Price	Robert	H.	LCDR	CVLG-25	1943	OCT	06	1944	JUN	12	
Prueher	Joseph	W.	CAPT	CVW-7	1983	JUN	25	1984	MAY	04	'Broadway Joe'
Prueher	Joseph	W.	CAPT	CVW-8	1987	JUL	31	1988	DEC	09	'Broadway Joe'
Pugh	Paul	E.	CDR	CVG-21	1957	FEB	05	1958	DEC	22	
Quirk	William	J.	CDR	CVW-17	1972	APR	07	1973	MAR	24	

Raby	John		CDR	CVG-9	1943	AUG	13	1943	OCT	24	
Radcliff	Roderick	T.	CAPT	RCVSG-50	1969	JUL	07	1970	JUL	10	
Raddatz	Robert	W.	CDR	CVSG-54	1960	MAY	18	1961	SEP	21	1st CAG
Radford Jr.	Cyrus	S.	LCDR	CVG-150	1945	JAN	22	1945	JAN	26	ACTING
Rainey Jr.	Daniel	L.	CDR	CVW-8	1984	AUG	22	1986	JAN	08	
Ramage	James	D. 'Jig Dog'	CDR	CVG-19	1952	DEC	10	1954	MAY	28	
Ramsey	Logan	C.	LCDR	YORKTOWN	1937	JUL	01	1938	JUN		Senior Squadron CO
Ramsey	Paul	H.	CDR	CVG-11	1942	OCT	10	1943	FEB	08	1st CAG
Raynor	Leon	T.	LT	CVLG-32	1945	AUG	22	1945	OCT	14	
Ready	John	K. 'Jack'	CAPT	CVW-11	1979	APR	16	1980	JUL	02	
Reed	Roy	E.	CDR	CVG-9	1958	MAY		1959	JUN	12	
Reed Jr.	William	H.	CDR	CVSG-56	1971	MAY	22	1972	MAY	26	
Rees	William	L.	LCDR	YORKTOWN	1940	MAY	20	1941	OCT	06	
Reinhardt	William	C.	CDR	CVG-7	1959	MAR		1960	MAR		
Rembert Jr.	John	P.	CDR	CVLG-40	1943	JUN	15	1944	JUN	01	1st CAG
Renfo	Carter	B.	CAPT	CVW-6	1984	JUL	19	1986	JAN	23	
Replogle	Thomas	H.	CDR	CVW-17	1970	MAR	31	1971	APR	03	
Reynolds	Marvin	D.	CDR	CVW-3	1973	FEB	28	1974	FEB	28	
Richardson	A.	Fields 'Spider'	CAPT	CVW-6	1989	APR	18	1991	JAN	04	
Richardson	David	C.	CDR	CVAG-13	1948	AUG	10	1949	AUG		CVG-13 1 SEP 48
Richardson	Phillip	D.	CAPT	CVWR-30	1973	JUN	24	1974	NOV	17	
Riendeau	Gerald	L.	CDR	CVW-1	1974	JUN	26	1975	JUL	19	
Riera	R.	Emmett 'Rebel'	CDR	CVBG-1	1947	MAY		1948	AUG	02	
Riera	R.	Emmett 'Rebel'	CDR	CVG-11	1944	NOV	13	1945	APR	09	
Ring	Stanhope	C.	LCDR	HORNET	1941	OCT	06	1942	JUN	30	1st CAG
Rippey	William	H.	CDR	CVSG-60	1963	JUN	14	1964	JUN	19	
Robb	James	A. 'Rookie'	CAPT	CVW-9	1992	FEB	27	1993	JUL	26	
Robb	Randolph	R.	CAPT	CVW-1	1995	JUL	23	1996	JUL	17	
Robinson	Richard	W.	LCDR	CVEG-25	1944	AUG	28	1945	APR	08	REFORM+CVEG/KIA near Okinawa(F6F-5)
Roche	John	R.	ENS	CVEG-35	1945	JAN	02	1945	JAN	07	REFORMED/ACTING
Rockwell Jr.	John	H. 'Rocky'	CDR	CVG-2	1955	JUN		1956	FEB		KIFA near Fallon/JUN 55 ASSUMED
Rodee	Walter	F.	LCDR	HORNET	1942	JUN	30	1942	OCT		LAST CAG
Rogers	David	N.	CAPT	CVW-14	1978	JUL	10	1979	DEC	22	
Rogers	Richard	S.	CDR	CVG-1	1951	OCT		1953	MAR		
Rogers Jr.	Thomas	S.	CAPT	CVW-6	1969	JUN	03	1970	SEP	02	
Romberg	Henry	A.	CDR	CVEG-36	1944	JUN					1st CAG
Rooney	Carl	W.	CDR	CVLG-46	1944	APR	15	1945	SEP	14	ONLY CAG?
Roudebush	Jack		CDR	CVLG-27	1943	OCT	15	1943	NOV	08	REFORMED/1st CAG?
Rouzee	George	M.	LCDR	CVLG-49				1945	NOV	11	LAST CAG?
Rowe	Gordon		LCDR	ScoutWing	1932	JUL	01				AS OF DATE
Rowney	James	V.	CDR	CVG-7	1956	JAN		1957	OCT	31	
Roy	James	C.	CAPT	CVW-17	1983	FEB	22	1984	MAR	23	
Ruddy Jr.	Joseph	A.	CDR	RANGER	1943	JUL	15	1944	MAY	06	CVG-4 3 AUG 43
Ruddy Jr.	Joseph	A.	LCDR	CVEG-29	1942	JUL	18	1943	APR	13	1st CAG
Ruddy Jr.	Joseph	A.	LCDR	CVLG-31	1943	MAY	01	1943	JUN		1st CAG/JUN 43?
Ruelow	Stanley	E.	CDR	CVG-8	1945	JAN	25	1945	NOV	23	LAST CAG
Rumble	Richard	E.	CDR	CVG-10	1963	AUG	05	1964	AUG	06	CVW-10 20 DEC 63
Rumsey II	Dexter	C.	CDR	CVG-6	1952	JUL	08	1952	AUG	04	ACTING
Rush	Max	R.	CDR	CVSG-56	1962	JUN	22	1963	JUN	17	
Russell	Hawley	'Monk'	CAPT	RCVG-12	1959	OCT	30	1960	JUN	02	
Russell	Hawley	'Monk'	CDR	CVAG-17	1947	JUN	20	1949	APR	11	CVG-17 1 SEP 48
Russell Jr.	Thomas	B.	CDR	CVW-14	1965	MAR	05	1966	MAR	08	
Rutherford	Ralph	B.	CAPT	CVW-5	1970	DEC	10	1971	DEC	10	
Rynd	Robert	W.	CDR	CVG-2	1951	JAN	07	1951	JUL	01	
Sampson	Richard	D.	LCDR	CVEG-40	1944	JUN	01	1945	APR	05	REFORM/CVEG 15 AUG/KIA Okinawa
Sanders	Charles	C.	CDR	CVG-15	1959	JAN		1960	FEB	12	
Sanderson	James	R.	CDR	CVW-3	1964	JUL	02	1965	JUL	19	
Saunders	David	M.	CDR	CVSG-52	1962	JUL	05	1963	JUL	29	
Schaeffer	Valentine	H.	LCDR	LEXINGTON	1937	JUN		1938	JUN		Senior Squadron CO/Informal CAG
Schaufelberger Jr.	Albert	A. 'Shoes'	CAPT	CVW-5	1968	AUG	24	1969	NOV	24	
Schecter	Gordon	E.	CDR	CVLG-45	1944	APR	01	1945	MAR	18	1st CAG/KIA over Japan in F6F-5
Schrader	Frederick	R. 'Fritz'	CDR	CVG-11	1944	AUG		1944	OCT	13	KIA off Formosa in F6F-5
Schriefer Jr.	Luther	F.	CDR	CVW-3	1980	JAN	04	1981	MAR	20	
Schroeder	Raymond	C. 'Chuck'	CDR	CVW-2	1978	JUL	08	1979	OCT	27	
Schumann Jr.	Roland	W. 'Bud'	CDR	CVG-7	1953	FEB		1954	MAR		
Schwab	Erwin	G.	CDR	CVG-11	1952	SEP	29	1954	FEB	01	
Schwartz	Sheldon	O. 'Lefty'	CDR	CVW-15	1970	JUL	10	1971	JUN	02	
Scott	James	H.	CAPT	CVW-21	1969	APR	11	1970	MAY	14	
Scott	Thomas	P.	CDR	CVW-17	1974	JUN	22	1975	SEP	09	

Searcy Jr.	Seth	S. 'Pete'	CDR	CVG-88	1944	AUG	18	1945	OCT	29	ONLY CAG?
Sease	Hugh	St. Clair	LCDR	TrpBmbWing	1931	JUL	01				AS OF DATE
Sedaker	Thomas	S.	CDR	CVG-19	1960	JUN	01	1961	MAR	17	
Seely	James	M.	CAPT	CVW-9	1974	JUL	14	1975	JUL	03	
Seidel	Melvin	L.	CDR	CVWR-20	1978	FEB	25	1979	MAY	26	
Seligman	Morton	T.	LCDR	RANGER	1937	OCT	01	1939	JUN		1st Formal CAG on 1 JUL 38
Sells	Warren	H.	CDR	CVW-21	1964	APR	29	1965	JUN	03	
Selmer	Robert	J.	CDR	CVG-2	1960	APR	08	1961	DEC	20	
Selmer	Robert	J.	CDR	CVG-6	1957	NOV		1958	JAN	21	ACTING
Severns	Robert	L.	CDR	CVSG-60	1962	JUL	31	1963	JUN	14	
Sevilla	Bernard		CDR	CVG-9	1960	SEP	02	1961	DEC	20	
Shaffer	James	B.	CDR	CVSG-58	1964	MAY	21	1965	FEB	19	
Shartel	Howard	A.	CDR	CVSG-52	1968	AUG	07	1969	APR	18	
Shawcross	William	H.	CDR	CVG-16	1961	FEB	13	1962	MAR	16	1st CAG
Sheets	Roger	E.	CDR	CVW-15	1972	APR	06	1973	FEB		
Shelton	Doniphan	B.	CDR	CVW-19	1964	MAR		1965	MAR	01	
Shepherd	Burton	H.	CDR	CVW-16	1966	DEC	19	1968	FEB	05	
Sherlock	James	C.	CAPT	CVW-7	1991	JAN	18	1992	JUL	29	
Sherman	Forrest	P.	LCDR	FitWing	1932	JUL	01				AS OF DATE
Sherman	John	W.	CAPT	CVW-9	1999	JUN	16				
Sherrill	Wallace	A.	CDR	CVG-85	1944	MAY	15	1945	SEP	24	ONLY CAG?
Shifley	Ralph	L.	CDR	CVG-8	1944	MAY		1945	JAN	01	
Shinn	Allen	M.	CDR	CVG-89	1944	OCT	22	1945	DEC	28	1st CAG
Shipman	James	L.	CAPT	CVW-9	1966	JAN	04	1967	JUL	13	
Short	Giles	E.	LCDR	ENTERPRISE	1938	JUL	01	1939	JUN		1st Formal CAG
Shrine Jr.	Bertram		CDR	CVSG-56	1969	MAY	20	1970	MAY	13	
Shryock	William	A.	CAPT	RCVG-4	1960	APR	28	1961	JUL	07	
Shryock	William	A.	CDR	CVG-15	1955	JAN		1957			PRIOR TO JUL 57
Shugart	Kenneth	L.	CDR	CVW-7	1967	MAY	16	1968	JUN	12	
Siegener	Adolf	L.	CDR	CVG-11	1949	JUL	01	1949	DEC	28	
Silber	Samuel	L. 'Sambo'	CDR	CVG-4	1945	JUL	04	1947	JUN		CVAG-1 15 NOV 46
Simmons	Donald	K. 'Sickle'	CAPT	CVWR-30	1987	AUG	30	1989	APR	15	
Simmons Jr.	George	C. 'Chuck'	CDR	CVG-3	1951	OCT		1952	JUL		ASSUMED PERIOD
Simpson Jr.	Edward	W.	LT	CVEG-35	1945	JAN	07	1945			ACTING
Sinton	William		LCDR	LEXINGTON	1937			1937	JUN		Senior Squadron CO
Sinton	William		LCDR	LEXINGTON	1938	JUL	01	1939	JUN	01	1st formal CAG
Sizemore	William	G.'Gene'	CAPT	CVW-1	1969	JUN	17	1970	MAY	01	
Slater	Thomas	S.	CAPT	CVW-15	1982	NOV	09	1984	MAY	13	
Sliney	James	G.	LCDR	CVLG-30	1943	DEC	21	1944	JUN	25	
Smevog	H.	Wayne	CDR	CVSG-56	1972	MAY	26	1973	JUN	30	LAST CAG
Smiley	Curtis	S.	CDR	YORKTOWN	1941	OCT	06	1942	APR	25	
Smith Jr.	Armistead	B. 'Chick'	CDR	CVG-7	1957	OCT	31	1959	MAR		
Smith	Bernard	J. 'Hawk'	CAPT	CVW-3	1981	MAR	20	1982	JUL	25	or 31 MAR 81?
Smith Jr.	Daniel	F.	CDR	CVG-20	1943	DEC	03	1945	FEB	24	
Smith	George	E.	CDR	RCVSG-51	1966	MAY		1967	MAR	31	
Smith	L.	Wayne	CDR	CVW-6	1968	FEB	21	1969	JUN	03	
Smith Jr.	Leighton	W. 'Snuffy'	CDR	CVW-15	1977	SEP	14	1978	NOV	22	
Smith	Lloyd	A.	LCDR	CVG-11	1944	OCT	13	1944	NOV	01	ACTING
Smith	Robert	E.	CAPT	CVW-11	1980	JUL	02	1981	AUG	15	
Smith	Robert	H.	CDR	CVG-15	1953	OCT		1954	MAY	01	OCT 53 ASSUMED
Smith	Wendell	K.	CDR	CVW-1	1966	JAN	19	1967	MAR		
Snider	George	W.	LCDR	CVLG(N)-52	1944	OCT	20	1945	FEB	12	1st CAG/CVGN 06 JAN 45
Snowden	Ernest	M.	CDR	CVG-16	1943	OCT	07	1944	AUG	24	
Snowden	Macon	S.	CDR	CVW-5	1965	JAN	12	1965	DEC	13	
Snyder	Jack	L.	CDR	CVW-5	1965	DEC	13	1966	NOV	21	
Snyder	James	M.	CDR	CVW-6	1971	JUL	28	1972	JUL	03	
Snyder	Thomas	A.	CDR	CVSG-54	1971	JUL	13	1972	JUL	01	LAST CAG
Southerland	James	J. 'Pug'	CDR	CVG-4	1949	FEB		1949	OCT	18	
Southerland	James	J. 'Pug'	CDR	CVLG-23	1945	APR	21	1945	JUL	15	
Southerland	Leonard	B.	CDR	CVG-16	1942	NOV	28	1944	OCT	07	1st CAG
Soverel	William	W.	LCDR	CVG(N)-55	1945	MAR	01	1945	DEC	11	ONLY CAG?
Sparks	Samuel	A.	CDR	CVSG-62	1961	SEP	25	1962	OCT	01	ONLY CAG
Spears	Paul	H.A.	CDR	CVSG-59	1961	MAY	12	1962	MAY	14	
Speelman	Thomas	W.	CDR	CVSGR-70	1972	APR	15	1973	OCT	27	
Speer	Paul	H.	CDR	CVW-11	1969	OCT	02	1970	JAN	21	
Speer	Paul	H.	CDR	CVW-14	1970	JAN	26	1970	JUN	12	
Spencer	Paul	E.	CDR	CVW-8	1964	NOV	30	1965	NOV	21	
Spencer	Ross	B.	LT	CVEG-60	1945	APR	19	1945	MAY	15	ACTING
Springer	Judson	H.	CDR	CVW-8	1978	JAN	31	1979	APR	10	
Sprouse	Haywood	G. 'Woodie'	CAPT	CVW-15	1981	JUL	22	1982	NOV	09	

Spruit	Robert	E.	CDR	CVW-16	1965	OCT	04	1966	SEP	30	
Staggs	William	R.	LCDR	CVEG-29	1943	APR	13	1944	JAN	05	
Stanley	Thomas	A.	CDR	CVSGR-80	1971	JAN	23	1972	AUG	20	
Stapler	Charles	R.	CDR	CVG-19X	1951	MAY	23	1951	JUN	10	1st CAG/KIA over Korea
Starr	Mark	R.	CDR	RCVSG-51	1965	JUL	23	1965	SEP	24	
Stear	David	S.	CDR	CVG-7	1961	JUL		1962	JUL	21	
Stebbins	Edgar	E. 'Steb'	CDR	CVG-5	1943	NOV	21	1944	JUN	25	
Stegg	Robert	J.	CDR	CVG-10	1961	MAY	22	1961	SEP	11	ACTING
Stephens	Franklin	T. 'Frank'	CDR	CVG-3	1963	JUL	02	1964	JUL	02	CVW-3 20 DEC 63
Stephenson	Paul	D.	CDR	CVW-21	1974	JUN	26	1975	DEC	12	LAST CAG
Stetson	John	B. 'Jock'	CDR	CVG-16	1962	MAR	16	1963	FEB	09	
Stevens	Jack	M.	CDR	RCVSG-50	1965	JUL	30	1966	AUG	10	
Stevens	Paul	F.	CAPT	RCVG-12	1961	OCT	31	1963	JAN	18	
Stockdale	James	B.	CDR	CVW-16	1965	FEB	18	1965	SEP	09	POW over NVN in A-4E
Stockton	Jackson	A. 'Jack'	CDR	CVSG-53	1961	OCT	12	1962	OCT	31	
Stollenwerck	William	M.	CDR	CVW-8	1969	SEP	12	1970	JUN	06	
Stone	Robert	S.	CDR	CVSG-58	1965	FEB	19	1966	MAY	31	LAST CAG
Stone	Troy	E.	CDR	CVSG-57	1966	JUL	08	1967	JUN	29	
Stout	Edward	N.	CDR	CVSG-54	1969	JUN	26	1970	JUN	23	
Strange	John	C.	LCDR	CVLG-50	1944	APR		1944			
Strean	Bernard	M. 'Smoke'	CDR	CVBG-75	1945	JUN	01	1946	OCT	29	1st CAG
Strean	Bernard	M. 'Smoke'	CDR	CVG-98	1944	SEP	26	1945	MAY	08	1st CAG
Streeper	Harold	P.	CDR	CVG-15	1961	JUL	06	1962	JUL	27	
Streit	John	B.	CDR	CVW-5	1973	AUG	03	1974	JUL	23	
Stubbs	Harry	K.	LCDR	CVG-99	1945	JUL	14	1945	SEP	06	LAST CAG?
Stufflebeem	John	D.	CAPT	CVW-1	1996	JUL	17	1997	DEC	31	
Stump	Felix	B.	LCDR	ScoutWing	1934	JUL	01				AS OF DATE
Stump	Felix	B.	LCDR	ScoutWing	1936	APR	01				AS OF DATE
Sult	George	H.	CDR	CVG-19	1957	SEP	20	1958	SEP	05	
Swanson	Charles	A.L. 'Cal'	CAPT	CVW-17	1969	JAN	18	1970	MAR	31	
Swanson	Leroy	V.	CDR	CVG-17	1952	JUN	13	1955	OCT	07	
Sweeney	John	R.	CDR	CVG-4	1952	OCT		1954	FEB		
Sweeney	John	R.	CAPT	RCVG-12	1960	JUN	02	1961	OCT	31	
Swenson	Richard	M.	LT	CVLG-24	1943	SEP	06	1943	OCT		ASSUMED ACTING CAG
Switzer	Wendell	G.	LCDR	RANGER	1937	JUL	01	1937	OCT	01	Senior Squadron CO
Switzer III	William	H.	CAPT	CVW-13	1987	JAN	16	1988	NOV	18	
Switzer III	William	H.	CAPT	CVW-15	1984	MAY	13	1986	FEB	16	
Swope	James	S.	CDR	CVG-15	1960	FEB	12	1961	JUL	06	
Sykes	James	B.	LCDR	SARATOGA	1937	OCT	01	1938	JUN		Informal CAG
Tallent	Hamlin	B. 'Ham'	CAPT	CVW-14	1995	DEC	15	1997	MAR	19	
Talley Jr.	George	C.	CDR	CVG-1	1961	OCT	09	1962	DEC	19	
Tate	Benjamin	C.	CDR	CVSG-58	1961	JUL	15	1962	JUN	02	
Tate	Jackson	R.	LCDR	YORKTOWN	1939	JUN	26	1940	MAY	20	
Taylor	Jeremy	D. 'Bear'	CDR	CVW-3	1976	MAY	17	1977	AUG	27	
Taylor	Jesse	D.	LCDR	CVG-93	1945	JUN	18	1946	APR	30	LAST CAG?
Taylor	Robert	J. 'Tails'	CAPT	CVW-11	1996	MAY	14	1997	OCT	09	
Taylor	Thaddeus	M.	CDR	CVW-16	1964	FEB	03	1965	FEB	18	
Teague	Forrest	S. 'Tooter'	CDR	CVW-11	1974	MAY	01	1975	JUN	03	
Teel	Richard	A.	CDR	CVG-153	1945	MAR	26	1946	OCT	16	1st CAG
Terry	Bert	D.	CDR	CVW-15	1976	AUG	07	1977	SEP	14	
Terry	Harold	L.	CDR	CVW-21	1967	OCT	06	1969	APR	11	
Thomas	James	A.	LCDR	CVG-10	1943	APR	07	1943	JUL	15	
Thomassy Jr.	Louis	E. 'Tim'	CDR	CVW-2	1981	JAN	15	1982	APR	07	
Thompson	Paul	V.	LCDR	CVEG-26	1945	JAN	17	1945	JUL	08	REFORMED 10 JAN 45/KIFA off Hawaii
Thrash	William	G.	LCOL	MCVG-5							ONLY CAG?/UNKNOWN EXISTENCE DATES
Tierney	John	M.	CDR	CVW-19	1965	MAR	01	1965	DEC	18	
Tillotson	F.	Lee 'Raider'	CDR	CVW-11	1981	AUG	15	1982	NOV	06	
Timm	Dwight	D.	CAPT	CVW-1	1975	JUL	19	1976	AUG	20	
Tinker	Charles	L. 'Chuck'	CAPT	CVWR-20	1976	FEB	28	1978	FEB	25	
Tissot	Ernest	E. 'Gene'	CDR	CVW-14	1967	JAN	23	1968	FEB	05	
Torrey Jr.	Philip	H.	CDR	CVG-9	1943	DEC	22	1945	FEB	16	KIA over Japan in F6F-5
Toy	Frank	E.	CDR	CVSGR-70	1969	APR	01	1970	DEC		1st CAG
Treanor Jr.	Thomas	S.	CAPT	CVW-7	1981	MAR	31	1982	MAR	31	
Tucker Jr.	Robert	E. 'Gene'	CAPT	CVW-17	1979	JUN	01	1980	JUL	10	
Tully	Claude	I.	CDR	CVW-6	1965	MAR	19	1966	APR	08	
Turk	Herman	L.	CDR	CVW-1	1971	MAY		1972	MAY	24	
Turner	Charles	H.	CDR	CVBG-75	1946	OCT	29	1947	NOV	19	CVBG-3 15 NOV 46
Turner	Frederick	C.	CDR	CVG-3	1962	JUL	18	1963	JUL	02	
Tuttle	Jerry	O.	CAPT	CVW-3	1974	FEB	28	1975	FEB	10	
Tylutki	Ray	C.	CDR	CVG-14	1956	FEB	20	1957	AUG	09	20 FEB 56 APPROX.

Tyson	James	J.	CAPT	CVW-14	1970	JUN	12	1971	MAY	10	
Updegraff	William	N.	LCDR	ScoutWing	1933	OCT	01				AS OF DATE
Upham	Frank	K.	CDR	CVG-4	1945	FEB		1945	MAY	12	
Upham	Frank	K.	CDR	CVG-98	1945	MAY	08	1946	OCT	01	
Urban Jr.	Henry		CDR	CVW-11	1966	DEC	12	1968	JAN	13	
Utter	Harmon	T.	CDR	CVG-83	1944	MAY	01	1945	SEP	24	1st CAG/ONLY?
Van Meter	Karl	S.	CDR	CVG-12	1953	JUN	30	1954	OCT	29	
Van Meter	Karl	S.	CDR	CVG-19	1951	MAY	19	1951	JUL	27	ACTING
Vaughn	Thomas	N.	CAPT	CVW-9	1996	MAY	30	1997	DEC	05	
Vejtasa	Stanley	W. 'Swede'	CDR	CVLG-44	1945	JUN	11	1945	SEP	18	LAST CAG
Vernon	Larry	J.	CDR	CVW-5	1982	JAN	20	1983	APR	27	
Vieweg	Frederic		LT	CVG-5	1945	MAY	08	1945	MAY	09	ACTING
Vogel Jr.	Raymond	W. 'Sully'	CDR	CVG-11	1950	JUL	01	1950	AUG	19	KIA over Korea in F4U-4B
Vogel Jr.	Raymond	W. 'Sully'	CDR	CVG-89	1944	OCT	02	1944	OCT	22	ACTING
Vogel Jr.	Raymond	W. 'Sully'	CDR	CVG-89	1945	DEC	28	1946	APR	27	LAST CAG
Voris	Roy	M. 'Butch'	CDR	CVG-5	1957	DEC	13	1958	DEC	10	
Vorse Jr.	Albert	O. 'Scoop'	CDR	CVG-80	1944	OCT	16	1945	MAR		
Vredenburgh III	James	B.	LCDR	CVLG-31	1943	JUN		1944	JAN		JUN 43?
Wagner Jr.	Theodore	A.	CDR	CVW-9	1964	FEB	17	1965	JAN	11	
Walker	Donald	P.	CDR	CVSG-54	1961	SEP	21	1962	AUG		
Walker	Grover	C. 'Bull'	CDR	CVW-3	1965	JUL	19	1966	JUL	06	
Wallace Jr.	Daniel	J.	LCDR	CVLG-31	1944	JUN	29	1945	MAR	05	REFORMED DEC 44/KIFA over Monterey
Wallace	Kenneth	R.	CDR	CVW-15	1968	MAY	24	1969	MAY	29	
Waples	John	M.	CAPT	CVW-2	1979	OCT	27	1981	JAN	15	
Ward	J D	'Jake'	CDR	CVW-15	1966	JUN	25	1967	FEB	03	J D IS TRUE NAME
Ware	Robert	M.	LCDR	CVG-17	1944	APR	24	1944	MAY	13	ACTING
Warren	John	L.	LT	CVG-4	1945	MAY	12	1945	MAY	18	ACTING
Watkins	George	C.	CDR	CVG-13	1961	AUG	21	1962	OCT	01	ONLY CAG
Watson	Dewitt	C.	LCDR	ScoutWing	1930	OCT	01				AS OF DATE
Watson Jr.	Thomas	C.	CDR	CVW-9	1972	MAY	20	1973	MAY	24	
Watts	Donald	L.	LCDR	CVG-102	1950	AUG	01	1950	AUG	23	ACTING
Webb	Steven	L.	CAPT	CVW-11	1990	JUL	19	1992	JAN	31	
Weber	Bruce	S.	CDR	CVG-17	1949	APR	11	1950	NOV	02	
Weber	Bruce	S.	LCDR	CVG-98	1944	AUG	28	1944	SEP	26	ACTING
Weber	Bruce	S.	LCDR	CVLG-31	1945	MAR	14	1945	OCT	24	LAST CAG
Wedell	Edward	V.	CAPT	RCVG-12	1959	JUL	03	1959	JUL	31	ACTING
Weeks	John	M.	CDR	RCVSG-50	1960	JUN	30	1961	JUN	05	1st CAG
Weideman	Craig	F.	CAPT	CVWR-20	1999	JAN	30				
Weinel	John	P.	CDR	CVG-5	1954	NOV	30	1956	JUL	06	
Weinel	John	P.	LCDR	CVLG-22	1945	MAR		1945	SEP	19	REFORMED/LAST CAG
Weller	Oscar	A.	LCDR	RANGER	1939	JUN		1940	JUN	01	
Werner	Ralph	L.	CDR	ATG-181	1955	JUL		1957	AUG	12	
Werner	Ralph	L. 'Bud'	CAPT	RCVG-4	1961	JUL	07	1963	JUN	20	
Westerman	William	R.	CDR	CVW-7	1977	JUN	03	1978	NOV	02	
Weymouth	Ralph		CDR	CVG-11	1950	AUG	19	1951	MAY	22	
Wheatley	Gary	F.	CDR	CVW-1	1977	SEP	23	1978	DEC	11	
Whidden	Wynn	V.	CDR	CVSG-56	1961	AUG	22	1962	JUN	22	
Whitby	Ralph	E. 'Rusty'	CDR	CVW-7	1978	NOV	02	1979	DEC	14	
White Jr.	A.	Hardin	CAPT	CVW-3	1989	JUN	09	1991	JUN	24	
White Jr.	Clarence	H.	CDR	CVG-95	1945	JAN	02	1945	OCT	31	ONLY CAG?
White	Donald	M.	CDR	CVG-2	1950	JAN	01	1951	JAN	07	
White	Donald	F.	LCDR	CVLG-23	1944	JUL	01	1945	FEB	17	POW over Japan in F6F-5
White Jr.	George	W.	CAPT	CVW-6	1978	MAY		1979	MAY	04	MAY 78 ASSUMED
Whitehead	Richard	F.	LCDR	ENTERPRISE	1939	JUN		1940	MAY		
Whitney	Laurens	A.	CDR	ATG-1	1952	SEP		1953	NOV	28	REFORMED/ASSUMED PERIOD
Wick	Homer	C.	LCDR	FitWing	1928	JUL	01				AS OF DATE
Wilks	Robert	E.	CDR	CVW-2	1985	OCT	28	1987	JAN	21	
Williams	E.	Royce	CDR	CVW-11	1965	DEC	31	1966	DEC	12	
Williams	MacPherson	B.'The Colonel'	CDR	CVG-3	1943	OCT	01	1944	DEC	16	Shot down near Manila in F6F-5
Williamson	Thomas	B.	LCDR	RANGER	1941	MAY	03	1942	APR	11	
Willson Jr.	Lawrence	E.	CDR	CVSG-57	1963	AUG	22	1964	SEP	18	
Wilson	Christopher	T. 'Boomer'	CAPT	CVW-2	1988	AUG	19	1990	FEB	21	
Wilson	Christopher	T. 'Boomer'	CAPT	CVWR-30	1984	JUL	07	1986	FEB	01	
Wilson Jr.	John	R. 'Smoke'	CDR	CVW-14	1974	NOV	21	1976	FEB	28	
Wilson	Richard	A.	CDR	CVW-14	1981	APR	21	1982	JUN	10	
Wingard Jr.	Walter	C.	CDR	CVG-14	1943	SEP	15	1944	NOV	02	1st CAG
Winnefeld	James	A.	CDR	CVSG-57	1969	JUL	09	1969	SEP	30	LAST CAG
Winston	Robert	A.	CDR	CVLG-31	1944	JAN		1944	JUN	29	
Winters Jr.	T.	Hugh	CDR	CVG-19	1943	AUG	15	1943	AUG	18	ACTING
Winters Jr.	T.	Hugh	CDR	CVG-19	1944	SEP	17	1945	JAN	24	

Wirt Jr.	William	O.	CDR	CVSG-55	1967	JUN	07	1968	JUL	10	
Wise	Aubrey	L. 'Al'	CAPT	CVW-7	1987	FEB	27	1989	APR	14	
Wiseman	E	Robert	CDR	CVSG-54	1962	AUG		1963	AUG		
Wissler	John	G.	CDR	CVW-3	1966	JUL	06	1967	JUL	04	
Wittenberg	Robert	R.	CAPT	CVW-1	1989	JUL	08	1990	SEP	05	
Wojcik	Thaddeus	P.	MAJ	MCVG-12						ONLY CAG?/UNKNOWN EXISTENCE DATES	
Wolf	Rexford	E.	CDR	CVW-17	1981	NOV	12	1983	FEB	22	
Wolf	Robert	L.	CDR	CVSG-60	1961	JUN	08	1962	JUL	31	
Wondergem	John	M.	CDR	RCVSG-50	1964	JUN	02	1964	JUL	06	ACTING
Wood	E.	Wetherill	LCDR	CVLG-27	1944	APR	26	1944	JUN	19	KIA off Marianas in F6F-3
Wood	George	O.	CDR	CVG-8	1954	APR		1955	MAY		APR 54 ASSUMED
Wood	Phillip	R.	CDR	CVW-11	1977	DEC	10	1979	APR	16	
Wood	Robert	B.	CDR	CVG-8	1952	AUG	11	1954	MAR	14	
Wood	Robert	B.	LCDR	CVG-92	1944	DEC	02	1945	JAN		ACTING
Wordell	Malcolm	T. 'Mac'	CDR	CVLG-44	1944	FEB	02	1945	JUN	11	1st CAG
Worthington	John	R. 'Jocko'	CAPT	CVW-9	1994	OCT	26	1996	MAY	30	
Wright	Donald	J.	CAPT	CVWR-30	1981	NOV	21	1983	MAY	21	
Wright	James	M.	CDR	CVG-80	1945	MAY		1946	SEP	16	REFORMED/LAST CAG
Wright	Timothy	W.	CAPT	CVW-6	1980	JUN	06	1981	AUG	06	
Wyatt	Charles	A.	CAPT	CVW-8	1997	JUL	01	1999	JAN	15	
Yakeley III	Jay	B. 'Spook'	CAPT	CVW-14	1989	NOV	17	1991	MAY	10	
Yates	Earl	P. 'Buddy'	CDR	CVG-8	1961	NOV	04	1962	NOV	05	
Yost	Donald	K.	LCOL	MCVG-4						ONLY CAG?/UNKNOWN EXISTENCE DATES	
Young	Edwin	J.S.	LCDR	CVG-9	1942	APR	10	1942	JUN	19	ACTING
Young	Howard	L. 'Brigham'	LCDR	ENTERPRISE	1941	APR		1942	MAR	21	
Zelibor	Thomas	E. 'Chain'	CAPT	CVW-3	1995	MAR	25	1996	APR	12	
Zerr	John	J. 'Swede'	CAPT	CVW-14	1983	SEP	09	1985	MAY	08	
Zettel	Marcus	A.	CDR	CVSG-52	1967	JUN	30	1968	AUG	07	
Zlatoper	Ronald	J. 'Zap'	CAPT	CVW-1	1982	JUL		1983	AUG		
Zlatoper	Ronald	J. 'Zap'	CAPT	CVW-15	1986	FEB	16	1987	MAY	08	
Zoehrer	Herbert	A.	CDR	CVSG-53	1970	JUN	12	1971	JUN	09	
Zortman	James	M.	CAPT	CVW-17	1996	JAN	26	1997	JUL	01	

Notes